APPLICATIONS OF CHLOROPHYLL FLUORESCENCE

Applications of Chlorophyll Fluorescene

in Photosynthesis Research, Stress Physiology, Hydrobiology and Remote Sensing

An introduction to the various fields of applications of the in vivo chlorophyll fluorescence also including the proceedings of the first International Chlorophyll Fluorescence Symposium held in the Physikzentrum, Bad Honnef, F.R.G., 6–8 June 1988

edited by

HARTMUT K. LICHTENTHALER

Botanisches Institut
(Plant Physiology and Biochemistry),
University of Karlsruhe,
Karlsruhe, F.R.G.

KLUWER ACADEMIC PUBLISHERS

DORDRECHT / BOSTON / LONDON

Library of Congress Cataloging in Publication Data

International Chlorophyll Fluorescence Symposium (1st : 1988 : Bad Honnef, Germany)
Applications of chlorophyll fluorescence : in photosynthesis research, stress physiology, hydrobiology, and remote sensing / edited by Hartmut K. Lichtenthaler.
p. cm.
"An introduction to the various fields of applications of the in vivo chlorophyll fluorescence ... proceedings of the First International Chlorophyll Fluorescence Symposium, held in the Physikzentrum, Bad Honnef, F.R.G., 6-8 June 1988."
Includes indexes.

1. Chlorophyll--Analysis--Congresses. 2. Fluorimetry--Congresses.
I. Lichtenthaler, Hartmut K. II. Title.
QK898.C5I57 1988
581.19'218--dc19 88-25223
CIP

IBSN-13: 978-94-010-7771-2 e-ISBN-13: 978-94-009-2823-7
DOI: 10.1007/978-94-009-2823-7

Published by Kluwer Academic Publishers,
P.O. Box 17, 3300 AA Dordrecht, The Netherlands.

Kluwer Academic Publishers incorporates
the publishing programmes of
D. Reidel, Martinus Nijhoff, Dr W. Junk and MTP Press.

Sold and distributed in the U.S.A. and Canada
by Kluwer Academic Publishers,
101 Philip Drive, Norwell, MA 02061, U.S.A.

In all other countries, sold and distributed
by Kluwer Academic Publishers Group,
P.O. Box 322, 3300 AH Dordrecht, The Netherlands.

Sponsor and organizer of the first International Chlorophyll Fluorescence Symposium, Bad Honnef, June 1988.

Dr. Volker Schäfer
WE-Heraeus Foundation
Hanau (F.R.G.)

Prof. Dr. Hartmut Lichtenthaler
University of Karlsruhe

The **chairmen** of the International Chlorophyll Fluorescence Symposium held at Bad Honnef from June 6 to 8, 1988:

From left to right: R.J. **Strasser**, Genève; H.R. **Bolhàr-Nordenkampf**, Vienna; D. **Walker**, Sheffield; A. **Melis**, Berkeley; Z. **Sesták**, Praha; G. **Bertolini**, Ispra; W. **Schmidt**, Konstanz; H.K. **Lichtenthaler**, Karlsruhe (Scientific Coordinator); G. **Guyot**, Montfavet; D. **Ernst**, Hannover; V. **Schäfer**, Hanau (Heraeus Foundation).

The participants of the first International Chlorophyll Fluorescence Symposium in front of the Physikzentrum, Bad Honnef:

G. Bertolini, R. Blaich, H.R. Bolhàr-Nordenkampf, C. Buschmann, D. Ernst, G. Guyot, J. Harbinson, A.R. Holzwarth, L. Kocsányi, G.H. Krause, H. Laasch, E. Lehoczki, H.K. Lichtenthaler, H. Maske, A. Melis, M. Methy, E. Nagel, E. Ögren, U. Rinderle, A. Rosema, V. Schäfer, W. Schmidt, G. Schmuck, H. Schneckenburger, U. Schreiber, Z. Sesták, U. Sieber, D. Siefermann-Harms, P. Siffel, J. Snel, S. Somersalo, R.J. Strasser, Z. Szigeti, M. Tevini, A.F. Theisen, P.R. van Hasselt, O. van Kooten, D. Walker, E. Weis.

PREFACE

This book is a general **introduction into in vivo chlorophyll fluorescence** and contains the contributions of the first International Chlorophyll Fluorescence Symposium held in the Physikzentrum Bad Honnef, F.R.G. from June 6 to 8, 1988. This Symposium was made possible by a generous support from the Wilhelm and Else Heraeus Foundation, Hanau, which is gratefully acknowledged. The book not only comprises all aspects of the applications of chlorophyll fluorescence in photosynthesis, stress physiology, hydrobiology and remote sensing, but also gives access to measuring techniques, data acquisition and earlier literature references. Thus it is far more than just a common proceedings book, it is a general introduction to all forms of application of the non-destructive in vivo chlorophyll fluorescence including the newest results.

In a **first chapter** the inverse correlation between in vivo chlorophyll fluorescence and photosynthetic quantum conversion and CO_2-assimilation is outlined, the origin and life-time of the chlorophyll fluorescence at room and liquid nitrogen temperatures are given as well as the induction kinetics (Kautsky effect) and the methodological approaches to register different forms of chlorophyll-fluorescence signatures.

In **chapter 2** the applications of chlorophyll fluorescence in stress physiology and stress detection of plants are pointed out, starting with a general stress concept of plants and various examples of how short-term and long-term stress effects (e.g. herbicides, water stress, mineral deficiency, air pollutants) as well as regeneration of the plants' vitality and photosynthesis when the stressor is removed, can be detected and quantified.

Chapter 3 deals with the application of chlorophyll fluorescence and delayed fluorescence (luminescence) in hydrobiology, limnology and oceanography including remote sensing of phytoplankton.

Chapter 4 contains the basic approach and theoretical considerations for a future remote sensing and stress detection of terrestrial vegetation via laser-induced chlorophyll fluorescence and also gives examples and models as to how reflectance and chlorophyll fluorescence can interfere with and complement each other.

In vivo chlorophyll fluorescence has hitherto found broad application in photosynthesis research, stress physiology, hydrobiology and limnology as well as in remote sensing of phytoplankton in oceanography. It will possibly also be applied in the remote sensing and stress detection of terrestrial vegetation, which is the research topic of several European laboratories which cooperate within the LASFLEUR programme (LAser-induced FLuorescence in EURope).

It is hoped that the Proceedings book of the first International Chlorophyll Fluorescence Symposium "Applications of Chlorophyll Fluorescence", which mediates between different research fields, will be a useful starting help for colleagues and graduate students who are not yet in the field as well as for the active researcher in planning new experiments. The book may help not only to avoid major mistakes and measuring artefacts but also to interpret the measured chlorophyll fluorescence signatures in the right way and find access to the application of newer techniques.

I wish to thank Dr. Volker Schäfer, W. & E. Heraeus Foundation for good cooperation and Ir. A.C. Plaizier, Kluwer Academic Publishers, Dordrecht for publishing the proceedings book, my colleagues for providing their newest results in time, my coworkers and graduate students of the Botanisches Institut Karlsruhe for assistance and last but not least my wife Regine for her valuable aid in editing the book.

Karlsruhe, June 1988

Hartmut K. Lichtenthaler
Scientific Coordinator of the
Chlorophyll Fluorescence Symposium

TABLE OF CONTENTS

CHAPTER 1. Applications in photosynthesis research

CHAPTER 2. Applications in stress physiology and environmental research

CHAPTER 1

Applications in photosynthesis research

THE PHOTOSYNTHETIC APPARATUS AND CHLOROPHYLL FLUORESCENCE. AN INTRODUCTION

G. Heinrich Krause[1] and Engelbert Weis[2], Institute for Biochemistry of Plants (1) and Institute for Ecological Plant Physiology (2), University of Düsseldorf, Universitätsstraße 1, D-4000 Düsseldorf 1, Germany (FRG)

Key words: fluorescence of chlorophyll *in vivo*, fluorescence quenching, light harvesting complex, photoinhibition of photosynthesis, photosystem

ABSTRACT

An introduction to the basic relationship between chlorophyll *a* fluorescence yield and functioning of the photosynthetic apparatus is given. The principles governing the fluorescence decline by photochemical and non-photochemical mechanisms are summarized. Several ways of non-photochemical fluorescence quenching, related to the energization of the thylakoid membrane (q_E), to the phosphorylation of the light harvesting complex of photosystem II (q_T), and to photoinhibition of photosynthesis (q_I) are discussed.

FLUORESCENCE AND PRIMARY REACTIONS OF PHOTOSYNTHESIS

Chlorophyll *a* fluorescence emitted from the chloroplast thylakoid membranes reflects the primary processes of photosynthesis, such as light absorption, excitation energy transfer and the photochemical reaction in photosystem (PS) II. However, since these events are integrated into overall photosynthesis including electron transport, proton transfer across the thylakoid membranes, photophosphorylation and CO_2 assimilation, the yield of chlorophyll fluorescence is influenced by numerous factors in a very complex manner. Fluorescence is used as a tool to investigate different aspects of photosynthesis. The usefulness of this tool depends on an exact interpretation of the phenomena observed. Therefore, it is important that in the complex network of relations between fluorescence emission and events taking place in the photosynthetic apparatus, single effects can be resolved and characterized.

In the following, the basic principles that govern the yield of fluorescence in the photosynthetic apparatus of higher plants will be summarized. (For more detailed information see reviews of Butler 1977, Krause and Weis 1984, Briantais et al. 1986.) The transition from the first excited singlet state to the ground state of chlorophyll molecules occurs in competing first order reactions; the most important are the photochemical reaction, radiationless (thermal) transition, and fluorescence emission, characterized by their respective rate constants K_P, K_D and K_F. At room temperature, most

---

Abbreviations: DCMU, 3-(3,4-dichlorophenyl)-1,1-dimethylurea; F_M, maximum fluorescence; F_O, initial fluorescence; $F_V = F_M - F_O$, maximum variable fluorescence; LHC II, light harvesting chlorophyll protein complex of photosystem II; PS, photosystem; PQ, plastoquinone; Q_A, Q_B, primary and secondary quinone-type electron acceptors of photosystem II.

H. K. Lichtenthaler (ed.) Applications of Chlorophyll Fluorescence, 3–11.

of the chlorophyll fluorescence emission originates from PS II. Only in the long-wavelenth region (around 730 nm), PS I emission seems to contribute significantly to the initial fluorescence, F_O (Holzwarth, personal communication). There is general agreement that fluorescence is emitted by the antenna pigments only, not from the reaction centers. Excitation energy that reaches the pigment P_{680} of PS II can 1) initiate the primary electron donation, 2) be converted to heat and 3) migrate back to the antennae, where fluorescence emission can take place. In contrast, there seems to be no back transfer of excitons from P_{700} in the reaction center of PS I.

In PS II, a further way for de-excitation of chlorophyll is the energy transfer to mostly non-fluorescent pigments of PS I. The fluorescence yield of PS II therefore is

$$\Phi_F = \frac{F}{I_a} = \frac{k_F}{k_F + k_D + k_T + k_P} \qquad \text{(equation 1)}$$

where F is the emitted fluorescence, I_a the absorbed light flux and K_T the rate constant of excitation energy transfer to PS I. However, this equation applies only, when all reaction centers are in the active, 'open' state. In the primary photochemical reactions, an electron, which is ultimately derived from oxidation of water, is transferred via the secondary donor Z, the pigment P_{680}, the primary acceptor pheophytin (Pheo) to the quinone-type acceptor Q_A (equation 2).

$$\begin{array}{ll} Z - P_{680} - \text{Pheo} - Q_A - Q_B & \text{(open)} \\ \quad\quad h\nu \downarrow & \\ Z - P^{*}_{680} - \text{Pheo} - Q_A - Q_B & \\ \quad\quad e^{\ominus} \downarrow & \text{(equation 2)} \\ Z - P_{680} - \text{Pheo} - Q_A^{\ominus} - Q_B & \text{(closed)} \\ \quad\quad \downarrow & \\ Z - P_{680} - \text{Pheo} - Q_A - Q_B^{\ominus} & \text{(open)} \end{array}$$

All components noted in equation 2 are known to be integral parts of the reaction center (Satoh 1988). In the state of reduced Q_A, the reaction center is 'closed', as no further 'useful' photochemical reaction can take place ($K_P = 0$).

It has been proposed by Klimov and Krasnovskii (1981) that light can induce a charge separation (forming P^{+}_{680} - Pheo^{-}) in the closed center, followed by fast recombination of this pair. However, because of the electric field formed by Q_A^{-}, the quantum yield of this process appears to be low (Schatz et al. 1987). Transfer of the electron from Q_A^{-} to the secondary quinone-type electron acceptor Q_B again opens the center.

If the fraction of open reaction centers is A and the fraction of closed centers 1-A, the fluorescence yield of PS II is

$$\Phi_F = A \cdot \frac{k_F}{k_F + k_D + k_T + k_P} + (1-A) \cdot \frac{k_F}{k_F + k_D + k_T} \qquad \text{(equation 3)}$$

(open units) (closed units)

When all centers are open (A = 1), a minimal fluorescence yield, Φ_{FO}, will be obtained. Maximum yield, Φ_{FM}, is observed, when all centers are closed (A = 0) (Butler and Kitajima 1975). Using equation 3, the potential yield of the photochemical reaction, Φ_{PO}, can be calculated as

$$\Phi_{P_O} = \frac{\Phi_{F_M} - \Phi_{F_O}}{\Phi_{F_M}} = \frac{F_V}{F_M} \qquad \text{(equation 4)}$$

where F_M is the maximal total fluorescence and $F_V = F_M - F_O$ the maximal variable fluorescence of the experimental signal. The F_V/F_M ratio (measured at 77K) has been found to be very uniform (0.832 ± 0.004) among leaves of many vascular plant species and ecotypes (Björkman and Demmig 1987), which indicates a constant and high potential efficiency of the PS II primary reaction (i.e., $K_P >> K_F + K_D + K_T$). Environmental stress factors that affect PS II cause a decrease in the F_V/F_M ratio (see below).

FLUORESCENCE RISE

When leaves, chloroplast or green algae are illuminated in the state of open reaction centers (A = 1), the closure of centers by reduction of Q_A in the primary photochemical reactions causes a rise in fluorescence emission from F_O to a 'peak', F_P, as depicted in Fig. 1 (Duysens and Sweers 1963; see equ.3). Since electrons are transferred from Q_A^- via Q_B to the plastoquinone (PQ) pool, the fluorescence rise reflects the reduction of these electron carriers. The fraction of reduced PQ is proportional to the area increment above the induction curve. In high light, F_P is equal to F_M, i.e., practically all Q_A becomes reduced. In the presence of the inhibitor 3-(3,4-dichlorophenyl)-1, 1-dimethylurea (DCMU), which blocks electron transport from Q_A to Q_B, the induction signal is simpler, as it reflects reduction of Q_A only (Fig. 1). In a given flux of exciting light then the fluorescence rise is strongly enhanced. The signal measured in the presence of DCMU can be used to analyse the reaction kinetics of Q_A reduction. Such analyses have revealed the existence of two kinetically different types of PS II, termed PS II_α and PS II_β (see Melis, this volume).

FLUORESCENCE AT LOW TEMPERATURE

Like the fluorescence rise at 20°C in the presence of DCMU, fluorescence induction at low temperatures, conveniently recorded at 77K, reflects photochemical reduction of Q_A only (Butler 1977), as other electron transport reactions are very slow. Measurements at 77K are applied for routine determinations of the F_V/F_M ratio (Björkman and Demmig 1987). In contrast to 20°C fluorescence, emission spectra at 77K exhibit distinct bands of PS I (at about 735nm) and PS II fluorescence (at about 695 and 685 nm). PS I fluorescence, however, is normally not influenced by photochemical events taking place in

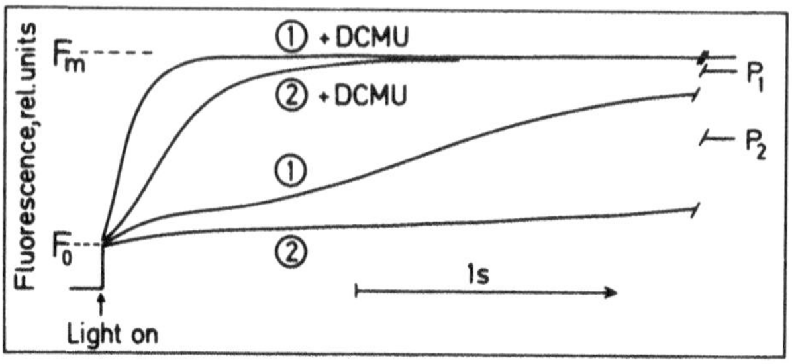

FIGURE 1. Fluorescence induction signals of isolated chloroplasts recorded at 20°C in the absence and presence of DCMU. Exciting light, 2.2 W m^{-2} (1) or 0.7 W m^{-2} (2).

the PS I reaction center. Variable fluorescence observed in the PS I band results from exciton transfer from PS II to PS I ('spillover') and reflects closure of PS II reaction centers (Butler 1977).

Fluorescence spectroscopy at 77K is frequently applied to study excitation energy distribution between PS II and PS I. However, care should be taken to consider distortion of spectra due to reabsorption of fluorescence by chlorophyll. Reabsorption is increased by light scattering effects. As predominantly PS II fluorescence (F_{695} and F_{685}) is reabsorbed, measured PS II band peaks are lowered when the chlorophyll concentration and light scattering are high. This usually applies to frozen leaves, which yield extremely low PS II emission, as compared to dilute chloroplast suspensions. Distortion of spectra from leaf material can be avoided by preparing a 'diluted leaf powder' (Weis 1985). A review on low-temperature fluorescence is given by Šestak in this volume. It should be noted that also at room temperature, any change in the physical state of a leaf may affect the ratio between short and long-wavelength emission.

FLUORESCENCE QUENCHING

The fluorescence rise in the induction signal at room temperature is followed by a fluorescence decline. This may occur in different phases, usually leading within one to several minutes to a low fluorescence yield in the steady state. According to equations 2 and 3, transfer of electrons from Q_A^- to Q_B (and the PQ pool), which re-opens the PS II centers, lowers the fluorescence yield. However, as already recognized by Duysens and Sweers (1963), the fluorescence decline can only partly be explained by reoxidation of Q_A^-. Today, besides this 'photochemical' quenching, various ways of 'non-photochemical' quenching, not related to Q_A^- reoxidation, are known. The extent of lowering of the fluorescence yield can be expressed by quenching coefficients ($0 \leq q \leq 1$), indicating the quenched proportion of maximal variable fluorescence F_V.

The following mechanisms of the decline in fluorescence yield can be distinguished:

1) Photochemical quenching (q_Q), related to the proportin of excitation energy 'trapped' by open centers. (Due to energy transfer between PS II units, q_Q is not directly proportional to oxidized Q_A.)

2) Non-photochemical quenching (q_N).

a) Energy-dependent quenching (q_E), related to the light-induced proton gradient across the thylakoid membrane.

b) Quenching by decrease of excitation energy distribution to PS II (q_T), related to phosphorylation of the light harvesting complex of PS II (LHC II).

c) Quenching related to photoinhibition of photosynthesis (q_I), possibly caused by transformation of PS II reaction centers.

d) Quenching by decrease in cation (Mg^{2+}) concentration, causing increased excitation energy transfer to PS I.

In intact systems, under most conditions, photochemical (q_Q) and energy-dependent quenching (q_E) are the major components of the fluorescence decline. In isolated chloroplasts or protoplasts, resolution of these components is possible by addition of DCMU (Krause et al. 1982, Horton and Hague 1988). This leads to immediate closure of PS II centers, causing fast reversion of q_Q, whereas a slower phase of relaxation indicates reversion of q_E. By use of the pulse amplitude modulation method (Schreiber et al. 1986), resolution of quenching is made possible also for intact leaves by application of light pulses that saturate Q_A reduction. The different mechanisms of non-photochemical quenching, q_E, q_T and q_I, (2a-c) may be distinguished by their different kinetics (Horton and Hague 1988) and will be briefly reviewed below. The cation effect on fluorescence (2d) does not seem to be of physiological significance for intact systems. However, its elaboration has considerably deepened our understanding of the organization and functioning of the photosynthetic apparatus (see Briantais et al. 1986).

Energy-dependent quenching (q_E)

It has been shown that the extent of q_E is linearly related to the intrathylakoid proton concentration (Briantais et al. 1979, Laasch 1987). In high light, when a high proton gradient is built up across the thylakoid membrane, about 80% of variable fluorescence can be quenched ($q_E = 0.8$). Low-temperature fluorescence analyses of this 'high energy state' have indicated (Krause et al. 1983) that the quenching is based on an increase in the rate constant of thermal deactivation of PS II, K_D. It is hypothesized that by low intrathylakoid pH, a structural change of unknown nature is induced that lowers the photochemical efficiency of PS II and transforms the trapped excitation energy to heat. Such mechanism is thought to function in a regulated manner, serving for protection against damaging effects of excess excitation energy (Krause and Laasch 1987, Weis and Berry 1987, Krause et al. 1988; see also Weis and Lechtenberg, this volume). As q_E responds to changes in the proton gradient within seconds, a fast response of the photosynthetic system to altered light conditions is possible. It has been experimentally shown that inactivation of isolated chloroplasts by high light (photoinhibition) is diminished in the presence of q_E. Antimycin A, which inhibits q_E without affecting the proton gradient (Oxborough and Horton 1987), causes increased photoinhibition in chloroplasts (Krause et al. 1988).

Quenching related to phosphorylation of the LHC II (q_T)

Phosphorylation of the LHC II has been postulated to balance excitation energy distribution between the two photosystems (for a review see Briantais et al. 1986). The phosphorylated LHC II apparently is detached from PS II and migrates from the grana regions to the stroma lamellae of the thylakoids. This diminishes excitation of PS II relative to PS I, and thereby lowers the fluorescence yield. It has been proposed that phosphorylated LHC II transfers excitons to PS I, but this concept is still disputed (Deng and Melis 1986). The activity of the protein kinase responsible for the phosphorylation is controlled by the redox state of the PQ pool and by the ΔpH across the thylakoid membrane (Fernyhough et al. 1984). The contribution of this mechanism to fluorescence quenching is relatively small ($q_{T\ max} \approx 0.2$), and its kinetics is slower than that of q_E (Krause and Behrend 1983, Horton and Hague 1988).

Quenching related to photoinhibition (q_I)

Photoinhibition of photosynthesis, occurring in excessive irradiation, is related to a quenching of variable fluorescence that develops within minutes to hours, depending on the PFD and acclimation state of the photosynthetic apparatus. Photoinhibition is promoted by various environmental stress factors that impair the normal utilization of photosynthetic energy in carbon metabolism. The quenching is expressed as a decrease in F_V/F_M in the induction signal recorded after a dark period of several minutes, following exposure to high light. The characteristics of 77K fluorescence show, like for q_E, that the q_I mechanism is based on an increased rate constant of thermal deactivation, K_D (Powles and Björkman 1982, Ögren and Öquist 1984, Barényi and Krause 1985, Demmig and Björkman 1987). This is in agreement with the hypothesis that photoinhibition is related to a transformation of PS II reaction centers to 'quenchers' (Cleland and Critchley 1985, Cleland et al. 1986). Such altered centers would still trap excitation energy with high efficiency, but are less capable of a photochemical reaction and dissipate an increased part of the energy as heat. Another hypothesis relates photoinhibition to primary inactivation and following degradation of the Q_B-binding protein (see Kyle 1987). Both hypotheses may find a common basis, as it is now known that the Q_B protein is an integral part of the reaction center (see Satoh 1988). An increase of thermal deexcitation of PS II antennae has also been considered as a cause of quenching (Demmig and Björkman 1987), and a correlation with reactions of the xanthophyll cycle (zeaxanthin formation from violaxanthin) was found (Demmig et al. 1987).

Photoinhibition and related fluorescence changes are partially or fully reversible within minutes to hours upon return to favourable conditions. Thus, at least the reversible photoinhibitory quenching should not be viewed as a damage but rather as a protective mechanism that allows for thermal dissipation of excess energy, in addition to the q_E effect. As discussed by Somersalo and Krause (this volume), reversible photoinhibition can be interpreted as a relatively fast response of the photosynthetic apparatus to changing light conditions that precedes long-term acclimation processes.

CONCLUSIONS

The relationship of chlorophyll fluorescence emission to the overall process of photosynthesis is indicated in the scheme of Figure 2. Light absorbed by the two photosystems can partly be transformed to heat. A small fraction of the energy (about 0.5 to 3%) is re-emitted in form of fluorescence, which may serve as a probe for various events that alter the pathway of energy conversion. Under optimal conditions in limiting light, most absorbed photons are utilized to drive electron transport, which induces the build-up of a ΔpH. Phosphorylation of the LHC II might optimize the energy distribution between PS II and I. The energy-rich products NADPH and ATP flow into carbon metabolism.

Conditions of excess light and limiting carbon metabolism probably induce regulatoy processes that increase energy conversion to heat in a controlled, non-destructive manner: An increase in the ΔpH, caused by limited utilization of ATP, would evoke the q_E mechanism. In addition, when q_E becomes light-saturated and then is insufficient for thermal deactivation, PS II reaction centers may be transformed to 'photoinhibited' centers that increase dissipation of excess energy. It should be noted that a controlled conversion of excitation energy can also be assumed for PS I (see Weis, this volume). De-excitation of PS II and 'photosynthetic control' via a high ΔpH would limit electron transfer to P_{700}. As a consequence its oxidized form, P^+_{700}, which is known to convert trapped excitation energy to heat, accumulates in high light.

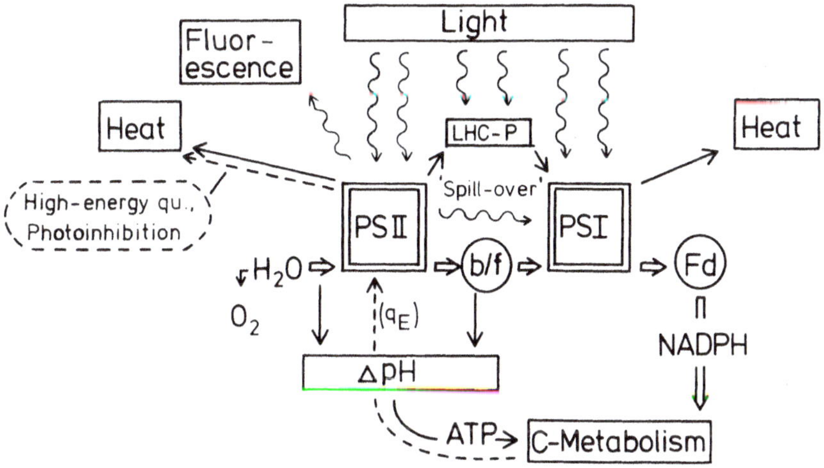

FIGURE 2. Scheme illustrating the relationships between fluorescence emission and other pathways of energy conversion in the photosynthetic apparatus.

REFERENCES

Barényi B and Krause GH 1985. Inhibition of photosynthetic reactions by light. A study with isolated spinach chloroplasts. Planta 163: 218-226.

Björkman O and Demmig B 1987: Photon yield of O_2 evolution and chlorophyll fluorescence characteristics at 77K among vascular plants of diverse origins. Planta 170: 489-504.

Briantais J-M, Vernotte C, Picaud M and Krause GH 1979. A quantitative study of the slow decline of chlorophyll a fluorescence in isolated chloroplasts. Biochim. Biophys. Acta 548: 128-138.

Briantais J-M, Vernotte C, Krause GH and Weis E 1986. Chlorophyll a fluorescence of higher plants: chloroplasts and leaves. In: Govindjee, Amesz J and Fork DJ. Light emission by plants and bacteria. Academic Press, New York.

Butler WL 1977. Chlorophyll fluorescence: a probe for electron transport and energy transfer. In: Trebst A and Avron M. Encyclopedia of plant physiology, new series vol. 5. Springer, Berlin.

Butler WL and Kitajima M 1975. Fluorescence quenching in photosystem II of chloroplasts. Biochim. Biophys. Acta 376: 116-125.

Cleland RE and Chritchley C 1985. Studies on the mechanism of photoinhibition in higher plants. II. Inactivation by high light of photosystem II in thylakoids and O_2 evolving particles. Photobiochem. Photobiophys. 10: 83-92.

Cleland RE, Melis A and Neale PJ 1986. Mechanism of photoinhibition: Photochemical reaction center inactivation in system II of chloroplasts. Photosynth. Res. 9: 79-88.

Demmig B and Björkman O 1987. Comparison of the effect of excessive light on chlorophyll fluorescence (77K) and photon yield of O_2 evolution of leaves of higher plants. Planta 171: 171-184.

Demmig B, Winter G, Krüger A and Czygan F-C 1987. Photoinhibition and zeaxanthin formation in intact leaves. Plant Physiol. 84: 218-224.

Deng X and Melis A 1986. Phosphorylation of the light-harvesting complex II in higher plant chloroplasts: effect on photosystem II and photosystem I absorption cross section. Photobiochem. Photobiophys. 13: 41-52.

Duysens LNM and Sweers HE 1963. Mechanism of the two photochemical reactions in algae as studied by means of fluorescence. In: Jap. Soc. of Physiol. Studies on microalgae and photosynthetic bacteria. Univ. of Tokyo Press, Tokyo.

Fernyhough P, Foyer CH and Horton P 1984. Increase in the level of thylakoid protein phosphorylation in maize mesophyll chloroplasts by decrease in the transthylakoid pH gradient. FEBS Lett. 176: 133-138.

Horton P and Hague A 1988. Studies on the induction of chlorophyll fluorescence in isolated barley protoplasts. IV. Resolution of non-photochemical quenching. Biochim. Biophys. Acta 932: 107-115.

Klimov VV and Krasnovskii AA 1981. Pheophytin as a primary electron acceptor in photosystem II reaction center. Photosynthetica 15: 592-609.

Krause GH and Behrend U 1986: ΔpH-dependent chlorophyll fluorescence quenching indicating a mechanism of protection against photoinhibition of chloroplasts. FEBS Lett. 200: 298-302.

Krause GH, Briantais J-M and Vernotte C 1983. Characterization of chlorophyll fluorescence quenching in chloroplasts by fluorescence spectroscopy at 77K. I. ΔpH-dependent quenching. Biochim. Biophys. Acta 723: 169-175.

Krause GH, Briantais J-M and Vernotte C 1982. Photoinduced quenching of chlorophyll fluorescence in intact chloroplasts and algae. Resolution into two components. Biochim. Biophys. Acta 679: 116-124.

Krause GH and Laasch H 1987. Energy-dependent chlorophyll fluorescence quenching in chloroplasts correlated with quantum yield of photosynthesis. Z. Naturforsch. 42c: 581-584.

Krause GH, Laasch H and Weis E 1988. Regulation of thermal dissipation of absorbed light energy in chloroplasts indicated by energy-dependent fluorescence quenching. Plant Physiol. Biochem. 26 (in press).

Krause GH and Weis E 1984. Chlorophyll fluorescence as a tool in plant physiology. II. Interpretation of fluorescence signals. Photosynth. Res. 5: 139-157.

Kyle DJ 1987. The biochemical basis for photoinhibition of photosystem II. In: Kyle DJ, Osmond CB and Arntzen CJ. Photoinhibition. Topics of photosynthesis, vol 9. Elsevier, Amsterdam.

Laasch H 1987. Non-photochemical quenching of chlorophyll a fluorescence in isolated chloroplasts under conditions of stressed photosynthesis. Planta 171: 220-226.

Ögren E and Öquist G 1984. Photoinhibition of photosynthesis in Lemna gibba as induced by the interaction between light and temperature. III. Chlorophyll fluorescence at 77K. Physiol. Plant. 62: 193-200.

Oxborough K and Horton P 1987. Characterisation of the effects of Antimycin A upon high energy state quenching of chlorophyll fluorescence (qE) in spinach and pea chloroplasts. Photosynth. Res. 12: 119-128.

Powles SB and Björkman 1982. Photoinhibition of photosynthesis: effect on chlorophyll fluorescence at 77K in intact leaves and in chloroplast membranes of Nerium oleander. Planta 165: 97-107.

Satoh K 1988. Reality of P-680 chlorophyll protein. -Identification of primary photochemistry in oxygenic photosynthesis. Physiol. Plant. 72: 209-212.

Schatz GH, Brock H and Holzwarth AR 1987. Picosecond kinetics of fluorescence and absorbance changes in photosystem II particles excited at low photon density. Proc. Natl. Acad. Sci. 84: 8414-8418.

Schreiber U, Schliwa U and Bilger W 1986. Continuous recording of photochemical and non-photochemical chlorophyll fluorescence quenching with a new type of modulation fluorometer. Photosynth. Res. 10: 51-62.

Weis E. 1985. Chlorophyll fluorescence at 77K in intact leaves: Characterization of a technique to eliminate artifacts related to self-absorption. Photosynth. Res. 6: 73-76.

Weis E and Berry JA 1987. Quantum efficiency of photosystem II in relation to 'energy'-dependent quenching of chlorophyll fluorescence. Biochim. Biophys. Acta 894: 198-207.

SOME ASPECTS OF THE RELATIONSHIP BETWEEN CHLOROPHYLL *a* FLUORESCENCE AND PHOTOSYNTHETIC CARBON ASSIMILATION.

D. A. Walker, Research Institute for Photosynthesis, University of Sheffield, Sheffield, S10 2TN, United Kingdom.

INTRODUCTION

When a photon strikes a chlorophyll molecule it becomes excited, an electron is raised from a ground state in the molecular orbital to an excited state. Blue photons, carrying higher energy, raise electrons to the excited singlet state 2, red photons raise electrons to excited state 1. These of course are the photons which are preferentially absorbed. Green photons are mostly not absorbed and it is the green light which is reflected (from and transmitted through) leaves which gives them their colour. Excited state 2 decays very rapidly to excited state 1 as electrons cascade down the energy gradient, losing energy by radiationless de-excitation. None of this energy is available for photosynthesis. It is the red, more long-lived, excited state created directly by a red photon (or indirectly by a blue photon) which is the starting point of photosynthesis and, therefore, of most of biological energy transduction. It is from here that electrons are transported through the two photosystems to reduce NADP and finally CO_2. In addition the creation of the red excited state is accompanied by the generation of the massive oxidising potential, the positively charged "holes" within PSII, which are refilled by electrons drawn from water. Otto Warburg believed photosynthetic energy transduction to be perfect. It is not. Oxidation is inevitably associated with reduction. Electrons which have been raised to an excited state will fall back into the ground state if there is no acceptor available. Much of the energy dissipated in these circumstances is dissipated as heat. A fraction, released as electrons drop more or less directly back to the ground state, brings about the emission of photons. This is fluorescence (Lavorell and Etienne 1977). Even this is not an entirely perfect process of energy transduction. The photons resulting from immediate de-excitation of the red excited state of chlorophyll *a* (the precise mechanism is still uncertain) are also red but, because some of the excitation energy is lost in other ways, fluorescence is longer wavelength red light (the Stokes Shift) than that which brings about excitation at peak absorption. Some fluorescence is immediately re-absorbed. Only a very small part of the light energy which enters the leaf is re-emitted in this way. Nevertheless this is the source of the inverse relationship which makes it such a useful signal (Walker 1988).

The broad inverse relationship which exists between chlorophyll *a* fluorescence and photosynthetic carbon assimilation has been recognised for many years. It was implicit in the early concept of "gainful employment". This embodied the notion that chlorophyll excitation energy was used to drive photosynthesis and that energy which was not, for any reason, used in this process, was re-emitted as fluorescence. In fact, this is only part of a larger story. The inverse relationship clearly holds but the relationship is complex. Excitation energy must be dissipated. What is not "gainfully" used in photosynthesis must be dissipated through other channels. Chlorophyll fluorescence, however, is only one minor channel of energy dissipation. Moreover, there are many factors, mostly still imperfectly understood, which can switch dissipation from one channel to another. In the present context, possibly the most important of these is thermal dissipation. In certain circumstances, it is clear that changes in the molecular environment within and about the thylakoid membrane can switch dissipation backwards and forwards between light and heat.

H. K. Lichtenthaler (ed.) Applications of Chlorophyll Fluorescence, 13–20.

We can begin to understand the complexity and appreciate the exciting features of the fluorescence signal (Horton 1985, Krause and Weis 1984) if we look at changes in fluorescence occasioned by experimental manipulation of the gas phase surrounding a leaf. Apart from anything else it is only then that we can begin to appreciate the immediacy of the interaction between carbon assimilation and fluorescence. Perhaps the simplest and most quoted example is the gas transient which occurs in going from to air to carbon-dioxide free air and back to air again. This is illustrated in Figure 1 (Walker 1988).

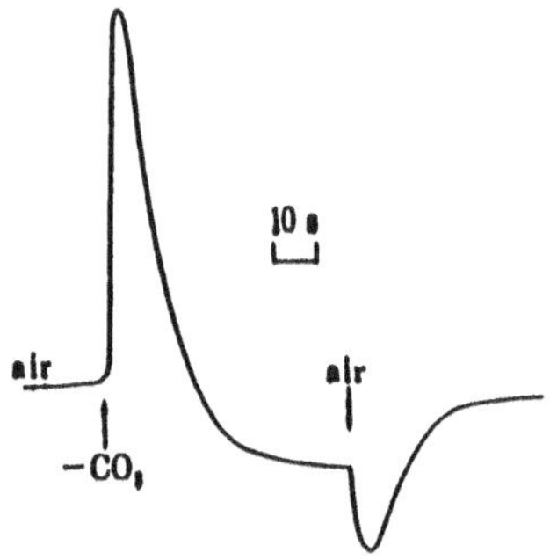

A "gas-transient"

Figure 1.

On an appropriate time scale for a consideration of carbon assimilation (i.e. seconds and minutes rather than milli or microseconds) fluorescence rises abruptly from the steady-state as soon as CO_2 is withdrawn. Within seconds, however, this abrupt increase is overtaken by a decline so that within 30 seconds, or so, the signal has fallen below the steady-state. A similar, but not identical, change in signal accompanies the restoration of CO_2 to the gas-phase.

This "gas transient" is interpreted as follows. Linear photosynthetic electron transport moves electrons from water through the two photosystems to ferredoxin (Figure 2). From ferredoxin these are passed to CO_2, via NADP, or to O_2 (the Mehler reaction or

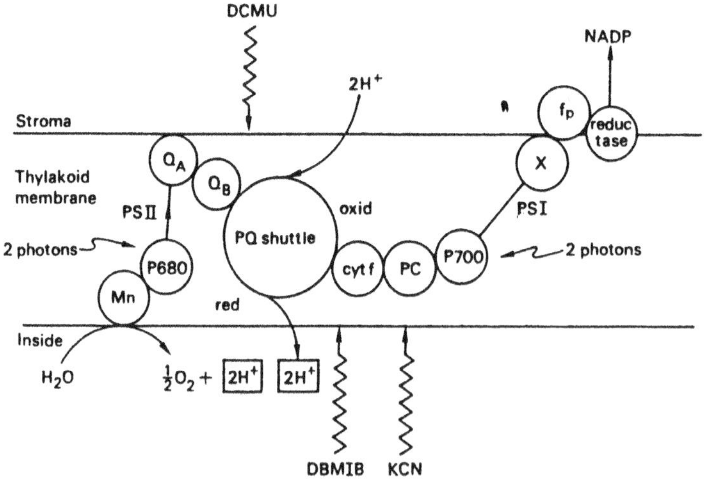

Figure 2. Organisation of electron carriers in the thylakoid membrane. Note the inward movement of protons from the stroma. The proton gradient, discharging through the ATPase (Figure 3) drives ATP formation from ADP and Pi. (Walker 1988)

pseudocyclic electron transport). If CO_2 is removed, NADP becomes rapidly reduced and the lack of NADP as a terminal electron acceptor affects other carriers in the chain. Once Q_A (the primary electron acceptor in PSII) is reduced (i.e. once its open electron traps are closed) it no longer "quenches" fluorescence and the fluorescence signal rises. Although Q_A is a quinone "Q" stands for "quencher" in this context. This was the name that Duysen and Swears (1963) first applied to a then unknown component in PSII which "quenches" fluorescence when it is oxidised but not when it is reduced. Such quenching is now referred to as q_Q, or photochemical quenching. If this were the only quenching mechanism the inverse relationship between fluorescence and PCA would be less complex than it is. It is immediately clear, however, that there must be other quenching mechanisms because the rise in fluorescence which follows removal of CO_2 is overtaken by a decline. In fact this decline does not threaten the concept of gainful employment. It is simply that the flux of energy dissipated through thermal channels is increased. Excitation energy which is not being used in PCA is always dissipated as heat as well as light (fluorescence); now the thermal fraction increases. This is a form of "non-photochemical" quenching and it is designated "q_E" because it is associated with the so called "high energy state" of the thylakoid membrane (Krause and Weis 1984). In the present context it is believed to indicate the fall in ATP consumption consequent upon the cessation of CO_2 fixation. In short, as CO_2 is withdrawn and PCA ceases, ATP is not longer consumed in this process and energy dissipation is switched from dissipation as fluorescence into thermal channels. The underlying mechanisms are still poorly understood. One aspect seems reasonably straightforward. During PET, protons are picked up at the outer side of the thylakoid membrane and put down at the inner side, within the thylakoid lumen. This is largely because of the chemical nature of plastoquinone, a major electron carrier between the two photosystems and the way in which it is organised in the membrane (Figure 3).

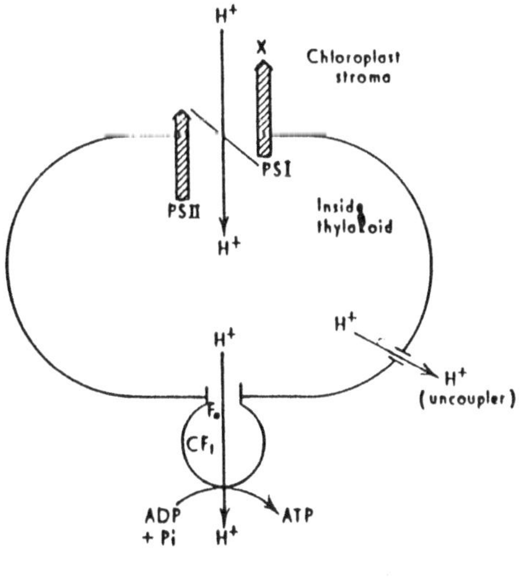

Figure 3. Photosynthetic electron transport linked to proton transfer into the thylakoid compartment. The proton gradient across the thylakoid membrane can be dissipated by transport through the coupling-factor (CFi), which facilitates ATP synthesis, or artificially by uncouplers. As the proton gradient builds up q_E increases (Walker 1988).

Chlorophyll excitation initiates electron transport. Plastoquinone offered an electron by PSII must simultaneously acquire a proton in order to proceed to the reduced state (plastoquinol). This proton is derived from the stroma of the chloroplast. The next carrier in the PET chain, cytochrome f, can accept the electron but not the proton which is discharged into the lumen. Accordingly, as electrons pass through the PET chain there is a consequent lateral (vectorial) movement of protons from the stroma, which becomes more alkaline, to the lumen, which becomes more acid. It is this ΔpH or proton gradient which drives the esterification of ADP as the protons discharge through the ATPase (Figure 3). If ATP consumption is constrained (as it is when CO_2 is withdrawn from the gas phase surrounding the leaf) ADP will no longer be available to discharge the proton gradient. Hence as long as electron transport persists (cyclic flow or Mehler reaction, or both) the ΔpH (the "high-energy" state) will increase. As it does, the thermal channel of dissipation opens wider and fluorescence is quenched (q_E quenching). What exactly opens the channel remains obscure. The establishment of a proton gradient involves swelling of the thylakoid compartment and protonation of the membrane and these structural and charge-related changes may be causal. (It should be noted, however, that they are *not invariably* related and that antimycin A can inhibit q_E without affecting ΔpH.)

For the present then we must accept a great many uncertainties and proceed with caution bearing in mind that there are also a number of additional quenching mechanisms (Horton 1985). We are still left, however, with a reasonable framework in which to begin to interpret fluorescence signals (Sivak *et.al*. 1984,5,6 and Walker 1988). As we have seen this working hypothesis offers an adequate explanation of the air to CO_2-free air transient. Removal of CO_2 causes reduction of NADP and Q_A. Accordingly q_Q relaxes and fluorescence rises. At the same time, ATP consumption is drastically diminished, ADP is no longer readily available to discharge the proton gradient, ΔpH increases and the initial rise in fluorescence is overtaken by a decline. The physical changes in the thylakoid membrane referred to above are also related to changes in light-scattering (at about 535nm) and have been used by Ulrich Heber (Heber 1969) as an indicator of ΔpH. The light-scattering signal during the air to CO_2-free air gas transient is consistent with the interpretation given here since it suggests a simple rise in ΔpH following removal of CO_2 and a corresponding decline following restoration of CO_2 (Figure 4)

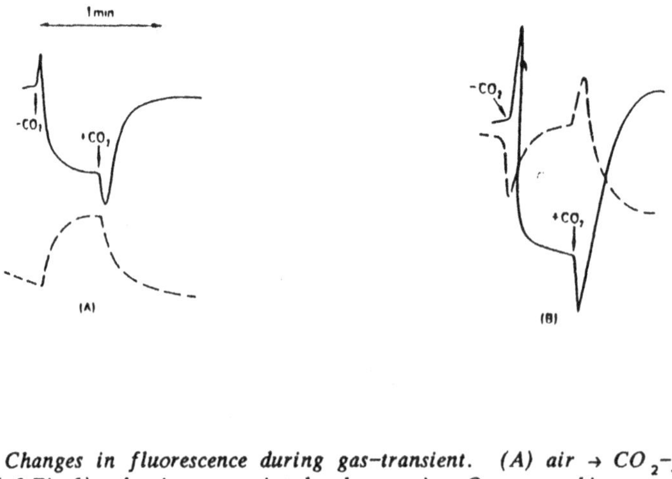

Figure 4. Changes in fluorescence during gas-transient. (A) air → CO_2-free air → air transient (cf.Fig.1) showing associated change in Q_E quenching as indicated by light-scattering. (B) as for (A) but involving higher CO_2 concentration and therefore more complex kinetics (after Walker 1988).

The complex nature of the relationship immediately becomes apparent, however, if the corresponding transient to and from higher CO_2 concentration is examined. The fluorescence excursion is superficially identical but the light-scattering displays an initial dip before it rises and *vice-versa*. It is not difficult to accommodate this dip within the hypothesis. The proton gradient is largely created by non-linear PET to NADP and if, in high CO_2, there is a larger pool of PGA (constituting a sink for ATP) it is likely that initial discharge of the proton gradient will exceed build-up. We are nevertheless immediately reminded that fluorescence is a rich, complex and ambiguous signal since it has at least two major quenching components which can move rapidly in either direction and act either in concert or in an anti-parallel fashion. Techniques (which are discussed elsewhere in these proceedings) now exist which allow these signals to be "de-convoluted" (Schreiber *et.al.* 1986). Figure 5 shows another gas-transient which is still most easily explained in terms of q_Q and q_E, as above. In this experiment each second gas transient involved a decrease in O_2 to 2% CO_2. This decrease in the availability of the alternative electron acceptor (oxygen) gives rise to an increase in the height of the fluorescence excursion. This can be experimentally diminished, in some circumstances, by feeding nitrate. This serves, to a small extent, as a replacement for CO_2 or O_2.

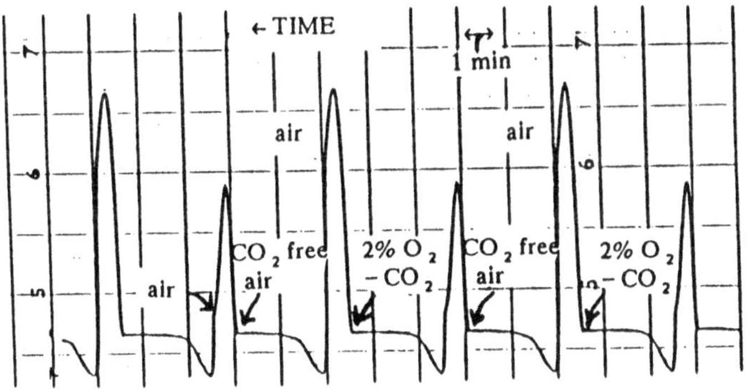

Figure 5. "Gas-transients" in spinach. This is a facsimile of a chart showing fluorescence transients inspinach induced by changes in the gas-phase. Air to CO_2-free air transients alternate with similar transients in which the O_2 was dropped from 21% to 2% at the same moment as the CO_2 was removed (Walker 1988).

THE M-PEAK

In Figure 6 the slower changes which follow re-illumination after a period of darkness, include a decline from a peak (P) to a terminal (T) value via a secondary maximum (M). Perhaps inevitably, this "M-peak" has meant different things to different observers. One approach (Horton, 1985) is to describe any secondary maximum, or maxima (because there may be several) as "M-peaks". This has the virtue of simplicity but it must be immediately emphasised that it can create a pitfall for the unwary if it is assumed that all "M peaks" are causally related in precisely the same way. A comparison of A and B in Figure 4 shows that fluorescence signals which are superficially very similar may mask underlying components which are quite different and the same may be true of "M-peaks" if they are not additionally defined. For example, re-illumination after a dark interval may give rise to a "CO_2-gulp" and an associated "O_2 burst" prior to the termination of induction. In some circumstances, the termination of induction may be followed by oscillations (see below). All three events (the gulp and burst, the termination of induction and oscillations) may give rise to secondary maxima or "M-peaks" but the underlying causes, the relative contributions of q_Q and q_E and their timing may differ substantially.

Figure 6 illustrates an "M-peak" of the sort which may sometimes be observed when a leaf is abruptly re-illuminated with low intensity light following a period of darkness long enough to permit appreciable depletion of metabolites and some dark-deactivation of enzymes but not so long that the leaf is fully "dark-adapted". This sort of "M-peak" was attributed (Walker, 1981) to the termination of induction (i.e. the onset of more rapid carbon assimilation after an initial lag). According to this view, Q_A will be rapidly reduced as soon as the leaf is illuminated because, although it must accept electrons from excited chlorophyll, it is denied the possibility of reoxidation by the remainder of the electron transport chain. This is because carbon assimilation has been "switched-off" in the dark and requires time (spent on light-activation of enzymes and building-up metabolites) to get going again. Some CO_2 assimilation will soon occur, however, and electrons will also be passed to O_2 as an alternative acceptor so that Q_A oxidation will become detectable within seconds and there will be a decline in fluorescence from the original peak value as q_Q quenching increases in parallel with O_2 evolution. This decline will be reinforced by q_E quenching as cyclic and linear electron transport lead to the establishment of a proton-gradient across the thylakoid membrane (Figures 2 and 3) which is not yet being discharged at its maximal rate. Such a discharge of protons through the ATPase will increase as the flux of metabolites through the Benson-Calvin cycle increases and ATP consumption by its PGA and Ru5P kinases makes more ADP available for photophosphorylation. In relatively low light, this new drain on the transthylakoid proton gradient will bring about a significant relaxation of q_E quenching (or a significant slowing of the rate of increase of q_E) and cause fluorescence to rise from S (Figure 6) towards M. Almost immediately further changes will combine to reverse this trend. Any decrease in the proton gradient will tend to accelerate electron transport as will the increasing availability of NADP reoxidised in PGA reduction. Associated increases in q_Q will once again push fluorescence downwards from M and q_Q will once again be reinforced by an increasing q_E caused by an increasing proton gradient. In higher light (Sivak *et al*, 1985), the drain on the proton gradient associated with the termination of induction will have a smaller impact on the fluorescence kinetics and the associated M-peak will be smaller.

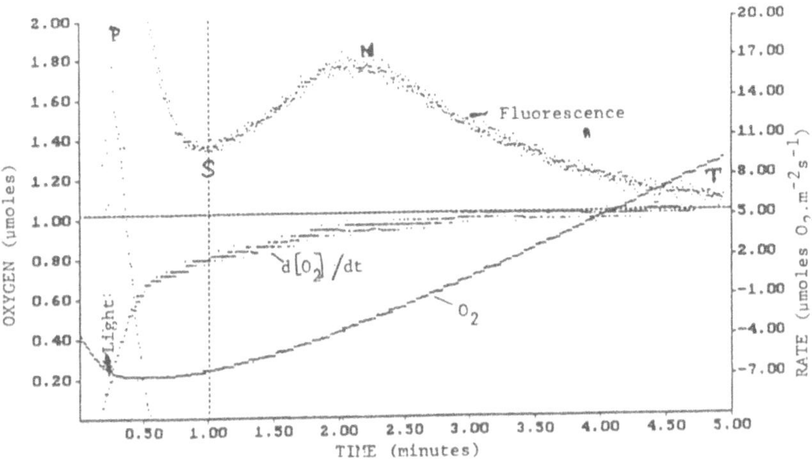

***Figure 6** "M-Peak" fluorescence kinetics following re-illumination, in very low light, after darkness. Induction has been re-established (i.e. the period of darkness has been long enough to allow metabolite pools to be partially depleted and enzymes to be*

inactivated). When the leaf is re-illuminated, therefore, it is not fully prepared for photosynthetic carbon assimilation at maximum rates, and there is an initial lag. As this lag terminates, some of the newly developed proton-gradient is discharged in ATP synthesis, q_E relaxes and fluorescence starts to rise again. As this happens, electron transport to CO_2 causes a simultaneous increase in the proton gradient so that it pushes fluorescence down from its "M-Peak". At higher light intensities (Sivak et al, *1985) the M-Peak is lost or modified - after Walker (1981).*

OSCILLATIONS.

PCA can be induced to oscillate in many, but not all, species provided that certain conditions prevail (Sivak and Walker 1984, Walker and Osmond 1986). High [CO_2], high light and low temperatures favour oscillation and the common feature seems to be orthophosphate (Pi) supply. Provided that cytosolic Pi is approaching limitation (either because conditions such as these have been selected or because it has been experimentally sequestered) almost any perturbation, such as a sudden increase in CO_2 or light, may initiate oscillatory behaviour. The inverse relationship between PCA and fluorescence then becomes immediately apparent but, except at the outset, the fluorescence peaks anticipate the PCA troughs and the fluorescence troughs anticipate the PCA peaks. Once extremely puzzling, the explanation of the anticipation is now based on out-of-phase quenching. It was shown, for example, that light-scattering also anticipates O_2 and if this, like q_E, is associated with changes in the proton gradient it seemed likely that while q_Q would be synchronous with O_2, q_E would come before and that this would also be true of F_v conditioned by the sum of q_Q and q_E quenching. "Q-analysis" (Schreiber *et.al.* 1986). supports this view (Figure 7)

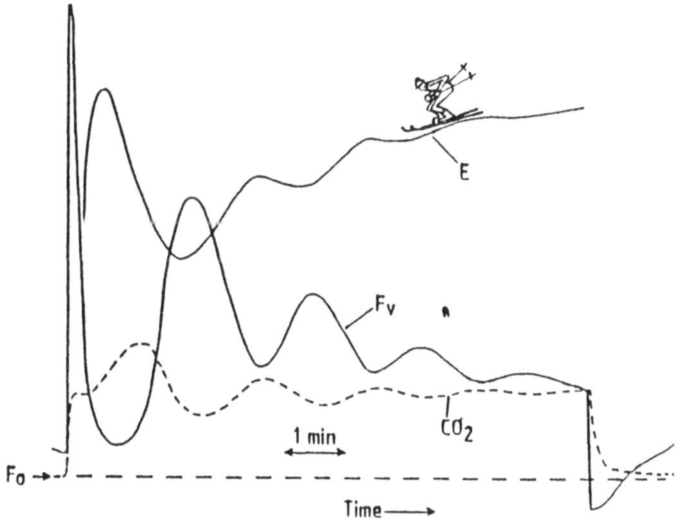

Figure 7. Oscillatory behaviour in spinach showing relationship between Carbon dioxide fixation, fluorescence and energisation.
Oscillations were induced by re-illumination in 0.55% CO_2, 2% O_2. The fall in fluorescence (Fv) anticipates the rise in CO_2 fixation and vice versa. Similarly, the increase in energisation (E), which is largely responsible for the fall in fluorescence anticipates Fv. "E" was defined by q-analysis and is the difference between maximal and variable fluorescence (E=qE.(Fv)m-(Fv)s (Sivak and Walker 1986).

REFERENCES

Duysens, W.L.N.M. and Sweers, H.E. 1963 Mechanism of two photochemical, reactions in algae as studied by means of fluorescence. In: *Studies on Microalgae and Photosynthetic Bacteria* (Jap. Soc. of Plant Physiol., eds) Tokyo, University of Tokyo Press. pp. 353–372.

Horton, P. 1985 Interactions between electron transfer and carbon assimilation. In: *Photosynthetic Mechanisms and the Environment*. (Barber, J. and Baker, N.R., eds). Topics in Photosynthesis Vol. 6. Elsevier Science Pubs., pp. 135–187.

Krause, G.H., and Weis, E. 1984 Review: Chlorophyll fluorescence as a tool in plant physiology. II. Interpretation of fluorescence signals. *Photosynthesis Research, Vol 5*. Martinus Nijhoff/Dr. W. Junk, The Hague, pp 139–157.

Lavorell, J. and Etienne, A.L. 1977 *In vivo* chlorophyll fluorescence. In: *Primary Processes in photosynthesis* (Barber, J., ed). Elsevier/North Holland Biomedical Press, Amsterdam pp 203–268.

Schreiber, U and Bilger, W 1986 Rapid assessment of stress effect on plant leaves by chlorophyll fluorescence measurements. NATO workshop Sesimbra Portugal Oct 1985. In press

Schreiber, U., Schliwa, U. and Bilger, W. 1986 Continuous recording of photochemical and non-photochemical chlorophyll fluorescence quenching with a new type of modulation fluorometer. *Photosynthesis Research 10*, 51–62.

Heber, U. 1969 Conformational changes of chloroplasts induced by illumination of leaves *in vivo*. *Biochim. Biophys. Acta 180*, 302–319.

Sivak, M.N. and Walker, D.A. 1984 New perspectives in the understanding of the regulation of photosynthesis and its relation to chlorophyll fluorescence kinetics through the study of oscillations. In: *Oscillations in physiological systems: Dynamics and control* (Proc. Symp., Oxford, September 1984) Inst. of Measurement and control, London. pp 91–96.

Sivak, M.N. and Walker, D.A. 1986 Photosynthesis *in vivo* can be limited by phosphate supply. *New Phytol. 102*, 499–512

Sivak, M.N., Heber, U. and Walker, D.A. 1985 Chlorophyll *a* fluorescence and light-scattering kinetics displayed by leaves during induction of photosynthesis. *Planta 163*, 419–423.

Sivak, M.N., Lea, P.J. and Walker, D.A. 1986 New developments in the measurement of photosynthesis *in vivo*. Implications for plant genetic engineering. *Biochemical. Soc. Trans. 14*, 63–64.

Walker, D.A. 1981 Secondary fluorescence kinetics of spinach leaves in relation to the onset of photosynthetic carbon assimilation. *Planta 153*, 273–278.

Walker, D.A. and Osmond, C.B. 1986 Measurement of photosynthesis *in vivo* using a leaf disc electrode: Correlations between light dependence of steady-state photosynthetic O_2 evolution and chlorophyll *a* fluorescence transients. *Proc. Roy. Soc.* B *227*, 267–280.

Walker, D.A. 1988 The use of the oxygen electrode and fluorescence probes in simple measurements of photosynthesis. Revised edition. Oxygraphics Limited, Sheffield. pp1–188

TIME RESOLVED CHLOROPHYLL FLUORESCENCE
What kind of information on photosynthetic systems does it provide?

Alfred R. Holzwarth
Max-Planck-Institut für Strahlenchemie, Stiftstr. 34-36, D-4330 Mülheim/Ruhr, FRG

1. Abstract

The use of fast time-resolved Chl fluorescence in photosynthesis research is reviewed. Examples include applications to the study of the functional organization of the photosynthetic apparatus of green algae and higher plants, physiological adaptation mechanisms, cell development and stress conditions. The concept of a decay-associated spectrum is introduced and the various parameters which can be obtained from such data are described.

2. Introduction

The remarkably complex organization of the antenna systems and reaction centers of photosynthetic organisms presents an outstanding challenge to researchers trying to unravel the various details of their structural and functional relationships. One of the techniques that has been successfully applied over the last few years to these problems is time-resolved chlorophyll (Chl) fluorescence in the picosecond and nanosecond time range (for reviews see Holzwarth 1986a,b, Holzwarth 1987, Karukstis and Sauer 1983, van Grondelle 1985, van Grondelle and Amesz 1986, Geacintov and Breton 1987). This contribution will restrict itself to applications on oxygen-evolving photosynthetic systems.

Fluorescence methods in general are highly advantageous in photosynthesis research due to their non-invasive nature, their high sensitivity and the fact that they can be applied easily to highly scattering samples like, e.g., intact leaves, needles, etc. Steady state fluorescence and/or quasi-steady state techniques like, e.g., fluorescence induction, have been used extensively in the past to probe a large variety of phenomena related to the energy absorption, energy distribution, and

H. K. Lichtenthaler (ed.) Applications of Chlorophyll Fluorescence, 21–31.

energy transformation in the photosynthetic apparatus (for reviews see Goedheer 1972, Lavorel and Etienne 1977, Bose 1982, Krause and Weis 1984).

The interpretation of both steady state and time-resolved fluorescence data depends on certain assumptions and hypotheses concerning the fate of the light energy absorbed by the photosynthetic antenna pigments. Not surprisingly, the conclusions drawn from such studies are to a considerable extent model dependent, i.e., they are a direct consequence of the various assumptions made on the structural and functional organization of the photosynthetic apparatus. It is therefore of utmost importance to rigorously test the different hypotheses. However, the validity of a particular model can not in general be tested by steady state fluorescence methods themselves. A much more powerful method is time-resolved fluorescence which is able to resolve the Chl excited state kinetics on the picosecond and nanosecond time-scale. This advantage of time-resolved methods over steady state techniques arises from the fact that the former provide a much larger set of parameters for the characterization of a particular system than is possible with the latter. It is the aim of this contribution to

i) describe the different kinds of information that can be obtained from time-resolved fluorescence and
ii) to provide some examples for their use.

3. Results and Discussion

3.1 Kinetic Analysis of fluorescence decays

The Chl fluorescence kinetics from intact systems is highly complex (Holzwarth 1986a, Holzwarth 1987), comprizing at least 3-4 kinetic components. This complex kinetics requires the most powerful data analysis techniques. We have recently introduced the so-called "global analysis" of kinetic data into the analysis of photosynthetic systems and have shown that its use is highly advantageous (Wendler and Holzwarth 1987, Holzwarth et al. 1987, Suter et al. 1988). In contrast to the conventional single decay analysis, which considers each fluorescence decay separately, the "global" method aims at testing a large set of fluorescence decays against a particular kinetic and organizational model. The different fluorescence decays comprizing one set may have been recorded under various conditions of, e.g., excitation and detection wavelength, light intensity, inhibitor concentration, etc. Thus the "global" method checks a particular kinetic model for its ability to fit a multi-dimensional data surface. In this way, the ability to differentiate between a variety of possible kinetic models is dramatically improved.

A sum of exponentials decay law is usually used to analyze Chl fluorescence decays. The time-course I(t) of the fluorescence intensity is this given by

$$I(t,\lambda_{ex},\lambda_{em}) = \sum_{i=1}^{n} A_i(\lambda_{exc},\lambda_{em}) \cdot \exp(-t/\tau_i) \qquad (1)$$

where λ_{exc} and λ_{em} are the excitation and detection wavelengths, respectively, and A_i $(\lambda_{exc}, \lambda_{em})$ is the amplitude factor for the kinetic component with lifetime τ_i, A_i and τ_i are fitting parameters which are determined using least-square techniques. It is characteristic of the "global analysis" method that the set of lifetimes τ_i is identical for a given set of fluorescence decays. Often the resulting amplitudes A_i are plotted vs. λ_{exc} or λ_{em} which is then called a decay-associated excitation or emission spectrum (DAS), respectively (Wendler and Holzwarth 1987, Holzwarth et al. 1987). The usefulness of such plots becomes apparent from eqn. 2 which provides the relationship between the amplitude A_i in a DAS and various molecular parameters of the antenne pigments. In many cases this relationship is simply given by

$$A_i(\lambda_{exc},\lambda_{em}) = C \cdot I_{exc} \cdot \varepsilon_{exc} \cdot N_{Chl} \cdot F(\lambda_{em}) \cdot k_{rad} \qquad (2)$$

where C is a proportionality constant, I_{exc} is the excitation intensity, ε_{exc} the absorption coefficient at the excitation wavelength λ_{exc}, N_{Chl} the size of the antenna pool (number of Chls) giving rise to decay component i, $F(\lambda_{em})$ the normalized fluorescence spectrum of this antenna pool and k_{rad} the radiative lifetime of the pigments.

Thus DAS provide the shape of the excitation (absorption) and emission spectra of a particular antenna pool or photosystem as well as relative antenna sizes of different photosystems. These are important static parameters characterizing a photosynthetic system. They cannot be obtained by steady state fluorescence. On the other hand, the dynamic parameters describing a system can be obtained from the lifetimes τ_i. These provide information on the energy transfer and charge separation kinetics and on the redox state of the reaction centers. Furthermore the combined set of parameters, both static as well as dynamic ones, provides detailed information on the functional organization and energy distribution in the photosynthetic apparatus.

For the purpose of comparison between the results from steady state

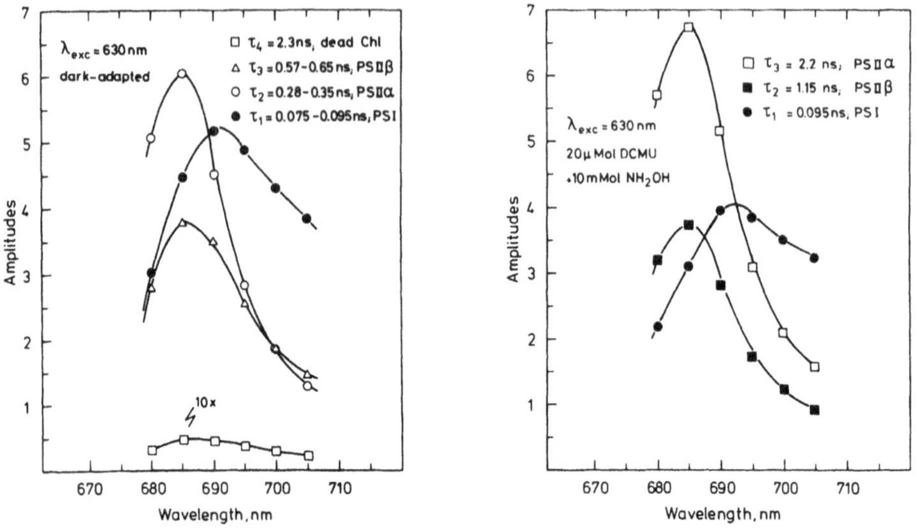

Fig. 1: (left) DAS spectra for dark-adapted cells of Scenedesmus at the F_o-level of fluorescence. (right) DAS spectra of Scenedesmus at the F_{max}-level of fluorescence. (taken from Wendler and Holzwarth 1987)

fluorescence and time-resolved fluorescence techniques we define the average lifetime τ_{av} as

$$\tau_{av} = \frac{\Sigma_i A_i \cdot \tau_i}{\Sigma_i A_i} \qquad (3)$$

The total fluorescence quantum yield ϕ_F is given by

$$\phi_F = \sum_{i=1}^{n} \phi_{Fi} = \sum_{i=1}^{n} A_i \cdot \tau_i = \tau_{av} \cdot \sum_{i=1}^{n} A_i \qquad (4)$$

where ϕ_{Fi} is the fluorescence yield due to the kinetic component with lifetime τ_i.

3.2 DAS of green algae and higher plant chloroplasts

In the following some examples for the application of DAS to photosynthetic systems are given. Fig. 1 shows the DAS of the green alga Scenedesmus obliquus with open (F_o) and closed (F_{max}) PS II reaction centers (Wendler and Holzwarth 1987). In each case three prominent lifetime components are observed. Two of the DAS have very similar spectral shapes with emission maxima around 685 nm both under F_o- and F_{max}-conditions. When closing PS II reaction centers their amplitudes remain approx. constant while their lifetimes both increase. The third component shows a red-shifted DAS with emission maximum around 690-695 nm. A slight decrease in the amplitude of this component is observed at F_{max} while the lifetime remains constant upon closing PS II reaction centers. Due to these properties the first two components have been assigned to PS II whereas the bathochromically shifted DAS is ascribed to originate from PS I.

Fig. 2 shows corresponding data for pea chloroplasts with open and closed PS II centers (Schatz and Holzwarth 1987). The spectra and lifetimes of most components are similar to those of the green algae (Fig. 1). A significant difference consists in the spectral shape of the PS I component which shows a spectrum with two emission bands at 685 and 730 nm, in contrast to the single-banded PS I spectrum of green algae (Holzwarth 1987).

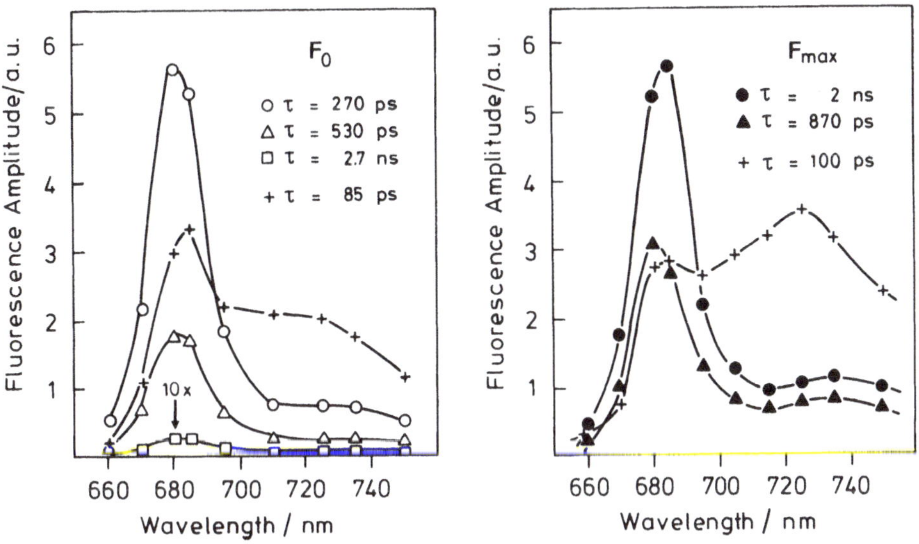

Fig. 2: (left) DAS spectra for dark-adapted pea chloroplasts at the F_0-level of fluorescence. (right) DAS spectra of pea chloroplasts at the F_{max}-level of fluorescence. (taken from Schatz and Holzwarth 1987)

The influence of a light state transition on the DAS of Scenedesmus under F_0-conditions is shown in Fig. 3 (compare with Fig. 1A). The most pronounced difference between these spectra occurs in the amplitudes of the two PS II components which change in a complementary fashion upon adaptation to different light conditions (Wendler and Holzwarth 1987). The most pronounced differences are observed in the dark-adapted state (Fig. 1A) as compared to the light II adapted state (Fig. 3). From a pronounced band at 650 nm the decay-associated excitation spectra of green algae reveal that the 2.0-2.4 ns at F_{max} originates from PS II units which are associated with the predominant part of the Chl a/b light-harvesting complex (Holzwarth et al. 1985, Holzwarth 1986a). An example for the corresponding excitation spectra under F_0-conditions is shown for Scenedesmus in Fig. 4. These spectra again reveal that the two PS II fluorescence components with lifetimes of approx. 300 ps and 550 ps show distinctly different excitation spectra. The corresponding PS II units have a different pigment composition in their antenna. At F_0 the shorter-lived PS II component is associated with most of the Chl a/b protein complex as is again revealed from the pronounced excitation band at 650 nm.

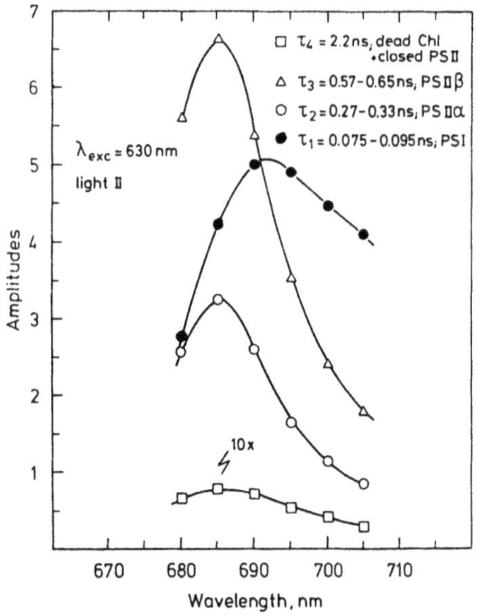

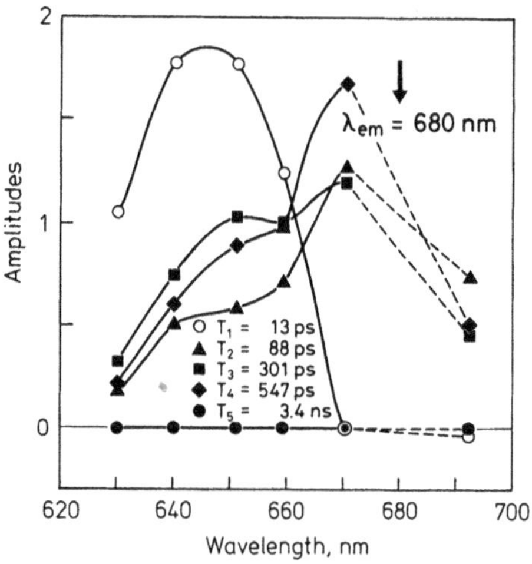

Fig. 3 (left): Decay-associated emission spectra for light-II-adapted cells of Scenedesmus (state II) at the F_0-level of fluorescence. Data from the same sample as in Fig. 1. (taken from Wendler and Holzwarth 1987)

Fig. 4 (right): Decay-associated excitation spectra for cells of Scenedesmus at the F_0-level of fluorescence. The detection wavelength was 680 nm.

Thus experimentally two PS II lifetime components are observed in green algae as well as in higher plants both at F_0- and F_{max}- conditions. The DAS reveal that these components have nearly identical emission spectra but different excitation spectra. Furthermore, for each of the PS II lifetime components at F_0 there exists a component at F_{max} with nearly identical amplitude. It has therefore been concluded that green algae and higher plants contain two different sets of PS II units called PS IIα and PS IIβ (Holzwarth et al. 1985, Wendler and Holzwarth 1987, Bittersmann et al. 1987). They differ in their antenna sizes, their antenna composition and in the photochemical and photophysical parameters of their respective reaction centers. PS IIα-centers, in accordance with other observations, are efficient centers for charge separation, oxygen evolution and plastoquinone reduction while PS IIβ centers have a low activity.

The lifetimes of all the fluorescence components described so far reflect the overall charge separation kinetics in these systems. The charge separation rate does not only depend on the type of reaction center considered but also on the size of the associated antenna. With increasing antenna size the charge separation time and thus the excited state lifetime increases (Schatz and Holzwarth 1986, Schatz et al. 1987, 1988). Energy transfer in the antenne occurs at a much faster rate and is reflected by the 13 ps component in Fig. 4. Table 1 gives an assignment of the origin of the various fluorescence decay components observed in green algae and higher plants as revealed by time-resolved Chl fluorescence over the recent years.

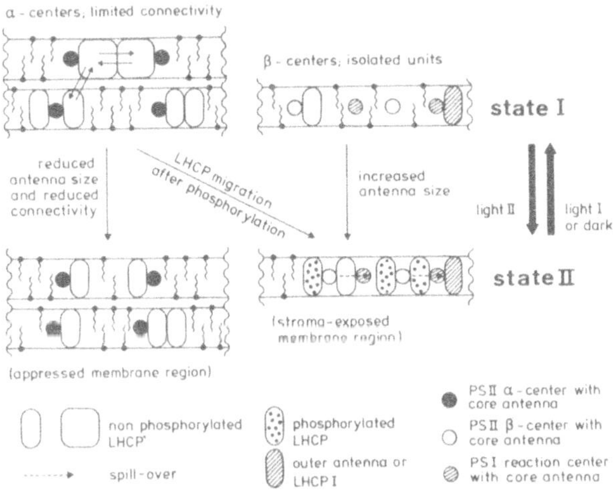

Fig. 5: Schematic model of the PS II antenna organization in states I and II. The regulation mechanism involves the reversible migration of phosphorylated LHCP from α-centers to β-centers. The model indicates that up to 30% of the total Chl in PS II may be mobile. (taken from Wendler and Holzwarth 1987)

From a large set of time-resolved fluorescence data a model for the functional energy distribution as well as its modification by light state transitions has been proposed (Wendler and Holzwarth 1987). This model is shown in Fig. 5. It takes into account both the heterogeneity of PS II, comprizing both PS IIα and PS IIβ units, the preferential association of PS IIα with the Chl a/b protein complex, and the attachment of phosphorylated Chl a/b protein to PS IIβ.

Table 1: Assignment of lifetime components in green algae and higher plants.

Component	Lifetime F_0	F_{max}	Interpretation
τ_1	10-15 ps	10-15 ps	Equilibration within antennae and between antennae and RC
τ_2	80-100 ps	80-100 ps	PS I charge separation time
τ_3	200-300 ps	2.-2.4 ns	charge separation in PS IIα units
τ_4	500-600 ps	1.-1.2 ns (0.8-1.0 ns for higher plants)	charge separation in PS IIβ units

In this way the relative absorption cross-section of PS IIα and PS IIβ units is regulated. This regulation is proposed to be the basic mechanism in light state transitions (Wendler and Holzwarth 1987). An important conclusion from these studies was that the direct absorption cross-section of PS I is not increased by a direct attachment of phosphorylated LHCP II to PS I units in state II. However, a small increase in PS I activity could be explained by an increased spill-over rate from PS IIβ to PS I in state II. These results show clearly that the PS II heterogeneity must not be ignored when interpreting steady state fluorescence data as is still often the case in published work. Severe misinterpretation of such data may be the consequence.

3.3 Applications of DAS to stress detection.

As a final example for the usefulness of DAS data in photosynthesis research some examples on photoinhibition and photobleaching in the cyanobacterium Anabaena variabilis are provided (Bittersmann et al. 1988, Nultsch et al. 1988). Fig. 6 compares the DAS of a control sample with that of cells that were submitted to photobleaching conditions. In the control sample the energy transfer in the phycobilisome antenna is fully functional, as can be concluded from the short lifetimes and the pronounced negative amplitude component.

In the sample exposed to high light intensity the modification of the antenna becomes readily apparent. The large amplitude of long lifetime components with emission maximum around 650 nm lets us conclude that the phycobilisome rods have been functionally detached from the phycobilisome under these conditions (Bitters-

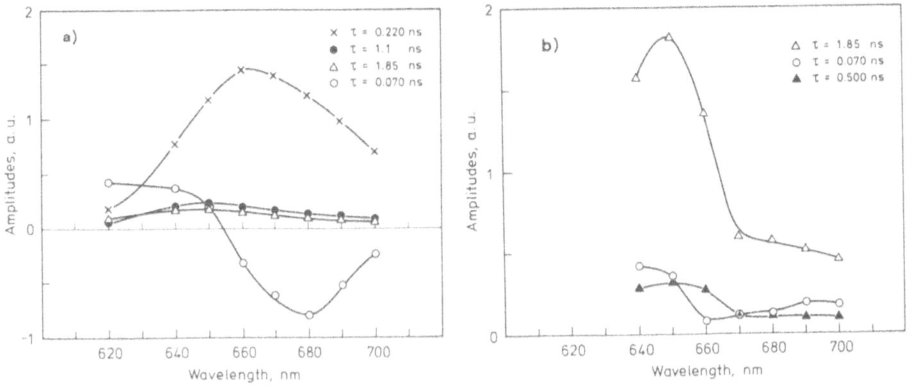

Fig. 6: Decay-associated emission spectra (λ_{exc} =580 nm) of cells of Anabaena variabilis grown under normal weak light conditions (a) and under strong white light conditions for 5 days (b). (taken from Bittersmann et al. 1988)

mann et al. 1988). On the basis of further recent time-resolved data it was possible to clearly distinguish between the rates and the mechanisms of photobleaching on the one hand and photoinhibition on the other hand. In agreement with other studies photoinhibition was shown to be a short-term effect which most probably modifies to acceptor side of the PS II reaction center (Nultsch et al. 1988).

4. Conclusion

The examples given in this short review reveal that time-resolved Chl fluorescence is a powerful tool to determine important parameters which characterize the state of the photosynthetic apparatus. In particular the measurement of DAS provides a wealth of information that can not be obtained otherwise. Evidently time-resolved fluorescence in the picosecond to nanosecond time-scale is not only useful as a tool for basic photosynthesis research. It is expected that this technique should be equally powerful in many areas of applied research. Because of its high information content it should be especially useful for. e.g.. the determination of the effects of stress conditions on the state of photosynthetic organisms. in ecophysiology and in other areas where environmental in fluences on the organization of the photosynthetic apparatus are studied. Not many studies have been carried out on these topics by time-resolved methods so far. However. the results of the available studies are highly encouraging and let us expect the increasing use of the various fast time-resolved methods in the future.

5. References

Bittersmann E, Holzwarth AR, Agel G and Nultsch W, 1988. Picosecond time-resolved emission spectra of photoinhibited and photobleached Anabaena variabilis. Photochem. Photobiol. 47: 101-105.

Bittersmann E, Senger H and Holzwarth AR, 1987. Synchronous cultures of the green alga Scenedesmus obliquus: Comparison of picosecond decay associated emission spectra and fluorescence induction kinetics. J. Photochem. Photobiol. 1: 247-260.

Bose S, 1982. Chlorophyll fluorescence in green plants and energy transfer pathways in photosynthesis. Photochem. Photobiol. 36: 725-731.

Geacintov NE and Breton J, 1987. Energy transfer and fluorescence mechanisms in photosynthetic membranes. In: CRC Crit. Rev. Plant Sci. 55: pp. 1-44. Chemical Rubber Company, Boca Raton.

Goedheer JC, 1972. Fluorescence in relation to photosynthesis. Ann. Rev. Plant Physiol. 23: 87-112.

Holzwarth AR, 1986a. Fluorescence lifetimes in photosynthetic systems. Photochem. Photobiol. 43: 707-725.

Holzwarth AR, 1986b. Excited state kinetics of chlorophyll antenna pigments. In: Encyclopedia of Plant Physiology: Photosynthesis III New Series 19, Staehelin LA and Arntzen CJ Eds., pp. 299-309, Springer, Berlin.

Holzwarth AR, 1987. Picosecond fluorescence spectroscopy and energy transfer in photosynthetic antenna pigments. In: Topics in Photosynthesis: The Light Reactions 8, Barber J Ed., pp. 95-157, Elsevier, Amsterdam.

Holzwarth AR, Wendler J and Haehnel W, 1985. Time-resolved picosecond fluorescence spectra of the antennna chlorophylls in Chlorella vulgaris. Resolution of Photosystem I fluorescence. Biochim. Biophys. Acta 807: 155-167.

Holzwarth AR, Wendler J and Suter GW, 1987. Studies on chromophore coupling in isolated phycobiliproteins. II. Picosecond energy transfer kinetics and time-resolved fluorescence spectra of C-phycocyanin from Synechococcus 6301 as a function of the aggregation state. Biophys. J. 51: 1-12.

Karukstis KK and Sauer K, 1983. Fluorescence decay kinetics of chlorophyll in photosynthetic membranes. J. Cell. Biochem. 23: 131-158.

Krause GH and Weis E, 1984. Chlorophyll fluorescence as a tool in plant physiology. 2. Interpretation of fluorescence signals. Photosyn. Res. 5: 139-157.

Lavorel J and Etienne AL, 1977. In vivo chlorophyll fluorescence. In: Primary Processes of Photosynthesis in Topics in Photosynthesis 2, Barber J Ed., pp. 203-268, Elsevier, Amsterdam.

Nultsch W, Bittersmann E, Holzwarth AR and Agel G, 1988. Distinction of photoinhibitory and photobleaching effects in Anabaena variabilis by picosecond time-resolved emission spectra and fluorescence induction. Photochem. Photobiol., submitted.

Schatz GH, Brock H and Holzwarth AR, 1987. Picosecond kinetics of fluorescence and of absorbance changes in photosystem II particles excited at low photon density. Proc. Natl. Acad. Sci. USA 84: 8414-8418.

Schatz GH, Brock H and Holzwarth AR 1988. A kinetic and energetic model for the primary processes in photosystem II. Biophys. J., in print.

Schatz GH and Holzwarth AR, 1986. Mechanisms of chlorophyll fluorescence revisited: Prompt or delayed emission from photosystem II with closed reaction centers?. Photosynth. Res. 10: 309-318.

Schatz GH and Holzwarth AR, 1987. Picosecond time resolved chlorophyll fluorescence spectra from pea chloroplast thylakoids. In: Progress in Photosynthesis Research 1, Biggins J Ed., pp. 67-69, Nijhoff, Dordrecht.

Suter GW, Klein-Boelting P, Bittersmann E, Stempfle W and Holzwarth AR, 1988. On the global analysis of time resolved spectra: Statistical error of decay times and amplitudes. Application to time resolved fluorescence. Chem. Phys., submitted.

van Grondelle R, 1985. Excitation energy transfer, trapping and annihilation in photosynthetic systems. Biochim. Biophys. Acta 811: 147-195.

van Grondelle R and Amesz J, 1986. Excitation energy transfer in photosynthetic systems. In: Light Emission by Plants and Bacteria, Govindjee, Amesz J and Fork DC Eds, 191-224, Academic Press, New York.

Wendler J and Holzwarth AR, 1987. State transitions in the green alga Scenedesmus obliquus probed by time-resolved chlorophyll fluorescence spectroscopy and global data analysis. Biophys. J. 52: 717-728.

PHOTOSYSTEM II HETEROGENEITY IN CHLOROPLASTS

A. Melis, G.E. Guenther, P.J. Morrissey, M.L. Ghirardi
Division of Molecular Plant Biology, 313 Hilgard Hall,
University of California, Berkeley, CA 94720 (USA)

Keywords - Fluorescence induction kinetics; photosystem II heterogeneity; chlorophyll antenna size; plastoquinone reduction; reaction-center repair.

Abstract - Two aspects of photosystem II (PSII) heterogeneity are discussed. (A) The PSII antenna heterogeneity refers to the existence of PSII centers with distinct photosynthetic unit size. Measurements of PSII antenna size in developing chloroplasts, in Chl b-less and Chl b-deficient mutants, and in mature higher plant chloroplasts indicated the existence of three distinct populations of PSII centers: $PSII_\alpha$ with a total of 210 (or more) chlorophyll (a+b) molecules; $PSII_\beta$ with a total of 120 Chl (a+b) molecules. In the absence of Chl b, a PSII complex containing only 50 Chl a molecules (PSII-50) was identified. A developmental relationship among PSII-50, $PSII_\beta$ and $PSII_\alpha$ is proposed. According to this hypothesis, the formation of the complete PSII unit involves the assembly of three modular complexes (PSII-50, LHC II-inner, and LHC II-peripheral). Addition of LHC II-inner to PSII-50 yields $PSII_\beta$. Addition of LHC II-peripheral to $PSII_\beta$ yields $PSII_\alpha$. The relative proportion of PSII centers with the three antenna configurations depends on the developmental stage of the chloroplast and on the availability of Chl b. Mature wild type chloroplasts contain $PSII_\alpha$ (75-80°/o of the total PSII) and $PSII_\beta$ (20-25°/o of the total PSII). (B) The PSII reducing side heterogeneity refers to the existence of PSII centers with impaired Q_A^- to Q_B electron transfer interaction (Q_B-nonreducing centers). The fraction of PSII-Q_B-nonreducing centers is small (20-25°/o of the total PSII) as tested in several chloroplast preparations. This steady state concentration of Q_B-nonreducing centers appears independent of the developmental stage of the chloroplast and also independent of the PSII photosynthetic unit size. In mature spinach chloroplasts, $PSII_\beta$ and PSII-Q_B-nonreducing centers constitute one and the same pool of PSII centers. It is proposed that PSII-Q_B-nonreducing centers represent newly synthesized and/or repaired PSII centers which have not yet established a functional interaction between Q_A^- and Q_B.

The concept of PSII heterogeneity was first introduced to explain the biphasic fluorescence induction kinetics observed upon illumination of higher plant chloroplasts suspended in the presence of PSII herbicides [Melis and Homann, 1975; 1976]. Figure 1A shows a typical fluorescence induction kinetic trace with isolated spinach thylakoids suspended in the presence of the herbicide DCMU (3-(3,4-dichlorophenyl)-1,1-dimethylurea). Qualitatively, the fast and sigmoidal portion of the fluorescence kinetic trace is followed by a slower exponential phase (shaded area). The

H. K. Lichtenthaler (ed.) Applications of Chlorophyll Fluorescence, 33–43.

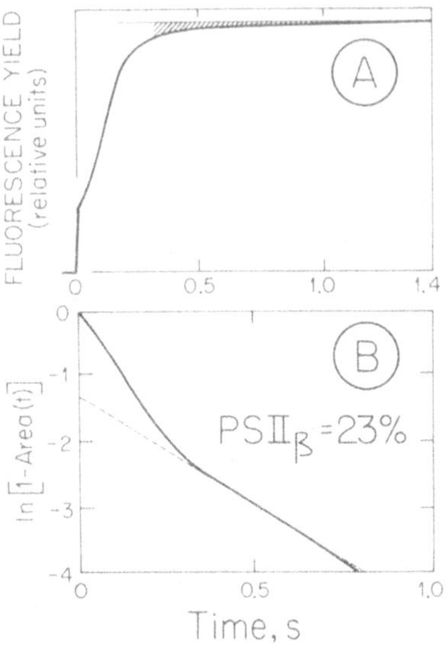

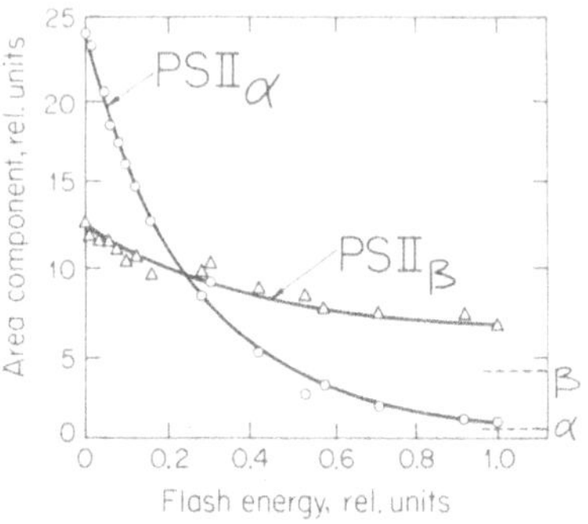

FIGURE 1. (A) Fluorescence induction trace with isolated spinach thylakoids suspended in the presence of 20 μM DCMU. The shaded area marks the portion of the fluorescence induction trace where the activity of $PSII_\beta$ is the only component remaining in the kinetics. (B) Semilogarithmic plot (first-order analysis) of the area over fluorescence induction, revealing the biphasic nature of PSII photoconversion. The relative amount of $PSII_\beta$ centers (23⁰/o of the total PSII) was determined from the value of the y-axis intercept with the slower linear phase (dashed line).

FIGURE 2. The flash saturation properties of $PSII_\alpha$ and $PSII_\beta$ in DCMU-poisoned spinach chloroplasts. The dashed lines represent the maximum changes obtained with a single saturating flash. The preilluminating flash was administered 3 ms before the onset of the actinic illumination. The half-peak duration of the flash was 8 μs. Identical results were obtained with subnanosecond flashes (not shown).

biphasic nature of the fluorescence induction curve is revealed clearly in a semilogarithmic plot of this area (Fig. 1B) where the fast sigmoidal phase ($0 < t \leq 0.27$ s) is followed by a slower linear phase ($0.27 \text{ s} < t$). A plot of the area over fluorescence is justified as this area is proportional to the amount of Q_A photoreduced [Bennoun/Li, 1973]. Analysis of the biphasic data revealed that the slow component accounted for 20-25⁰/o of the total phenomenon [Melis, 1985; McCauley/Melis, 1987]. Based on the relaxation of the fluorescence yield in the dark and on the flash saturation properties of the fluorescence yield, it was proposed that biphasic fluorescence induction kinetics reflect the function of two distinct populations of PSII centers, termed $PSII_\alpha$ and $PSII_\beta$, respectively [Melis and Homann, 1976]. Deconvolution of the biphasic kinetics indicated that rates of $PSII_\alpha$ conversion were 2-3 times faster than $PSII_\beta$ rates. This was originally attributed to a low quantum yield of photochemistry at $PSII_\beta$ [Melis/Homann, 1976]. It was reasoned that a quantum yield of photochemistry at $PSII_\beta$ equal to 0.4 ±0.1 would explain the results. However, later work by several investigators (reviewed by

Black et al, 1986) has revealed additional properties of the PSII heterogeneity phenomenon in chloroplasts and has led to a revision of the initial interpretation.

PSII ANTENNA HETEROGENEITY

It was recognized that rate differences between the two kinetic components (α and β) could simply reflect different rates of light absorption by the respective photosystem populations ($PSII_\alpha$ and $PSII_\beta$). Such interpretation required a different absorption cross section of light-harvesting pigments for $PSII_\alpha$ and $PSII_\beta$. This hypothesis was tested by the flash saturation curve of PSII fluorescence and by the flash saturation of the α and β-components in DCMU-poisoned chloroplasts [Melis/Duysens, 1979]. In this experiment, a short flash of varying intensity was fired a few ms before the onset of the continuous actinic illumination and the subsequent registration of the fluorescence induction curve. The effect of the preilluminating flash was to reduce the primary quinone acceptor Q_A of $PSII_\alpha$ and of $PSII_\beta$ to an extent depending on the energy of the flash and on the absorption cross section of the two photosystems. Figure 2 shows the flash saturation properties of $PSII_\alpha$ and $PSII_\beta$. The half-saturation flash intensity for $PSII_\alpha$ was approximately 3 times lower than that of $PSII_\beta$. This result was in fairly good agreement with the difference observed in the initial rates of the α and β components under conditions of continuous illumination (Fig. 1) and constituted evidence in support of the hypothesis that $PSII_\alpha$ has an absorption cross section about 3 times larger than that of $PSII_\beta$. Further support was provided by the findings of Thielen and Van Gorkom [1981a] who showed that both $PSII_\alpha$ and $PSII_\beta$ operate with a quantum yield of $\geq$ 90°/o, thereby implying that the slower rate of the β-phase is due to a smaller antenna size and not due to a lower quantum yield of photochemistry.

Independent work with isolated thylakoid membranes resulted in the development of a kinetic method for the quantitation of the functional Chl antenna size of PSI and PSII [Thielen/Van Gorkom, 1981; Melis/Anderson, 1983]. Results from this analysis revealed that in spinach chloroplasts $PSII_\alpha$ contains a total of 250 ±40 Chl (a+b) molecules in its light-harvesting antenna. By comparison, the antenna size estimate for $PSII_\beta$ was 120 ±20 Chl (a+b) molecules.

Intermittent light plastids and Chl b-deficient mutants

Chloroplasts developing under intermittent illumination and Chl b-deficient mutants are known to have a smaller photosynthetic unit size because of the limited availability of chlorophyll [Argyroudi/Akoyunoglou, 1970; Armond et al, 1976; Somerville, 1986]. Leaves from intermittent light plants and Chl b-deficient mutants show lower Chl content per leaf area and higher Chl a/Chl b ratios than their wild type counterparts grown under physiological conditions. Analysis of intermittent light (ImL) plastids and of Chl b-deficient mutants from several plant species revealed a substantially enhanced concentration of $PSII_\beta$ in the thylakoid membrane of these chloroplasts [Melis/Thielen, 1980; Thielen/Van Gorkom, 1981; Melis, 1984; Abadia et al, 1985; Ghirardi/Melis, 1988]. This is qualitatively depicted in the results of Fig. 3, where the variable portion of the fluorescence induction curve of ImL pea plastids and of the Chl b-deficient Su/su tobacco mutant is shown. The shaded areas mark the portion of the variable fluorescence yield contributed by

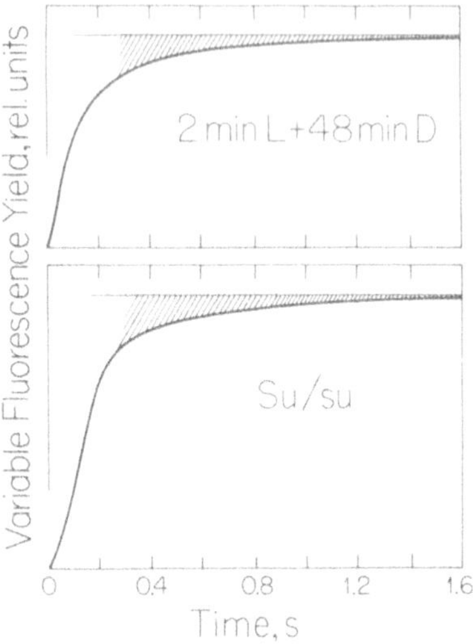

FIGURE 3. The variable portion of the fluorescence induction kinetics in isolated and DCMU-treated thylakoids (upper) from pea plants grown under intermittent illumination (2 min light and 48 min dark), and (lower) from Chl b-deficient Su/su mutants of tobacco. The Chl a/Chl b ratio of the ImL peas was 6.0 whereas that of the Su/su mutant was 5.0. Note the enhanced amplitude of the slower component in the fluorescence induction curve (shaded areas).

$PSII_\beta$. It was estimated that, both in the ImL plastids and in the Chl b-deficient Su/su tobacco mutant, $PSII_\beta$ accounted for 60-65% of the total PSII in the thylakoid membrane [Melis, 1984; Melis/Thielen, 1980]. Unlike the results obtained with mature wild-type chloroplasts, where $PSII_\alpha$ was the dominant form of PSII (Fig. 1), it appeared that $PSII_\beta$ is the dominant form of PSII in these developing plastids. Quantitation of the amount of $PSII_\alpha$ and $PSII_\beta$ in developing chloroplasts revealed a negative correlation between the relative concentration of $PSII_\beta$ and the acquisition of Chl b in the thylakoid membrane [Thielen and Van Gorkom, 1981a]. This result suggested a developmental relationship between $PSII_\beta$ and $PSII_\alpha$ in which $PSII_\beta$ served as a precursor to $PSII_\alpha$. In the absence of sufficient quantities of Chl a/b LHC II the formation of $PSII_\alpha$ units is retarded. Then, $PSII_\beta$ units with a substantially smaller photosynthetic unit size accumulate in the thylakoid membrane [Percival et al, 1984; Ghirardi et al, 1986].

The case of the Chl b-less mutant of barley

The results from the research summarized above suggested that whenever the acquisition of Chl b was retarded, either because of ImL conditions or because of mutations, $PSII_\beta$ complexes with an antenna size of 120 ±20 Chl (a+b) molecules accumulated in the thylakoid lamellae. In the case of the chlorina f2 mutant of barley [Highkin, 1950; Highkin/Frenkel, 1962] which totally lacks Chl b, the phenomenon of PSII antenna heterogeneity was not manifested. Fluorescence induction kinetics obtained with thylakoid membranes from the Chl b-less chlorina f2 mutant were single exponential functions of time reflecting the presence of a uniform PSII population of very small photosynthetic unit size [Percival et al, 1984; Ghirardi et al, 1986].

Measurements of the PSII antenna size in the chlorina f2 mutant suggested the presence of only 50 Chl a molecules [Ghirardi et al, 1986]. Hence, it was concluded that these mutant chloroplasts lacked the differentiation of PSII into $PSII_\alpha$ and $PSII_\beta$. Instead, they possessed a

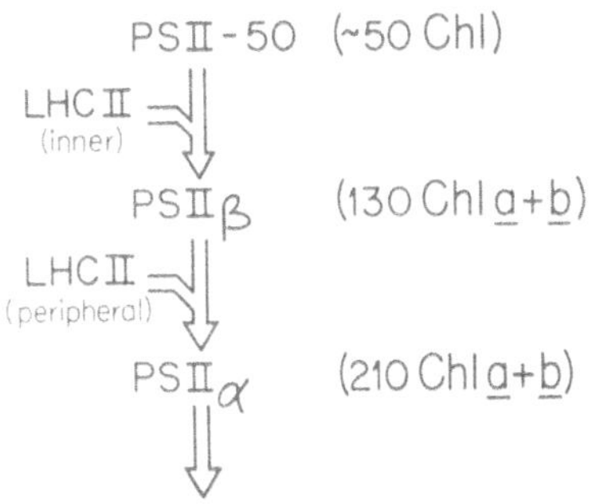

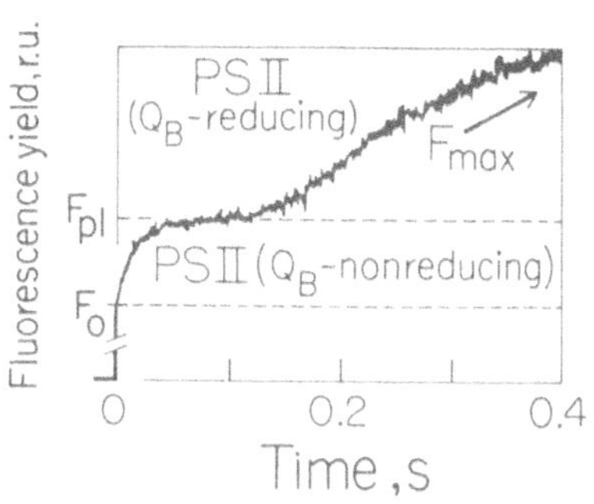

FIGURE 4. Schematic defining a step-wise process in the development of the light-harvesting antenna of PSII proposing the addition of two LHC II complements in the developing PSII antenna. The PSII-50 configuration contains 50 Chl a molecules, it lacks LHC II and is evident in the Chl b less chlorina f2 mutant of barley. The $PSII_\beta$ configuration features 130 Chl (a+b) molecules and is obtained upon addition of the LHC II-inner to the PSII-50 complex. The $PSII_\alpha$ configuration features a minimum of 210 Chl (a+b) molecules and is obtained upon addition of the LHC II-peripheral to $PSII_\beta$ units.

FIGURE 5. Fluorescence induction kinetics obtained with a spinach leaf disc under in vivo conditions. The exponential fluorescence yield increase from F_0 to F_{pl} reflects the kinetics of Q_A^- accumulation in PSII-Q_B-nonreducing centers. The main portion of the fluorescence yield increase from F_{pl} to F_{max} reflects the kinetics of Q_A^- accumulation in PSII-Q_B-reducing centers. The lag observed in the kinetics of the latter is due to the time needed for the reduction of the plastoquinone pool.

stable and functional form of PSII (PSII-50) in their thylakoid membrane. The Chl b-less barley mutant provided an additional example of the interplay between PSII heterogeneity in the thylakoid membrane and the Chl a/b LHC II as a functional component of the photosynthetic unit.

Measurements of PSII antenna size with ImL developing chloroplasts [Melis, 1984], with Chl b-deficient mutants [Abadia et al, 1985; Ghirardi et al, 1986, 1988; Percival et al, 1984; Thielen/Van Gorkom, 1981a, b] and with mature chloroplasts under various experimental conditions [Andersson et al, 1987; Horvath et al, 1987; Sunby et al, 1986; Melis/Anderson, 1983], all support the notion of PSII complexes with distinct Chl antenna sizes. For example, there appears to be a lack of PSII units with an antenna size between 50 Chl a (PSII-50) and 120 Chl (a+b) ($PSII_\beta$). Similarly, there is a lack of PSII reaction centers with an antenna size between 120 ($PSII_\beta$) and 210 Chl molecules ($PSII_\alpha$). It is proposed that, in the process of photosynthetic unit development, pigment molecules are not inserted continuously, in a linear fashion, to preexisting LHC II polypeptide subunits in the thylakoid lamella. Rather, two distinct and fully assembled LHC II complexes, containing the necessary polypeptides and the associated pigment molecules, are involved in the apparent transition from PSII-50 to $PSII_\beta$ and from $PSII_\beta$ to $PSII_\alpha$. Figure 4 presents a hypothetical schematic of a step-wise process in the development of the PSII unit in which the addition of LHC II-inner (also referred to as tightly bound LHC II [Staehelin, 1986]) augments the antenna of PSII-50 by about 80 Chl (a+b) molecules and yields $PSII_\beta$ with an antenna size of 130 Chl (a+b) molecules. In a second step, the addition of LHC II-peripheral increases the antenna size of PSII by at least another 80 Chl (a+b) molecules to yield $PSII_\alpha$, with an antenna size of 210 (or more) Chl (a+b) molecules.

Interconversion of $PSII_\alpha$ to $PSII_\beta$

If the formation of the complete PSII unit involves the assembly of modular complexes (PSII-50, LHC II-inner, LHC II-peripheral), it would be expected that certain physiological conditions may cause partial or complete disassembly of the modular PSII unit. To date, the dissociation of the LHC II-peripheral from $PSII_\alpha$, leading to the formation of $PSII_\beta$ and free LHC II, has been documented. Phosphorylation/dephosphorylation of the LHC II-peripheral apparently leads to a reversible dissociation of this complex from PSII, resulting in the conversion of $PSII_\alpha$ into $PSII_\beta$ and free LHC II [Kyle et al, 1982; Larsson/- Andersson, 1985; Andersson et al, 1987]. In addition, moderate heat treatment of chloroplast thylakoids [Sunby et al, 1986] and the homogeneous catalytic hydrogenation of thylakoid membrane lipids [Horvath et al, 1987], both lead to a dissociation of the LHC II-peripheral and to the ensuing conversion of $PSII_\alpha$ into $PSII_\beta$. Physiological conditions for the dissociation of LHC II-inner from $PSII_\beta$ have not been described in the literature.

The reversible dissociation of LHC II-peripheral from $PSII_\alpha$ is in contrast to the "tightly bound" nature of LHC II-inner and it suggests different subunit composition between the two LHC II complements and also differences in the nature of their association with PSII [Thielen et al, 1981; Staehelin, 1986]. However, elucidation of these properties will have to await further research advances in this field.

PSII REDUCING SIDE HETEROGENEITY

Evidence has been presented in the literature suggesting that a number of PSII reaction centers in chloroplasts, although photochemically competent, are unable to transfer electrons efficiently from Q_A^- to Q_B [Thielen/Van Gorkom, 1981b; Lavergne, 1982; Melis, 1985; Graan/Ort, 1986]. Following the nomenclature of Lavergne [1982], these centers are termed PSII-Q_B-nonreducing in order to distinguish them from the PQ reducing PSII centers (Q_B-reducing). It is possible to measure the relative concentration of PSII-Q_B-nonreducing centers _in vivo_, i.e., in intact leaf discs or algal cells in the absence of herbicides [Melis, 1985]. The method is based on the inability of these centers to transfer electrons from Q_A^- to Q_B, hence, upon illumination, their photochemical activity is revealed by an initial fluorescence yield increase from F_0 to the intermediate plateau F_{pl} [Forbush/Kok, 1968]. An example of this phenomenon is given in Fig. 5 where the fluorescence induction trace of a dark adapted spinach leaf disc is presented. The initial fluorescence yield increase from F_0 to F_{pl} precedes the main variable fluorescence rise to F_{max}. The relative fluorescence yield of the transition from F_0 to F_{pl} directly reflects the kinetics of photoreduction of Q_A in PSII-Q_B-nonreducing centers. The amplitude of this transition provides a measure of the relative concentration of PSII-Q_B-nonreducing centers. The fluorescence yield increase from F_{pl} to F_{max} is due to the progressive accumulation of the chemical species Q_A^- in PSII centers with efficient electron transfer to plastoquinone (PSII-Q_B-reducing). The lag in the fluorescence transition from F_{pl} to F_{max} reflects the time needed for electrons to accumulate in the plastoquinone pool [Malkin, 1971; Malkin/Michaeli, 1971].

Work from several laboratories has shown that fluorescence transients from F_0 to F_{pl} and from F_{pl} to F_{max} (Fig. 5) are manifested both with intact leaves and algal cells _in vivo_, and with isolated thylakoid membranes, as first demonstrated by Forbush and Kok [1968]. Work with isolated thylakoids revealed that artificial PSII electron acceptors, such as potassium ferricyanide, fully quenched the F_{pl} to F_{max} component of the fluorescence yield because of rapid removal of electrons from the plastoquinone pool [McCauley/Melis, 1987]. However, artificial electron acceptors failed to exert any effect on the F_0 to F_{pl} transition, suggesting that these electron acceptors cannot activate electron transport through PSII-Q_B-nonreducing centers [Forbush/Kok, 1968; McCauley/-Melis, 1987]. These observations opened the way for the study of the properties of PSII-Q_B-nonreducing centers from the kinetics of the F_0 to F_{pl} transition, unhindered by any overlap from the larger F_{pl} to F_{max} component [Melis, 1985].

Figure 6A compares the fluorescence induction kinetics of isolated thylakoid membranes suspended either in the presence of ferricyanide (FeCN) or in the presence of DCMU. The faster and sigmoidal α-component (due to $PSII_\alpha$) dominated the fluorescence induction kinetics in the presence of DCMU. In the presence of ferricyanide, the fluorescence induction kinetics (corresponding to the transition from F_0 to F_{pl}) reflected the accumulation of Q_A^- in PSII-Q_B-nonreducing centers. The latter was a monophasic exponential function of time occurring with kinetics identical to those of the β-component in the presence of DCMU (Fig. 6B). Further detailed study with spinach _in vivo_ and _in vitro_ suggested an interrelationship between $PSII_\beta$ and PSII-Q_B-nonreducing: (a) both displayed exponential fluorescence induction kinetics of equal rates (b) the fluorescence yield amplitude controlled by $PSII_\beta$ (Fv_β) was identical to that controlled by PSII-Q_B-nonreducing (F_0 to F_{pl}) and (c) the fraction of Q_A corresponding to $PSII_\beta$ was equal to the amount of Q_A photoreduced during the F_0 to F_{pl} transition [Melis, 1985]. On the basis of these observations, it was suggested that in mature spinach chloroplasts $PSII_\beta$ and PSII-Q_B-nonreducing constitute one and the same pool of PSII centers. A summary of the two aspects of PSII heterogeneity is presented in the schematic of Fig. 7 [Guenther et al, 1988].

Physiological Significance of PSII Heterogeneity

It is possible that PSII-Q_B-nonreducing centers are in a state of repair of the PSII unit in which the reaction center has replaced the damaged 32 kDa Q_B-binding protein. This hypothesis is consistent with the observation that PSII-Q_B-nonreducing centers are localized in stroma-exposed regions of the thylakoid lamella, they lack the LHC II peripheral, and have not yet established a functional interaction between Q_A^- and Q_B. The hypothesis is also consistent with the rapid turnover of the reaction-center 32 kDa protein [Matoo et al, 1981] and with the evidence that newly synthesized 32 kDa polypeptides first "assemble" in stroma thylakoids and subsequently move to the grana [Matoo/Edelman, 1987]. According to this model, a sequence of events is postulated following damage to the PSII reaction center: (a) uncoupling of the LHC II-peripheral from the damaged PSII unit in the grana partition

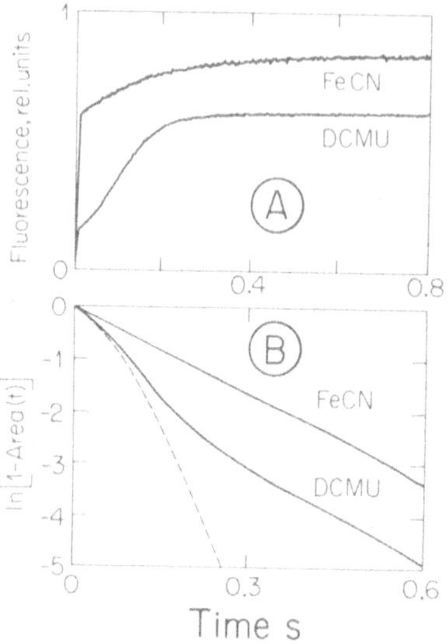

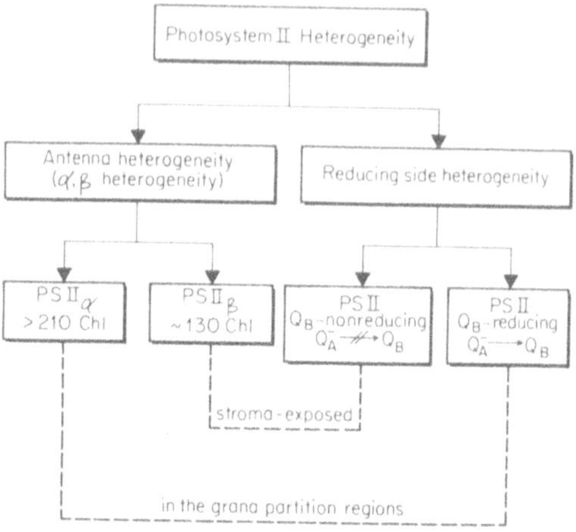

FIGURE 6. (A) Fluorescence induction traces with isolated spinach thylakoid membranes suspended in the presence of 20 μM DCMU or 2 mM potassium ferricyanide (FeCN). Compare the sigmoidal and biphasic trace in the presence of DCMU with the monophasic and exponential trace obtained in the presence of FeCN. The apparatus gain was set to 4x for the FeCN trace compared to 1x for the DCMU sample.

(B) Semilogarithmic plot of the area over the fluorescence induction of the traces shown above. Note the similarity between the slopes of the monophasic linear trace (FeCN) and the slow β component in the presence of DCMU. The dashed line shows the kinetics of $PSII_\alpha$ photoconversion.

FIGURE 7. Schematic delineating the two aspects of photosystem II heterogeneity discussed in this work. In wild type mature chloroplasts from higher plants and green algae there is a PSII antenna heterogeneity (α,β heterogeneity). This reflects the presence of a dominant $PSII_\alpha$, which is localized in the membrane of the grana partition region and contains 210 or more Chl (a+b) molecules per reaction center. Photosystem II_β is localized in stroma-exposed lamellae and lacks the LHC II-peripheral, resulting in a smaller antenna size of about 130 ±20 Chl (a+b) molecules. A reducing side heterogeneity reflects the presence of a small pool of PSII centers impaired in the Q_A-Q_B electron-transport process (Q_B-nonreducing). In wild type mature chloroplasts, $PSII_\beta$ and PSII-Q_B-nonreducing may constitute one and the same pool of PSII centers [Melis, 1985]. However, this strict relationship does not hold in developing chloroplasts. In ImL and Chl b-deficient chloroplasts, PSII-Q_B-nonreducing is a subpopulation of $PSII_\beta$.

region, (b) movement of the damaged $PSII_\beta$-like center from the grana partition region to stroma-exposed regions, (c) replacement of the damaged 32 kDa reaction-center protein complex resulting in the formation of a photochemically competent center in which the secondary Q_A^- to Q_B electron transfer interaction has not yet been established (PSII-Q_B-non-reducing), (d) activation of the Q_A^- to Q_B interaction followed by association with LHC II-peripheral and incorporation in the grana partition region in the form of a newly functional $PSII_\alpha$ unit. The above sequence of events is speculative, nevertheless, it may provide the basis of further experimentation by which to test the proposed model.

The hypothesis presented above explains the consistent presence of $PSII_\beta$ and/or PSII-Q_B-nonreducing centers in all mature higher plant and green algae examined to date. It must be stressed, however, that $PSII_\beta$ is defined solely by the smaller chlorophyll antenna size (lack of LHC II-peripheral), without reference to electron-transport properties on the reducing side of the reaction center. As such, $PSII_\beta$ and PSII-Q_B-nonreducing are not always synonymous. There are several examples of $PSII_\beta$ centers that are active in the process of electron transport to plastoquinone. Those include chloroplasts in the developing stage (like ImL plastids and Chl b-deficient mutants). These chloroplasts cannot acquire the LHC II-peripheral and, therefore, display enhanced concentrations of $PSII_\beta$ in the thylakoid membrane. However, under these conditions, a significant fraction of $PSII_\beta$ is active in the process of PQ reduction [Akoyunoglou, 1977; Melis/Thielen, 1980; Thielen/van Gorkom, 1981; Melis, 1984; Abadia et al, 1985; Ghirardi et al, 1986; 1988]. Additionally, evidence has been presented to support the notion that $PSII_\beta$ centers generated as a result of the phosphorylation of the LHC II-peripheral [Steinback et al, 1982; Farchaus et al, 1982; Larsson et al, 1986], or upon moderate heat treatment of thylakoid membranes [Sunby et al, 1986], remain active in the process of electron transport to plastoquinone. The above observations suggest that PSII-Q_B-nonreducing centers are a subset of $PSII_\beta$ so that the concentration of Q_B-nonreducing centers is equal to, or less than that of $PSII_\beta$ ([Q_B-nonreducing] $\leq$ [$PSII_\beta$]).

ALTERNATIVE EXPLANATIONS

An alternative model to explain the biphasic fluorescence induction kinetics was proposed by Hodges and Barber [1986]. They postulated a type of PSII heterogeneity which does not result from different antenna sizes but instead, it involves different degrees of DCMU inhibition among PSII centers. According to these authors, the slower phase of the fluorescence induction kinetics of chloroplasts in the presence of DCMU is attributed to "leakage" of electrons from Q_A^- (of $PSII_\beta$) to the PQ pool, presumably due to inefficient inhibition by the herbicide. This would result in a slower rate of Q_A^- accumulation in these centers. Their hypothesis was based on two observations: (a) excessive amounts of DCMU eliminated the slow exponential phase of the fluorescence induction curve [Schreiber/Pfister, 1982] and (b) prior addition of dithionite to isolated thylakoids, in sufficient amounts for the chemical reduction of the plastoquinone pool, also eliminated the slow exponential phase of the fluorescence induction [Hodges/Barber, 1983]. This alternative interpretation, however, has not addressed two related methodological shortcomings: (a) excessively high amounts of DCMU cause chaotropic effects in the thylakoid membrane [Ramanujam/Bose, 1983; Horvath et al, 1984], resulting in a preferential disorganization of stroma-exposed lamellae, and therefore, in the preferential inactivation of $PSII_\beta$ centers by herbicides; and (b) as shown by Klimov et al [1977], in the presence of strongly reducing agents like dithionite, the chloroplast fluorescence shows a steep rise followed immediately by a pronounced decline to a low fluorescence yield value. It is clear that in the presence of dithionite, the true F_{max} level will not be attained, effectively rendering the results of any kinetic analysis questionable. Moreover, the apparent lack of electron transfer from Q_A^- to Q_B in $PSII_\beta$ (Q_B-nonreducing centers) is incompatible with the hypothesis of "leakage" of electrons from the reducing side of this PS in the presence of DCMU.

In summary, the hypothesis of a PSII antenna heterogeneity ($PSII_\alpha$, $PSII_\beta$) and of a PSII reducing side heterogeneity (Q_B-reducing, Q_B-nonreducing centers) sufficiently explains a score of related observations. It provides an understanding of the supramolecular assembly and organization of the photosynthetic unit of PSII and also provides insight into the mechanism of repair of damaged PSII reaction centers (turnover of the 32 kDa polypeptide).

REFERENCES

Abadia, J., Glick, R.E., Taylor, S.E., Terry, N. and Melis, A. (1985) Plant Physiol. 79, 872-878

Akoyunoglou, G. (1977) Arch. Biochem. Biophys. 183, 571-580

Andersson, B., Sunby, C., Larsson, U.K., Maenpaa, P. and Melis, A. (1987) In, Progress in Photosynthesis Research (J. Biggins, ed), Vol. II, 669-676, Martinus Nijhoff Publishers, Dordrecht

Argyroudi-Akoyunoglou, J.H. and Akoyunoglou, G. (1970) Plant Physiol. 46, 247-249

Armond, P.A., Arntzen, C.J., Briantais, J.M. and Vernotte, C. (1976) Arch. Biochem. Biophys. 175, 54-63

Bennoun, P. and Li, Y.S. (1973) Biochim. Biophys. Acta 292, 162-168.

Black, M.T., Brearley, T.H. and Horton, P. (1986) Photosyn. Res. 8, 193-207

Farchaus, J.W., Widger, W.R., Cramer, W.A. and Dilley, R.A. (1982) Arch. Biochem. Biophys. 217, 362-367

Forbush, B. and Kok, B. (1968) Biochim. Biophys. Acta 162, 243-253

Ghirardi, M.L. and Melis, A. (1988) Biochim. Biophys. Acta 932, 130-137

Ghirardi, M.L., McCauley, S.W. and Melis, A. (1986) Biochim. Biophys. Acta 851, 331-339

Graan, T. and Ort, D.R. (1986) Biochim. Biophys. Acta 852, 320-330

Guenther, J.E., Nemson, J.A. and Melis, A. (1988) Biochim. Biophys. Acta, in press

Highkin, H.R. (1950) Plant Physiol. 25, 294-306

Highkin, H.R. and Frenkel, A.W. (1962) Plant Physiol. 37, 814-820

Hodges, M. and Barber, J. (1983) FEBS Lett. 160, 177-181

Hodges, M. and Barber, J. (1986) Biochim. Biophys. Acta 848, 239-246

Horvath, G., Droppa, M. and Melis, A. (1984) Photobiochem. Photobiophys. 7, 249-256

Horvath, G., Melis, A., Hideg, E., Droppa, M. and Vigh, L. (1987) Biochim. Biophys. Acta 891, 68-74

Klimov, V.V., Klevanik, A.V., Shuvalov, V.A. and Krasnovsky, A.A. (1977) FEBS Lett. 82, 183-186

Kyle, D.J., Haworth, P. and Arntzen, C.J. (1982) Biochim. Biophys. Acta 680, 336-342

Larsson, U.K. and Andersson, B. (1985) Biochim. Biophys. Acta 809, 396-402

Larsson, U.K., Ogren, E., Oquist, G. and Andersson, B. (1986) Photobiochem. Photobiophys. 13, 29-39

Lavergne, J. (1982) Photobiochem. Photobiophys. 3, 257-285

Malkin, S. (1971) Biochim. Biophys. Acta 234, 415-427

Malkin, S. and Michaeli, G. (1971) in Proc. Second International Congress on Photosynthesis (Forti, G., Avron, M. and Melandri, A., ed.), Vol. I, pp. 149-152

Matoo, A.K. and Edelman, M. (1987) Proc. Natl. Acad. Sci. USA 84, 1497-1501

Matoo, A.K., Pick, U. Hoffman-Falk, H. and Edelman, M. (1981) Proc. Natl. Acad. Sci. USA 78, 1572-1576

McCauley, S.W. and Melis, A. (1987) Photochem. Photobiol. 46, 543-550

Melis, A. (1984) J. Cell Biochem. 24, 271-285

Melis, A. (1985) Biochim. Biophys. Acta 808, 334-342

Melis, A. and Anderson, J.M. (1983) Biochim. Biophys. Acta 724, 473-484

Melis, A. and Duysens, L.N.M. (1979) Photochem. Photobiol. 29, 373-382

Melis, A. and Homann, P.H. (1975) Photochem. Photobiol. 21, 431-437

Melis, A. and Homann, P.H. (1976) Photochem. Photobiol. 23, 343-350

Melis, A. and Thielen, A.P.G.M. (1980) Biochim. Biophys. Acta 589, 275-286

Percival, M.P., Webber, A.N. and Baker, N.R. (1984) Biochim. Biophys. Acta 767, 582-589

Ramanujam, P. and Bose, S. (1983) Photochem. Photobiol. 37, 77-80

Schreiber, U. and Pfister, K. (1982) Biochim. Biophys. Acta 680, 60-68

Sommerville, C.R. (1986) Ann. Rev. Plant Physiol. 37, 467-507

Staehelin, L.A. (1986) In, Encyclopedia of Plant Physiology, Photosynthesis III, Vol. 19, pp. 1-83, Springer-Verlag, Berlin

Steinback, K.E., Bose, S. and Kyle, D.J. (1982) Arch. Biochem. Biophys. 216, 356-361

Sunby, C.A., Melis, A., Maenpaa, P. and Andersson, B. (1986) Biochim. Biophys. Acta 851, 475-483

Thielen, A.P.G.M. and Van Gorkom, H.J. (1981a) Biochim. Biophys. Acta 635, 111-120

Thielen, A.P.G.M. and van Gorkom, H.J. (1981b) In, Photosynthesis, Proceedings of 5th International Congress (Akoyunoglou, G., ed,), Vol. II, pp. 57-64, Balaban International Science Services, Philadelphia, PA

Thielen, A.P.G.M., Van Gorkom, H.J. and Rikjgersberg, C.P. (1981) Biochim. Biophys. Acta 635, 121-131

FLUORESCENCE PROPERTIES OF ISOLATED CHLOROPHYLL-PROTEIN COMPLEXES

Dorothea Siefermann-Harms
Kernforschungszentrum Karlsruhe, Institut für Genetik und Toxikologie, Postfach 3640, D-7500 Karlsruhe, F.R.G.

Key words: Chlorophyll-protein complexes; Photosystem I; Photosystem II; Chlorophyll-proteolipid particles; Light-harvesting chlorophyll-a/b-protein complex II (LHC-II); Fluorescence emission; Fluorescence excitation; Energy transfer; Limonene; Photooxidation.

ABSTRACT

Three examples are presented to illustrate the use of steady-state fluorescence in systems of isolated chlorophyll-protein complexes:

- The chlorophyll-protein subunits of Photosystem I and II have been isolated and their fluorescence maxima established, due to the combined effort of many workers. The available data are reviewed. They characterize the individual chlorophyll-protein complexes and add to the understanding of fluorescence emitted by intact thylakoids.

- Various lipids are able to incorporate separated chlorophyll-protein complexes into proteolipid particles. Fluorescence emission spectra of the separated complexes and of the proteolipid particles are used to reveal interruption and restoration, respectively, of energy transfer between the different chlorophyll-protein complexes.

- The pigments of the light-harvesting Chl-a/b-protein complex (LHC-II) are stable under strong white light, but they are readily photo-oxidised in the presence of limonene or Triton X-100. The fluorescence excitation spectrum of LHC-II reveals differences in the mode of action of limonene and Triton X-100. Evidence for the protective role of LHC-II apoprotein is obtained from these and other experiments.

INTRODUCTION

The fluorescence spectrum of higher plant chloroplasts at room temperature reveals one major broad emission band with maximum at 685 nm and a minor band near 740 nm. Upon cooling to 77 K, the fluorescence intensity increases about twentyfold, and two additional bands appear at 696 and 735 nm (Govindjee and Yang, 1966). The 685, 696 and 735 nm bands respond differently to temperature changes. Up to 170 K the 735-nm band

H. K. Lichtenthaler (ed.) Applications of Chlorophyll Fluorescence, 45–54.

increases slowly followed by a steep increase between 170 and 70 K. In contrast, the 685- and 695-nm bands are quite stable upon cooling from 220 to 120 K. At further cooling a steep increase of the 695-nm and a flat increase of the 685-nm band are observed (Satoh and Butler, 1978a).

Evidence from intact thylakoid membranes has accumulated over the years suggesting that the 685- and 695-nm emission bands originate from chlorophylls located in Photosystem II whereas the emission band at 735 nm originates from chlorophylls of Photosystem I. Based on their studies with developing bean leaves, Strasser and Butler (1977) calculated emission spectra for Photosystem I, Photosystem II and a light-harvesting Chl-a/b protein and correlated the maxima at 735, 695 and 685 nm to these thylakoid components.

The isolation of Chl-protein complexes from thylakoid membranes allowed to substantiate our understanding of thylakoid fluorescence. The fluorescence and other properties of the Chl-proteins will be summarized in the following. Furthermore, the application of fluorescence measurements in studies on the reassembly of separated Chl-protein complexes and on mechanisms protecting protein-bound pigments against photooxidation is presented.

FLUORESCENCE PROPERTIES OF ISOLATED CHLOROPHYLL-PROTEIN COMPLEXES

After gentle solubilisation of thylakoid membranes three different fractions may be separated, i. e. the Photosystem I supracomplex, a Chl-protein supracomplex with Photosystem II activities and a light-harvesting Chl-a/b-protein complex LHC-II that lacks photochemical activities (Satoh and Butler, 1978b). According to Andersson and Anderson (1980), the Photosystem I supracomplex is located in stroma thylakoids whereas the latter two fractions are the predominant proteins of grana membranes. As expected from this compartimentation, LHC-II is usually associated with Photosystem II but can approach Photosystem I under special conditions (Kyle et al,1983).

Tables I and II show the subunit composition of Photosystem I and Photosystem II. The P-700 bearing core complex of Photosystem I consists of two big polypeptides that might or might not be separated by electrophoretic procedures. The P-680 bearing reaction center complex of Photosystem II consists of two smaller polypeptides D1 and D2 in addition to cytochrome b-559 and shows strong analogy to the reaction center complex of purple bacteria (Nanba and Satoh, 1987).

Both, the core complex of Photosystem I and the reaction center complex of Photosystem II are associated with different light-harvesting proteins. In contrast to LHC-II, that is mobile within the thylakoids, the light-harvesting proteins CP 43, CP 47 and LHC-I represent integral parts of Photosystem II and Photosystem I, respectively. These integral light-harvesting proteins carry the chlorophyll species that at 77 K emit the long-wavelengh fluorescence at 695 nm of Photosystem II and at 730 nm of Photosystem I (Table I and II).

Table I: Components and fluorescence properties of Chl-protein complexes organized in Photosystem I

Species	Chl/P-700	Non-peptice components (mol / 100 mol Chl-a)				Polypeptides No. observed in size class (kDa)			Fluorescence emission Maxima (nm) at		Refs
		P-700	Chl-b	β-car	Xanth	≤ 21[a]	21-24[b]	60-70[c]	RT	77K	
Photosystem I supracomplex PS I-200											
Spinach	230	0.51	14.3	nr	nr	8	3	1	nr	735	(1)
	210	0.54	12.9	14.5	12.4		nr		nr	nr	(2)
Photosystem I supracomplex PS I-110											
Pea	110	1.0	9.1	nr	nr	6	3-4	2	690	736	(3,4)
Photosystem I core complex PS I-65											
Spinach	50-60	nr	nr	nr	nr	6	0	1	nr	725 (in D-M) 681 (in T-X)	(5)
	77	1.4	6.6	20.7	2.3	nr	nr	nr	nr	nr	(6)
Pea	65	1.6	6.2	nr	nr	6	0	2	690	722	(3,4)
Photosystem I core complex PS I-40											
Spinach	nr	nr	< 6	nr	nr	0	0	1	nr	721 (in D-M) 671 (in SDS)	(5)
Pea	ca. 40	2.5	< 6	nr	nr	6	0	2	nr	694 (in T-X)	(3)
	ca. 40	2.5	< 6	nr	nr	6	0	2	nr	721 (in D-M)	(5)
Light-harvesting Chl-a/b-protein LHC-I											
Spinach		0	22.2	nr	nr	0	1	0	nr	730	(7)
		0	33	11	30		nr		nr	nr	(6)
Pea		0	21.3	nr	nr	0	3-4	0	nr	730	(4)
		0	nr	nr	nr	1	1	0	nr	732	(8)

[a] : Polypeptides not associated with pigments; [b] : Polypeptides of LHC-I, pigment-bearing under non-denaturing conditions; [c] : Polypeptide(s) bearing P-700 and core antenna pigments under non-denaturing conditions.

Abbreviations: Chl, chlorophyll; β-car, β-carotene; Xanth, xanthophylls; D-M, dodecyl maltoside; SDS, sodium dodecyl sulfate; T-X, Triton X-100; nr, not reported.

References: (1) Ortiz et al (1985), (2) Siefermann-Harms (1984), (3) Mullet et al (1980), (4) Haworth et al (1983), (5) Nechushtai et al (1986), (6) Damm et al (1987), (7) Lam et al (1984), (8) Tapie et al (1984).

Table II: Components and fluorescence properties of Chl-protein complexes organized in Photosystem II (Spinach)

Chl/RC[a]	Non-peptide components (mol / 100 mol Chl-a)						Polypeptides[b] (kDa) intrinsic							extrinsic			O_2 evolution	Fluorescence emission Maxima (nm) at			Energy transfer $^1Car^* \rightarrow$ Chl-a	Refs
	Phe-a	Chl-b	β-car	Xanth	PQ-9	b-559	9	26	27	32	34	43	47	18	24	33		RT	77K	4K		
Purified Photosystem II grana preparation (Photosystem II Particles)																						
260[1]	nr	46.5	nr	nr	nr	nr											yes	685	685 ,695	695	near 100 %	(1)
220[1]	nr	50.0	nr	nr	3.2	1.8	+	+	+	+	+	+	+	+	+	+	yes					(2)
	nr	51.6	5.1	29.3	nr	nr																(3)
Photosystem II core complex																						
45[1]	nr	nr	nr	nr	nr	nr	+			+	+	+	+		+	+	yes	685	(685),695	695	35 %	(1)
55[1]	4.2	0	20.0	0	4.0	nr	+			+	+	+	+			+	yes					(4)
50[2]	4.2	0	20.0	0	3.7	nr	+			+	+	+	+				no					(5)
Reaction center complex of Photosystem II (D1-D2 complex)																						
4.1[3]	49	nr	nr	nr	nr	nr	+			+	+						no	679	nr	684	20 %	(6)
5.1[3]	39	0	19.6	0	0	25.5	+			+	+											(7)
3.8[3]	53	0	21.0	0	21.0	nr	+			+	+											(8)
Light-harvesting Chl-a-protein CP 43																						
	0	0	nr	0	0	0						+					no	nr	685	nr	nr	(9)
	0	0	11	0	0	0						+										(8)
Light-harvesting Chl-a-protein CP 47																						
	0	0	nr	0	0	0							+				no	681	nr	693	40 %	(6)
	0	0	10	0	0	0							+									(8)
Light-harvesting Chl-a/b-protein LHC-II																						
	0	77	1.3	53	nr	0		+	+								no	681	681	nr	100 %	(10)

a : Chlorophyll content per primary quinone acceptor Q_A ([1]), P-680 ([2]), or 2 pheophytin-a ([3]).
b : Polypeptides bearing pigments in their non-denatured state: 26, 27, 32 (D1), 34 (D2), 43, 47 kDa. 9 kDa: Apoprotein of cytochrome b-559.

Abbreviations: Chl, chlorophyll; Car, carotenoids; β-car, β-carotene; Phe, pheophytin; PQ-9, plastoquinone 9; RC, reaction center; Xanth, xanthophylls; nr, not reported.

References: (1) Van Dorssen et al (1987a), (2) Murata et al (1984), (3) Siefermann-Harms (1984) - data from combined LHC + PS-II complexes, (4) Yamada et al (1987), (5) Omata et al (1984), (6) Van Dorssen et al (1987b), (7) Nanba and Satoh (1987), (8) Akabori et al (1988), (9) Nakatani et al (1984), (10) Ref. 3 and unpublished results.

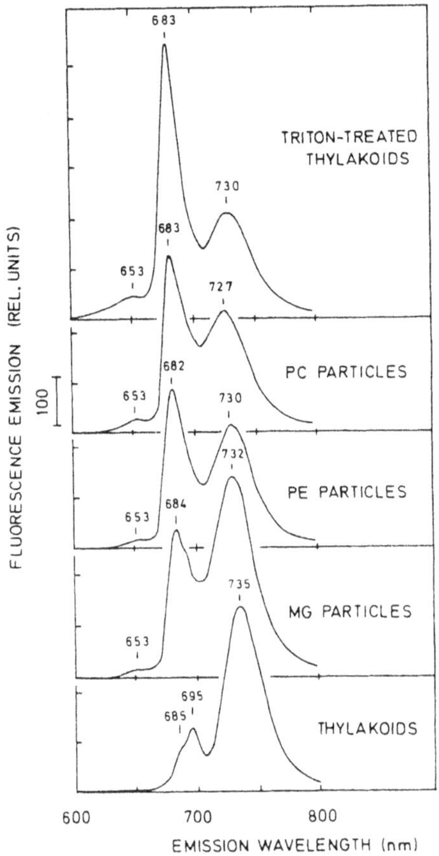

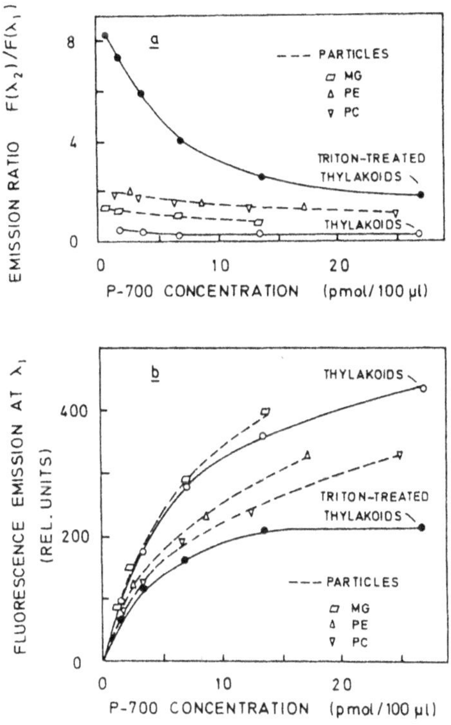

Fig. 1: 77 K fluorescence emission spectra for excitation at 470 nm of spinach thylakoids, Chl-proteolipid particles prepared with MG, PE or PC, respectively, and Triton-treated thylakoids. The 100 µl samples contained 6.1 µg Chl.

Fig. 2: Effect of Chl-protein concentration on the properties of 77 K fluorescence excited at 470 nm in thylakoids, Triton-treated thylakoids and Chl-proteolipid particles. P-700 content was used as a relative measure for Chl-protein concentration. (a) Variations in the ratio of fluorescence intensities of emission bands near 730 nm (λ_1) and near 684 nm (λ_2). (b) Variations in fluorescence intensity of the emission band near 730 nm (λ_1).
(from Siefermann-Harms et al, 1987)

MG: Monogalactosyldiacylglycerol; PE: phosphatidylethanolamine; PC; phosphatidylcholine

CHARACTERIZATION OF REASSEMBLED CHLOROPHYLL-PROTEOLIPID PARTICLES USING 77 K FLUORESCENCE EMISSION

Conditions required to reassemble physically separated Chl-protein complexes and polar lipids into proteolipid particles have been studied by several authors (Larkum and Anderson, 1982; Siefermann-Harms et al. 1982, 1987; Murphy et al,1984; Sprague et al,1985; Murphy, 1986). In most of the studies, the proteolipid particles were characterized based on their fluorescence emission at 77 K that differed from that of the mixture of solubilized Chl-protein complexes and was more or less comparable to that of intact thylakoids.

In Fig.1 the 77 K fluorescence emission spectra of intact thylakoids (bottom) and solubilized Chl-protein complexes (top) are compared with those of different proteolipid particles. When the Chl-protein complexes of the thylakoids are physically separated during membrane solubilization, the 730-nm emission band of Photosystem I is decreased and the 685- and 695-nm bands of Photosystem II are superimposed by a strong emission band at 681-683 nm of LHC-II. The changes reflect interruption of energy transfer from LHC-II to the photosystems. When the solubilized complexes are incorporated in proteolipid particles the emission band at 683 nm becomes less intense and the 730-nm band is increased, depending on the lipid used. The data suggest that energy transfer fom LHC-II to Photosystem I is restored at different degrees.

The fluorescence emission spectra of physically separated Chl-protein complexes and of Chl-protein complexes assembled in supra-molecular structures are affected differently by sample dilution. When solubilized membranes are diluted, their 683-nm emission band is enhanced and the fluorescence ratio at 683 versus 730 nm is strongly increased in spite of the decrease of fluorescence at 730 nm (Fig.2, Triton-treated thylakoids). Evidently, the distance between solubilized molecules affects LHC-II emission: its increase leads to lower reabsorption of emitted light by neighbour molecules. In case of thylakoids and proteolipid particles no such dilution effect is observed (Fig.2). This may easily be explained: since the position of Chl-protein complexes in membranes and proteolipid particles is fixed, their distance is little affected by sample dilution.

Fig.1 and 2 illustrate the use of 77 K fluorescence in studies of Chl-proteolipid particles. For further details on preparation and properties of proteolipid particles formed with different lipids see Siefermann-Harms et al (1987).

MODE OF LIMONENE ACTION IN THE DE-STABILIZATION OF LHC-II - AN ANALYSIS BASED ON FLUORESCENCE EXCITATION MEASUREMENTS

A few years ago we showed that the pigments of LHC-II, i.e. chlorophylls and carotenoids, are organized in close proximity and buried inside the LHC-II apoprotein (Siefermann-Harms and Ninnemann, 1982). Evidence for such organization came from the efficient energy transfer from excited carotenoids to Chl-a that requires dense packing of the pigments, and from the stability of LHC-bound Chl-a under strongly acidic conditions. Recently I observed that LHC-bound pigments are highly stable under photo-oxidative conditions (aerobiosis + white light at 1000 W/m^2;

Table III: Effect of limonene and Triton X-100 on the stability of LHC-II - bound Chl-a under strong white light or low pH, and on excitation energy transfer

		LHC-II	LHC-II + limonene	LHC-II + Triton X-100
Photosensitivity	% Chl-a destroyed under white light at 1000 W/m^2 after			
	15 min:	7 %	30 %	70 %
	30 min:	11 %	50 %	92 %
Acid sensitivity	% Chl-a destroyed during incubation at pH 1.25 for			
	5 min:	15 %	71 %	100 %
	15 min:	22 %	100 %	100 %
Excitation energy transfer	Efficiency of energy transfer $^1car^* \longrightarrow$ Chl-a:			
		100 %	100 %	< 10 %

LHC-II was isolated from spinach according to Siefermann-Harms and Ninnemann (1982). The reaction mixtures for photosensitivity (at pH 7.8) and acid sensitivity (at pH 1.25) measurements contained LHC-II equivalent to 1.9 µg Chl a+b/ml and, as stated, 0.22 mM R(+)-limonene or Triton X-100 at a Chl/Triton ratio of 1/300 (w/w). Conditions for excitation energy transfer measurements are given in Fig. 3.

Samples contained LHC-II (0.19 µg Chl/ml) in 50 mM Tricine buffer pH 7.8. a,b: Spectra monitored in the presence of 0.22 mM limonene - they were identical to those without limonene. c,d: Spectra monitored in the presence of Triton X-100 (Chl/Tr = 1/300). The excitation spectra for emission at 680 nm (F-680) and at 650 nm (F-650) are corrected for equal quantum number.

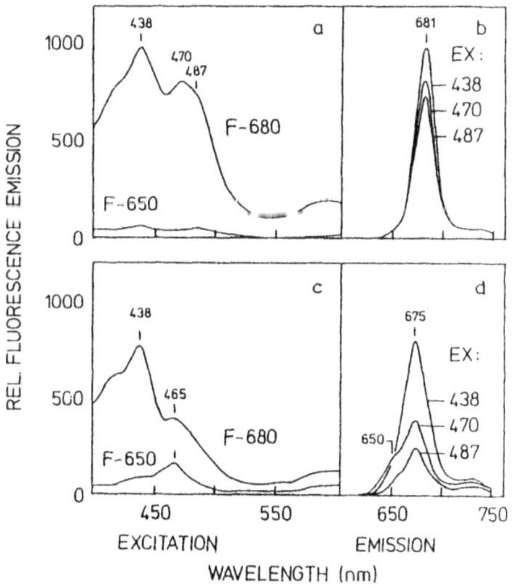

Fig. 3: Effect of limonene and Triton X-100 on fluorescence excitation (a,c) and emission (b,d) spectra of LHC-II at room temperature (from Siefermann-Harms, 1988)

Siefermann-Harms, 1988). As shown in Table III, both, the terpene limonene and the detergent Triton X-100 eliminate the photostability of LHC-bound Chl-a as well as its acid stability.

The role of carotenoids in the protection of Chl-a from photo-oxidation is well established (for review see Siefermann-Harms, 1987). Since Triton X-100 destroys the densely packed organization of chlorophylls and carotenoids in LHC it might be speculated that Triton acts on the photostability of LHC simply by removing the carotenoids. However, the mode of limonene action demonstrates that the presence of carotenoids in the LHC is not sufficient to protect Chl-a from photo-oxidation. As shown in Fig.3, limonene, in contrast to Triton X-100, does not modify the fluorescence properties of LHC. As for untreated LHC, the fluorescence excitation spectrum in the presence of limonene agrees with the absorbance spectrum of the complex, indicating that the energy absorbed by carotenoids or by Chl-b is transferred to Chl-a at an efficiency of 100%. Consequently, limonene does not disturb the organization of LHC-bound pigments.

As discussed elsewhere (Siefermann-Harms, 1988) these and additional data suggest that the apoprotein of LHC-II participates in protecting LHC-II bound pigments from the destructive effect of light plus O_2.

ACKNOWLEDGEMENTS

The author's work was supported in part by the Deutsche Forschungsgemeinschaft and by Project Europäisches Forschungszentrum für Maßnahmen zur Luftreinhaltung (PEF).

REFERENCES

Akabori K, Tsukamoto H, Tsukihara J, Nagatsuka T, Motokawa O and Toyoshima Y, 1988. Disintegration and reconstitution of Photosystem II reaction center core complex. I. Preparation and characterization of three different types of subcomplex. Biochim. Biophys. Acta 932: 345-357.

Andersson B and Anderson JM, 1980. Lateral heterogeneity in the distribution of chlorophyll-protein complexes of the thylakoid membranes of spinach chloroplasts. Biochim. Biophys. Acta 593: 427-440.

Damm I, Knoetzel J and Grimme LH, 1987. On the protective role of carotenoids in the PS I reaction centre and LHC I complexes of the thylakoid membrane. In: Biggins J. Progress in Photosynthesis Research. Proc. VIIth Internatl. Congr. on Photosynthesis. Vol. 2: 351-354. Martinus Nijhoff, Dordrecht, Boston, Lancaster.

Govindjee and Yang L, 1966. Structure of the red fluorescence band in chloroplasts. J. General Physiol. 49: 763-780.

Haworth Ph, Watson JL and Arntzen CJ, 1983. The detection, isolation and characterization of a light-harvesting complex which is specifically associated with Photosystem I. Biochim. Biophys. Acta 724: 151-158.

Kyle DJ, Staehelin LA and Arntzen CJ, 1983. Lateral mobility of the light-harvesting complex in chloroplast membranes controls excitation energy distribution in higher plants. Arch. Biochem. Biophys. 222: 527-541.

Lam E, Ortiz W and Malkin R, 1984. Chlorophyll a/b proteins of Photosystem I. FEBS Lett. 168: 10-14.

Larkum AWD and Anderson JM, 1982. The reconstitution of a photosystem II protein complex, P-700-chlorophyll a-protein complex and light-harvesting chlorophyll a/b-protein. Biochim. Biophys. Acta 679: 410-421.

Mullet JE, Burke JJ and Arntzen CJ, 1980. Chlorophyll proteins of Photosystem I. Plant Physiol. 65: 814-822.

Murata N, Miyao M, Omata T, Matsunami H and Kuwabara T, 1984. Stoichiometry of components in the photosynthetic oxygen evolution system of Photosystem II particles prepared with Triton X-100 from spinach chloroplasts. Biochim. Biophys. Acta 765: 363-369.

Murphy DJ, 1986. Reconstitution of energy transfer and electron transfer between solubilised pigment-protein complexes from thylakoid membranes. The role of acyl lipids. Photosynth. Research 8: 219-233.

Murphy DJ, Crowther D and Woodrow IE, 1984. Reconstitution of light harvesting chlorophyll-protein complexes with Photosystem 2 complexes in soybean phosphatidylcholine liposomes. FEBS Lett. 165: 151-155.

Nakatani HY, Ke B, Dolan E and Arntzen CJ, 1984. Identity of the Photosystem II reaction center polypeptide. Biochim. Biophys. Acta 765: 347-352.

Nanba O and Satoh K, 1987. Isolation of a Photosystem II reaction center consisting of D-1 and D-2 polypeptides and cytochrome b-559. Proc. Natl. Acad. Sci. USA 84: 109-112.

Nechushtai R, Nourizadeh SD and Thornber JP, 1986. A reevaluation of the fluorescence of the core chlorophylls of Photosystem I. Biochim. Biophys. Acta 848: 193-200.

Omata T, Murata N and Satoh K, 1984. Quinone and pheophytin in the photosynthetic reaction center II from spinach chloroplasts. Biochim. Biophys. Acta 765: 403-405.

Ortiz W, Lam E, Chollar S, Munt D and Malkin R, 1985. Topography of the protein complexes of the chloroplast thylakoid membrane. Studies of Photosystem I using a chemical probe and proteolytic digestion. Plant Physiol. 77: 389-397.

Satoh K and Butler WL, 1978a. Competition between the 735 nm fluorescence and the photochemistry of Photosystem I in chloroplasts at low temperature. Biochim. Biophys. Acta 502: 103-110.

Satoh K and Butler WL, 1978b. Low temperature spectral properties of subchloroplast fractions purified from spinach. Plant Physiol. 61: 373-379.

Siefermann-Harms D, 1984. Evidence for a heterogenous organization of violaxanthin in thylakoid membranes. Photochem. Photobiol. 40: 507-512.

Siefermann-Harms D, 1987. The light-harvesting and protective functions of carotenoids in photosynthetic membranes. Physiol. Plantarum 69: 561-568.

Siefermann-Harms D, 1988. Physiologische und biochemische Untersuchungen zur Montanen Vergilbung von Fichten. - Teil I: Die Rolle von Apoprotein und Pigmentanordnung beim Schutz der Pigmente des Licht-sammelnden Chl-a/b-Proteins vor Photo-oxidation. KfK-PEF, in press.

Siefermann-Harms D and Ninnemann H, 1982. Pigment organization in the light-harvesting Chl-a/b-protein complex of lettuce chloroplasts. - Evidence obtained from protection of the chlorophylls against proton attack and from excitation energy transfer. Photochem. Photobiol. 35: 719-731.

Siefermann-Harms D, Ross JW, Kaneshiro KH and Yamamoto HY, 1982. Reconstitution by monogalactosyldiacylglycerol of energy transfer from light-harvesting chlorophyll-a/b-protein complex to the photosystems in Triton X-100-solubilized thylakoids. FEBS Lett. 149: 191-196.

Siefermann-Harms D, Ninnemann H and Yamamoto HY, 1987. Reassembly of solubilized chlorophyll-protein complexes in proteolipid particles - comparison of monogalactosyldiacylglycerol and two phospholipids. Biochim. Biophys. Acta 892: 303-313.

Sprague SG, Camm EL, Green BR and Staehelin LA, 1985. Reconstitution of light-harvesting complexes and Photosystem II cores into galactolipid and phospholipid liposomes. J. Cell Biol. 100: 552-557.

Strasser RJ and Butler WL, 1977. Fluorescence emission spectra of Photosystem I, Photosystem II and the light-harvesting chlorophyll a/b complex of higher plants. Biochim. Biophys. Acta 462: 307-313.

Tapie P, Choquet Y, Breton J, Delepelaire P and Wollman F-A, 1984. Orientation of Photosystem-I pigments. Investigation by low-temperature linear dichroism and polarized fluorescence emission. Biochim. Biophys. Acta 767: 57-69.

Van Dorssen RJ, Plijter JJ, Dekker JP, Den Ouden A, Amesz J and Van Gorkom HJ, 1987a. Spectroscopic properties of chloroplast grana membranes and of the core of Photosystem II. Biochim. Biophys. Acta 890: 134-143.

Van Dorssen RJ, Breton J, Plijter JJ, Satoh K, Van Gorkom HJ and Amesz J, 1987b. Spectroscopic properties of the reaction center and of the 47 kDa chlorophyll protein of Photosystem II. Biochim. Biophys. Acta 893: 267-274.

Yamada Y, Tang X-S, Itoh S and SAtoh K, 1987. Purification and properties of an oxygen-evolving Photosystem II reaction-center complex from spinach. Biochim. Biophys. Acta 891: 129-137.

LOW TEMPERATURE FLUORESCENCE SPECTRA OF CHLOROPLASTS: METHODICAL ASPECTS AND POSSIBLE APPLICATIONS

P. ŠIFFEL* and Z. ŠESTÁK**

Institute of Experimental Botany, Czechoslovak Academy of Sciences,
Department of Physiology of Photosynthesis,
* CS-370 05 České Budějovice, Branišovská 31, Czechoslovakia
** CS-160 00 Praha 6, Na Karlovce 1a, Czechoslovakia

Key words: chlorophyll-protein complexes; chlorophyll in vivo; chlorophyll synthesis; chlorophyll degradation; Phaseolus vulgaris L.

ABSTRACT

The main advantages of measuring fluorescence emission and excitation spectra at low (77 K) instead of room (293 K) temperature are spectra with narrower and higher, better distinguishable bands, and the possibility to study rapid changes in the structure of photosynthetic apparatus. The sample for measuring correct fluorescence spectra must not contain a large amount of pigments (only pale thin leaves or chloroplast preparations are appropriate), must be properly mounted (chloroplasts on a translucent filter) and rapidly frozen, and not affected by substances causing degradation of pigment-protein complexes (not even mild detergents). Combination of emission and excitation spectra with their second derivatives and with calculated difference spectra enables to determine the presence of individual chlorophyll-protein complexes, their synthesis and degradation in thylakoids, chloroplasts and whole leaves.

INTRODUCTION

Spectra of whole leaves are often flattened or distorted due to a high concentration of densely pigmented chloroplasts in many layers of cells. A situation more similar to that in intact cells is obtained by measuring fluorescence spectra of isolated chloroplasts or thylakoid membranes.

By measuring fluorescence spectra of chloroplasts at room temperature only few, rather low broad bands appear. The shape of the spectra much improves by measuring at low temperature ($<$100 K). The most frequently used freezing initiator is liquid nitrogen (77 K); for special purposes liquid helium (4 K) is used.

There are two main reasons for measuring fluorescence spectra of chloroplasts at a low temperature: (a) The sharpening of spectral bands at low temperature and thus enhancement of their resolution; this in many cases enables the detection of bands not measurable at room temperature; (b) freezing to low temperature stops biochemical and some photochemical reactions which in turn gives a possibility to study the changes in the structure of photosynthetic apparatus, the duration of which is shorter than the time of measurement.

H. K. Lichtenthaler (ed.) Applications of Chlorophyll Fluorescence, 55–61.

METHODICAL RECOMMENDATIONS

By measuring at liquid nitrogen temperature care should be taken to reduce undesirable side effects which affect the shape of the spectrum. The distortions of the spectra might arise from the optical arrangement of the used measuring set (luminescence of filters, reflections, etc.). In available commercial instruments these artefacts are mostly restricted. Critical points in measuring low temperature fluorescence spectra are the sample mounting and concentration dependent reabsorption of excitation and emitted radiation.

Incorrect sample mounting may bring about a destruction of chloroplast membrane integrity and thus change the fluorescence emission spectra. At 77 K emission spectra are measured of chloroplasts contained in a plexiglass cuvette, groove of metal block, or glass capillary, or of chloroplasts adsorbed on cheese-cloth, filter paper or membrane filter (for their description see Harnischfeger 1977). Another way is to use frozen subcellular particles mixed with quartz particles to decrease chl concentration (Weis 1985). The three former methods have the disadvantage of a slow cooling rate, which allows the formation of large crystals of water and high ion concentration during freezing, and often results in the disturbance of thylakoid membranes and thus in a modified shape of fluorescence emission spectra, e.g. in the ratio of emission peaks F685/F740 (Harnischfeger 1977, 1979). Sufficiently high cooling rates are ensured only by using chloroplasts adsorbed on a wet support that enables direct contact of liquid nitrogen with chloroplasts. The effect of rate of chloroplast cooling on the fluorescence emission spectra was not confirmed by Butler and Strasser (1977) but the methods using the plexiglass cuvette or metal block did not ensure in our laboratory a satisfactory reproduction of fluorescence emission spectra measured during ontogeny of leaves.

By measuring spectra of chloroplasts adsorbed on a surface of some cloth or cellulose support it is desirable to know the amount of chlorophyll (chl) in chloroplasts adsorbed per unit surface area. The easiest solution is to filter a suspension of chloroplasts with a known concentration through a known surface of membrane filter.

The immersion of a part of sample holder into liquid nitrogen should be as simple as possible, in order to prevent liquid nitrogen bubbling, and must prevent the sample turning. One way is to mount the sample into a window of a holder made of metal of good thermal conductivity (e.g. copper), and immersion of the lower part of such a holder into liquid nitrogen. Another way is to fix the membrane filter with chloroplasts by freezing it to a plexiglass or quartz plate. The first way cannot ensure precise reproduction of sample temperature and thus is not convenient for fine measurement of emission spectra.

In order to prevent the freezing damage, various cryoprotective agents such as glycerol, glycol, dimethylsulphoxide and others have been added to chloroplast suspensions. Protective agents show a high viscosity in aqueous solutions, and thus support the formation of amorphous, non-crystalline structure. However, the real effect of cryoprotective agents on the thylakoid membrane remains obscure: they are added in high concentrations and thus osmotic pressure and inner impurities might affect the spectra in addition to direct effects, e.g. reduction of the extent of the State 1/State 2 transitions and quenching of photosystem (PS) 2 fluorescence, especially of antenna molecules (Butko and Szalay 1986). Thus the cryoprotective agents can be recommended only when absolutely neccessary, i.e. for polarization measurements.

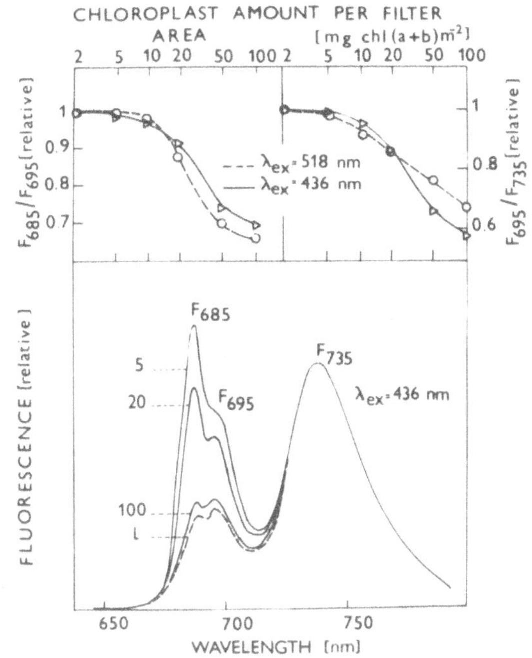

Fig.1. The dependence of shape of the low temperature (77 K) fluorescence emission spectrum of chloroplasts isolated from primary leaves of Phaseolus vulgaris L. (bottom) and ratios of intensities of individual emission bands (top) on the amount of chloroplasts [expressed as mg(chl a+b) m-2] per unit area of the membrane filter. Figures at curves give the amounts of chloroplasts, L is emission spectrum of the initial leaf.

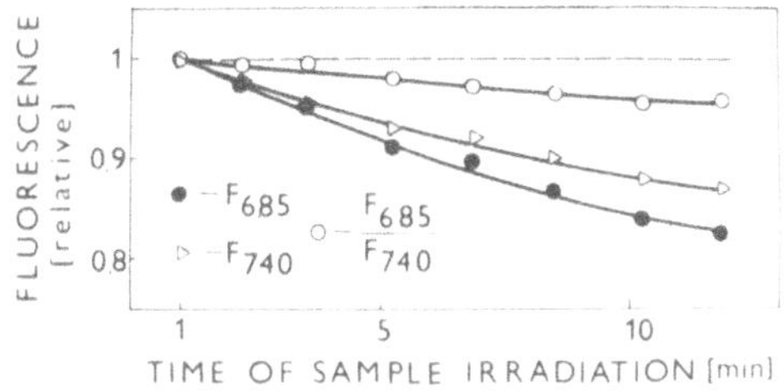

Fig.2. Changes in intensity of fluorescence emission at 685 nm (F685) and 740 nm (F740) and their ratio in the course of illumination on the membrane filter of chloroplasts from primary leaves of Phaseolus vulgaris L. Means of three measurements at 77 K related to the value after 1 min illumination.

The distortion of both the fluorescence emission and excitation spectra occurs also at a high pigment concentration. Pigments act as an inner filter and thus cause a wavelength-dependent absorption of excitation radiation or fluorescence emission in the sample. By increasing pigment concentration fluorescence excitation spectra flatten and in fluorescence emission spectra the intensity of short-wavelength part of the spectra relatively decreases (Fig. 1). The limiting pigment concentration, below which the fluorescence spectra do not change, depends on the method used. For chloroplasts adsorbed on the surface of membrane filter we did not find any concentration dependent change in the fluorescence emission spectra below 5-7 mg (chl) m-2. This amount is roughly one hundredth of the chl concentration usually present in leaves. This surface chl concentration must be further lowered for measuring of fluorescence excitation spectra of samples prepared from greening or yellow leaves. Such leaves contain a great amount of pigments, which do not transfer excitation energy to fluorescent species, but have a strong absorption band in the blue region.

Of course, we cannot eliminate the reabsorption that occurs within a single chloroplast. Nevertheless, Szalay et al. (1967) show that the contribution of this effect is not larger than 5-6 %.

Instead of isolated chloroplasts a suspension of blended leaves soaked into filter paper or membrane filter can also be used for fluorescence measurements, but the reproducibility of results is rather low. The surface concentration of chl can be roughly given by the amount of plant material and volume of the medium used for blending. The advantage of this method is it quickness: ca. 1 min from the homogenization to sample freezing.

To achieve a high reproduction of chloroplast fluorescence spectra care should be taken in: the choice of plant material including plant age (cf. Šesták and Šiffel 1988) and cultivation conditions, procedure of chloroplast isolation (avoid even mild detergents), conditions of sample preparation and measuring. Laboratory illumination might affect the State 1-State 2 transitions, and thus change the shape of fluorescence emission spectra. Emission spectrum (at 293 or 77 K) of chloroplasts depends on the time interval between the beginning of sample excitation and start of emission measurement (Fig. 2). Concentration of ions in the isolated chloroplasts may vary a little in dependence on ion concentrations in the used leaves, even if the ions concentration in the isolation medium is stable. Thus one should prefer a medium with the divalent ions concentration high enough to sature its effect on fluorescence emission spectrum.

SPECTRAL BANDS THAT CAN BE DISTINGUISHED ON FLUORESCENCE SPECTRA OF CHLOROPLASTS

Using the low temperature measurements connected with curve deconvolution and/or differentiation in the fluorescence emission spectra about 6 components have been distinguished (for review see Govindjee et al. 1986). These components have been related to those found in low temperature absorption spectra, and to photosystems and pigment-protein complexes. Nevertheless, some of them may be artefacts induced by screening or structural changes during freezing (Meister 1977).

The Soret (blue region) band of chloroplast absorption spectra is formed by many components with greatly overlapping bands; most of them cannot be separated even by the low temperature derivative absorption spectroscopy but some of them are more pronounced in the fluorescence excitation spectra. Only the bands of chl a around 431 and 438 nm, of chl b around 472 nm, and of carotenoids at around 459, 483, 494, 508 and 517 nm can be clearly distinguished in the second derivation of the spectrum (Fig. 3).

The room temperature fluorescence emission spectrum (Fig. 4) of chloroplasts has a form similar to that of chl in organic solvents, but it is shifted towards longer wavelengths by about 15 nm. Its shows a peak at about 685 nm and a shoulder at ca. 730 nm. At 77 K the ratio of both main peaks dramatically changes and a more complicated structure of chloroplast emission is revealed; three emission peaks localized at 685, 695 and around 735 nm are clearly observed. The second derivative reveals a further band at 681 nm and a complicated structure of the long-wavelength band. Two main components of the long-wavelength band localized at 715-725 and 735-745 nm belong to the surrounding of the reaction centre of PS 1 and to light-harvesting complex (LHC) 1, respectively. The emissions of LHC 2 and PS 2 also contribute to the long-wavelength band. The bands at 685 and 695 nm originate from the surrounding of the reaction centre of PS 2 and from recombination luminescence of the reaction centre of PS 2, respectively.

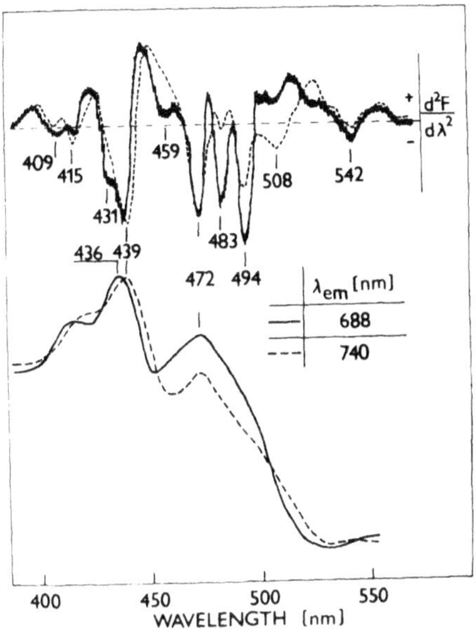

Fig.3. Fluorescence excitation spectra of chloroplasts isolated from primary leaves of Phaseolus vulgaris L. and measured at 77 K on a membrane filter. Excitation spectra for the Soret band were measured as emission at either 688 or 740 nm and their second derivatives were recorded without correction for the spectrum of the excitation halogene source.

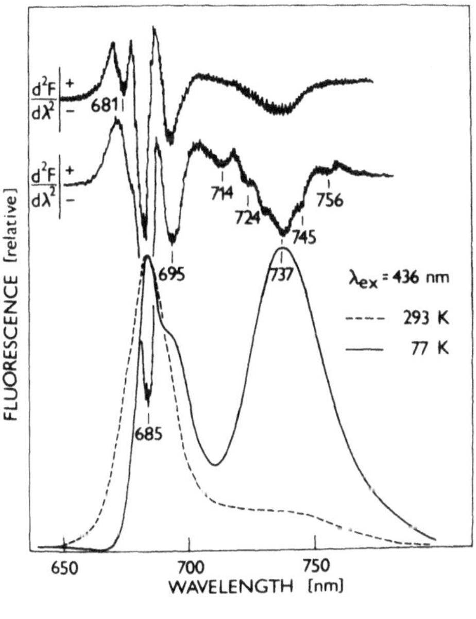

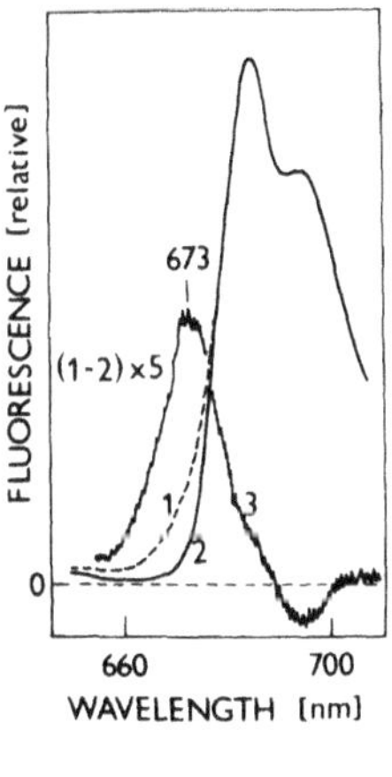

Fig.4.(left) Fluorescence emission spectra of chloroplasts isolated from primary leaves of 15 d-old plants of Phaseolus vulgaris L. and measured at 293 K (- - -) or 77 K (——) after excitation at 436 nm. The derivatives of the low temperature spectra were produced using 2 nm (a) or 5 nm (b) numerical steps of the calculator; only the larger step enables detection of the hidden shoulders.

Fig.5.(right) Fluorescence emission spectrum of whole yellowing primary leaf of Phaseolus vulgaris L. excited by 436 nm or 504 nm (2); their difference spectrum (5x multiplied) shows a peak at 674 nm, the short-wavelength maximum of chlorophyll a .

The shoulder at 681 nm belongs to the emission of LHC 2 (for references to individual bands see Govindjee et al. 1986).

In addition to these intensive components of fluorescence emission spectra, minor bands originating from intermittent products of chl synthesis or degradation can be observed. These bands, due to their very low intensity, may be observed only in the spectral range where they are not overlapped by strong emission of the antenna chl. Determination of these bands requires the use of low temperature differential measurement, which is based on the subtraction of two emission spectra excited at both the absorption maximum and minimum of the measured pigment. The differential measurements reveal, e.g., the band of chl a at about 673 nm (Fig. 5), the band of chl b at about 653 nm and bands at 632 and 641 nm belonging to protochlorophyllide (Lebedev et al. 1986, Šiffel et al. 1987). Similar emissions of chl a and b are observed during greening of etiolated leaves (for references see Šiffel et al. 1987) and the intensity of appropriate bands correlates with the rate of chl accumulation in the course of development of primary bean leaves (Šiffel et al. 1987), and thus they can be regarded as characteristic for chl synthesis. In addition to these four bands an emission of pheophytin can be observed near 670 nm in the leaves of plants darkened for several days.

POSSIBLE APPLICATION OF LOW TEMPERATURE SPECTRA

Low temperature spectroscopy is a non-destructive method often used for identification of isolated pigment-protein complexes, chl synthesis, degradative processs occurring in the photosynthetic apparatus under natural or artificial conditions (high irradiance, extreme pH, degradative enzymes, polluting gases, etc.).

The low temperature absorption and fluorescence spectra can also be used for a rough estimation of changes in the content of some pigment-protein complexes. The excitation spectra of LHC 1 and 2 have a much higher intensity between 460 and 490 nm than the excitation spectra of reaction centres complexes. Thus the ratio of chloroplast excitation spectra intensity at both the 435 and 472 nm maxima, i.e. E472/E435, can be used as a measure of relative content of LHC 1 + LHC 2. Absorption spectra are not convenient for this purpose, because they may include in the blue region pigments not connected to the pigment antenna.

The emission above 730 nm belongs mainly to LHC 1 and thus also the absorption around 708 nm belongs mainly to this complex. Hence these two bands may serve as an indicator of changes in relative concentration of LHC 1.

The changes in intensity of the LHC 2 emission band localized at 681 nm may display displacement of LHC and reaction centres of PS 2 that occurs at acid action on isolated chloroplasts (Lebedev et al. 1986) or during natural yellowing of leaves.

The short wavelength emissions of both chl and protochlorophyllide serve as a probe of chl synthesis in mature and senescing leaves under natural physiological conditions.

REFERENCES

Butko P and Szalay L, 1986. Effect of glycerol on the low-temperature fluorescence spectra of green algae. Acta Biochimica et Biophysica Hungarica 21: 391-395.

Butler WL and Strasser RJ, 1977. Does the rate of cooling affect fluorescence properties of chloroplasts at -196 °C. Biochimica et Biophysica Acta 462: 283-289.

Govindjee, Amesz J and Fork DC, 1986. Light emission by plants and bacteria. Academic Press, Orlando - San Diego - New York - Austin - Boston - London - Sydney - Tokyo - Toronto.

Harnischfeger G, 1977. The use of fluorescence emission at 77 oK in the analysis of the photosynthetic apparatus of higher plants and algae. Advances in Botanical Research 5: 1-52.

Harnischfeger G, 1979. Connection between the rate of cooling and fluorescence properties at 77 K of isolated chloroplasts. Biochimica et Biophysica Acta 546: 348-355.

Lebedev NN, Šiffel P and Krasnovskiĭ AA, 1985. Detection of protochlorophyllide forms in irradiated green leaves and chloroplasts by difference fluorescence spectroscopy at 77 K. Photosynthetica 19: 183-187.

Lebedev NN, Šiffel P, Pakshina EV and Krasnovskiĭ AA, 1986. The effect of acidification on absorption and fluorescence spectra of French bean chloroplasts and kinetics of pheophytin formation. Photosynthetica 20: 124-130.

Meister A, 1977. Messung von Absorptionspektren in vivo. Kulturpflanze 25: 141-154.

Šesták Z and Šiffel P, 1988. Changes in chloroplast fluorescence during leaf development. This volume.

Šiffel P, Lebedev NN and Krasnovskiĭ AA, 1987. Detection of short-wavelength chlorophyll a emission in green leaves. Photosynthetica 21: 23-28.

Szalay L, Török M and Govindjee, 1967. Effect of secondary fluorescence on the emission spectrum and quantum yield of fluorescence in chlorophyll a solutions and algal suspensions. Acta Biochimica et Biophysica Academica Scientiarum Hungaricae 5: 261-262.

Weis E, 1985. Chlorophyll fluorescence at 77 K in intact leaves: Characterization of a technique to eliminate artifacts related to self-absorption. Photosynthesis Research 6: 73-86.

NEW WAYS OF ASSESSING PHOTOSYNTHETIC ACTIVITY WITH A PULSE MODULATION FLUOROMETER.

U. SCHREIBER, C. NEUBAUER and C. KLUGHAMMER
Lehrstuhl Botanik I, Universität Würzburg,
Mittlerer Dallenbergweg 64, D-8700 Würzburg, FRG

ABSTRACT

The special features of a new type of modulation fluorometer (PAM Fluorometer) are outlined in comparison to conventional measuring systems. Fluorescence is excited by 1 µs pulses of a light-emitting-diode and the pulse signal is processed via a selective window amplifier, yielding an exceptionally high dynamic range. The system is also capable of measuring P700 changes around 820 nm, using a modified emitter-detector unit. Due to its tolerance of large non-modulated background signals, the PAM Fluorometer is well suited for measurements in full sunlight and for quenching analysis by the saturation pulse method (determination of photochemical and non-photochemical quenching components). Photochemical quenching is shown to correspond closely to O_2-evolution rate. However, as is examplified by the effect of ANT-2p, under special conditions an increase in photochemical quenching may go along with a decrease in quantum yield. Cyclic electron flow around PS II is suggested to be stimulated upon loss of PS II donor side capacity.

INTRODUCTION

The yield of chlorophyll fluorescence is influenced in multiple ways by the state of the photosynthetic apparatus. Chlorophyll may be visualized as a built-in fluorescence indicator, embedded in the thylakoid membrane: It directly participates in the primary events of energy absorption, transfer and transformation, and indirectly senses the secondary changes induced at the membrane level in the course of photosynthetic reactions. In this way fluorescence carries information on virtually all aspects of photosynthesis (for recent reviews, see Briantais et al. 1986, Renger and Schreiber 1986). To obtain specific information, selective measuring techniques are required, which allow to differentiate between various forms of fluorescence quenching. Furthermore, for interpretation and calibration of the fluorescence signals, it is essential to be able to measure other photosynthesis signals in parallel.

Recently a new type of pulse modulation fluorometer (PAM Fluorometer) was introduced displaying the properties required for a thorough analysis of fluorescence in comparison with other photosynthesis signals (Schreiber 1986, Schreiber et al. 1986, Schreiber and Bilger 1987, Neubauer and Schreiber 1987, Schreiber and Neubauer 1987). The analytical power of this tool has been further increased by the development of a P700 probe which can be operated in conjunction with the fluorometer (Schreiber et al. 1988).

In the present report, an overview is given on the properties of the PAM Fluorometer in comparison to conventional measuring systems. As an outstanding application, the so-called saturation pulse quenching analysis will be described.

H. K. Lichtenthaler (ed.) Applications of Chlorophyll Fluorescence, 63–69.

THE PAM MEASURING SYSTEM

In Fig. 1 the main features of the PAM Fluorometer are schematically summarized in comparison to other types of fluorometers. In most conventional fluorometers, actinic and excitation light are identical and a separation of fluorescence from stray actinic light is achieved by appropriate combinations of optical filters. The resulting DC-signal is strong. However, there are limitations with respect to applications where changes in actinic light intensity are essential, as e.g. for quenching analysis. Also there is no way of distinguishing fluorescence from other signals, like that caused by the red component of ambient day light. These limitations are overcome by use of a modulated excitation beam. In conventional modulation fluorometers a chopper creates symmetrical light-dark periods and the resulting fluorescence signal is processed by a lock-in amplifier. In the PAM Fluorometer a different approach was taken: Fluorescence is excited by 1 μs pulses of light from a light-emitting diode (LED), with relatively long dark periods between individual pulses. A so-called "selective window amplifier" is used, which is active only during the 1 μs sampling periods. Hence, any external disturbance not coinciding with the 1 μs sampling periods will be eliminated by this amplifier. Consequently the PAM Fluorometer displays an exceptionally large dynamic range, i.e. actinic light may exceed the integrated measuring light intensity by more than a factor of 10^6. In particular, contrary to conventional lock-in systems, transient artefacts caused by rapig changes of background signals (like fluorescence excited by actinic light) are prevented. This provides for reliable monitoring of kinetic changes at high time resolution. The resulting measuring system is well suited for a thorough analysis of fluorescence quenching by the "saturation pulse method" and for the recording of rapid induction and relaxation kinetics.

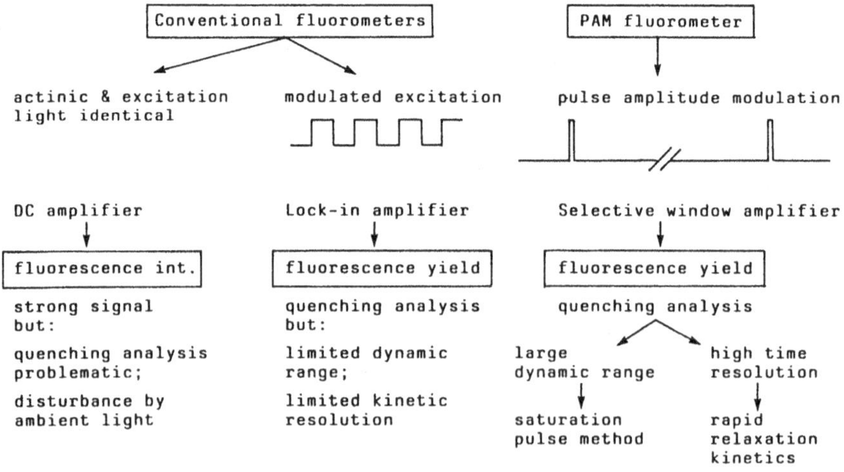

FIGURE 1 Scheme comparing main features of the PAM Fluorometer with those of conventional fluorometers.

With modulated fluorometers the obtained signal represents fluorescence yield, i.e. although the overall fluorescence intensity may vary by many orders of magnitude (depending on the actinic intensity), the fluorescence specifically excited by the modulated measuring beam can only vary between a minimal value F_o and a maximal value F_m. Measuring fluorescence yield in this way is very similar to measuring absorbance changes, as e.g. of cyt f or P700. Actually, it is possible to record absorbance changes of P700 with the PAM Fluorometer in a quite satisfactory way, by using a modified emitter-detector unit (different LED and filters) which simply plugs in to the main control unit (Schreiber et al. 1988). The basic equivalence of fluorescence yield and absorbance measurements is demonstrated in Fig. 2 for the case of P700. To excite fluorescence, a 650 nm measuring light is employed while 820 nm light is used to monitor absorbance of $P700^+$ around 820 nm. Fluorescence yield depends on the redox state of the primary PS II acceptor Q. It is minimal after dark adaptation (F_o= 1 rel. unit) when Q is fully oxidized and it becomes maximal ($F_m \sim 5xF_o$) when Q is fully reduced by actinic illumination which drives PS II. The transmission at 820 nm depends on the redox state of P700: In the dark, with P700 being reduced, transmission is maximal, and it decreases by about 0.8% in an intact leaf when P700 becomes fully oxidized by actinic illumination driving PS I. In both cases, with fluorescence and P700, it is essential to have two well defined limits between which the signal can vary. Knowing these limits, it is possible to assess the redox state of PS II and PS I at any given situation.

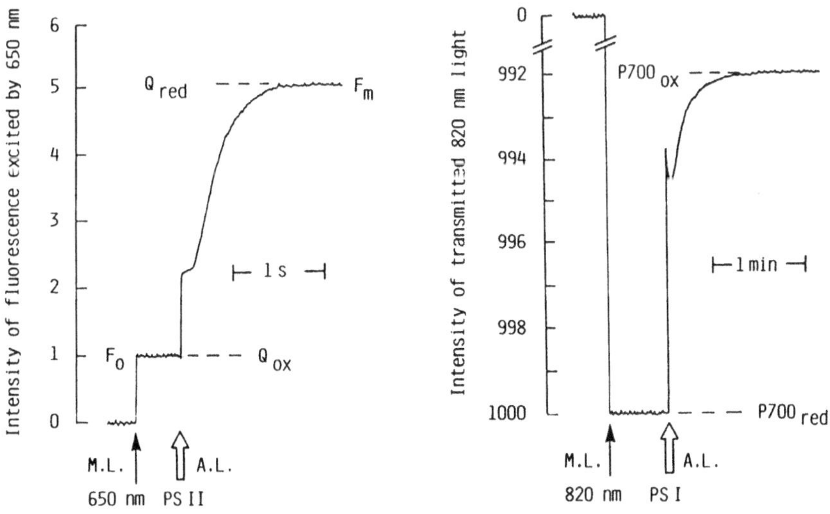

FIGURE 2 Equivalence of measuring fluorescence yield and 820 nm transmission changes with the PAM Fluorometer. For better comparison the transmission scale is inverted. See text for further explanations.

THE SATURATION PULSE METHOD.

Two basic types of fluorescence quenching may be distinguished, photochemical and non-photochemical quenching. Photochemical quenching results primarily from charge separation at PS II centers, while non-photochemical quenching can have a number of different causes, including membrane energization (energy dependent quenching), state I-state II shift (transfer quenching) and PS II donor side limitation (donor side dependent quenching) (see Horton and Hague 1988, Schreiber and Neubauer 1987). For the interpretation of fluorescence changes, it is essential to know what kind of quenching is involved. A low fluorescence yield can be as well expression of efficient electron transport (high photochemical quenching) as it can reflect strong damage to the photosynthetic apparatus (e.g. strong donor side dependent quenching following heat treatment). A clear-cut distinction between these two contrasting possibilities has become possible by the saturation pulse method (Bradbury and Baker 1981, Quick and Horton 1984, Dietz et al. 1985, Schreiber et al. 1986). The principle of this method is simple: A brief pulse of very strong light is applied to fully reduce all PS II acceptors, leading to transient elimination of photochemical quenching. The remaining quenching with respect to a dark-adapted control is non-photochemical.

In practice, it requires a measuring system with exceptional dynamic properties like the PAM Fluorometer to reliably measure fluorescence yield during a saturation pulse and also to be able to monitor the F_o-level. It is important to apply a sufficiently strong pulse (preferentially 500 ms of 2000 W/m^2) to assure full Q-reduction also under conditions which are favorable for high electron flow rates (activated Calvin cycle, elevated temperatures, high CO_2). On the other hand, only very low measuring light can be tolerated for determination of F_o (preferentially less than 10^{-2} W/m^2), particularly when Q-reoxidation (via PQ or backreaction) is slowed down (low temperature, electron transport inhibition, stress induced damage).

An application of particular diagnostic value is the recording of induction curves with repetitive (e.g. every 30 s) illumination by saturation pulses (Schreiber et al. 1986, Schreiber and Bilger 1987, Bilger et al. 1987). Fig. 3 shows the result of a comparative measurement of photochemical quenching, q_P, via the saturation pulse method and of oxygen evolution rate, as measured with a leaf disc oxygen electrode. Conditions were chosen such as to induce pronounced oscillations (see Walker et al. 1983). It is apparent that every oscillation of O_2-evolution rate is reliably reflected in a corresponding oscillation of q_P. On closer inspection, and when dO_2/dt and q_P are plotted against each other, small systematic deviations from a linear correlation are revealed. As has been discussed in detail elsewhere (Schreiber and Bilger 1987) electron transfer to molecular oxygen can be one cause for such deviations. Mehler reaction and photorespiration are known to play important roles during induction (Radmer and Kok 1976) and under conditions of closed stomates (Heber et al. 1988), respectively.

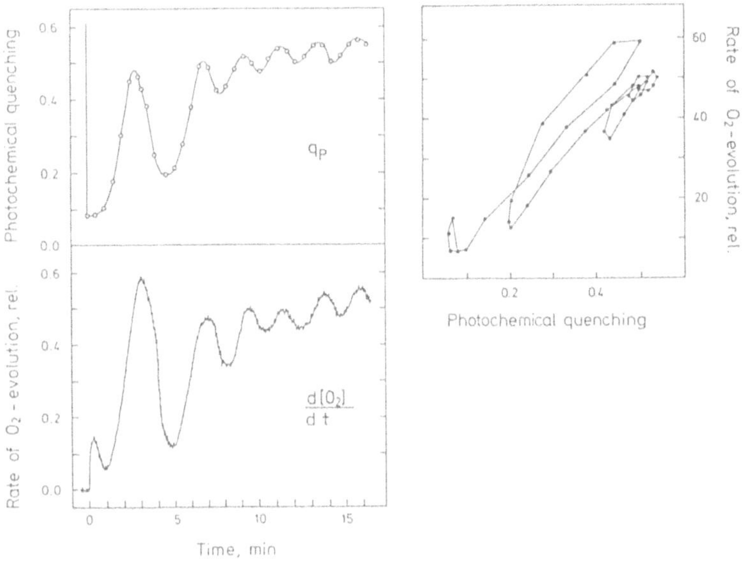

FIGURE 3 Comparison of photochemical quenching and the rate of O_2-evolution during induction of photosynthesis in barley. Simultaneous measurement of fluorescence and O_2 in a modified Hansatech leaf disc electrode. q_P was determined by repetitive application of saturation pulses. dO_2/dt was obtained by analog differentiating circuitry. In the plot of q_P vs. dO_2/dt each point corresponds to a saturation pulse, with consecutive points being connected. At time zero, continuous illumination with 150 W/m^2 white light was started; saturation pulses (2000 W/m^2) were applied every 30 s. Temperature, 15°C; saturating CO_2.

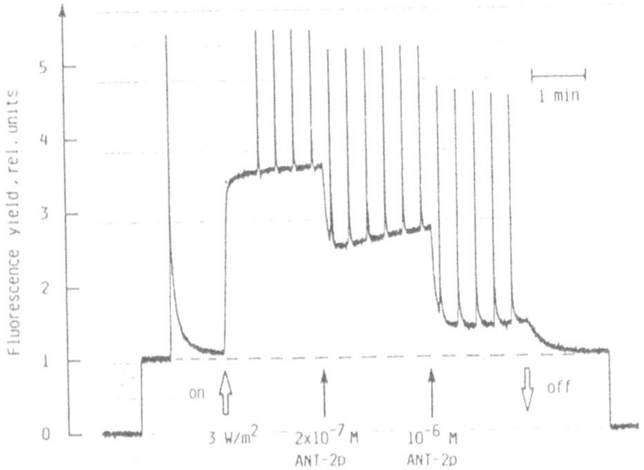

FIGURE 4 Photochemical and non-photochemical quenching induced by ANT-2p in isolated chloroplasts, as determined by repetitive application of saturation pulses. Class D spinach chloroplasts; 15°C; presence of 10^{-7}M nigericin.

Weis and Berry (1987) have shown that another mechanism causing deviations between the rate of linear electron flow and q_P is closely related to the extent of non-photochemical quenching. Actually, these researchers derived an empirical formula which allows calculation of electron transport rates on the basis of the quenching coefficients, q_P and q_{NP}, determined by the saturation pulse method. At present, it is not clear how non-photochemical quenching is linked to the lowering of quantum yield of open PS II reaction centers.

We have presented data which suggest that cyclic electron flow around PS II may play a major role under conditions of membrane acidification and donor side limitation of PS II (Schreiber and Rienits 1987, Schreiber and Neubauer 1987). The significance of this point may be further illustrated by the experiment shown in Fig. 4. The ADRY reagent ANT-2p induces non-photochemical and photochemical fluorescence quenching in isolated chloroplasts, with ΔpH formation and energy-quenching being prevented by nigericin. Parallel measurements of O_2-evolution and Hill reaction reveal a strong suppression of PS II quantum yield (data not shown). The essential point is that this is accompanied by a major increase in q_P. As linear electron flow is clearly suppressed, the only logical conclusion is that cyclic electron flow is stimulated. To be expressed as photochemical quenching, the postulated cycle must become saturated, at least in part, during a saturation pulse.

ACKNOWLEDGEMENT

This work was supported by the Deutsche Forschungsgemeinschaft (Forschergruppe Ökophysiologie and SFB 176).

REFERENCES

Bilger W, Schreiber U and Lange OL (1987) Chlorophyll fluorescence as an indicator of heat induced limitation of photosynthesis in *Arbutus unedo*. In: Tenhunen et al. Plant response to stress. pp 391-399, Berlin, Springer Verlag.

Bradbury M and Baker NR(1981) Analysis of the slow phase of the in vivo chlorophyll fluorescence induction curve. Changes in the redox state of photosystem II electron acceptors and fluorescence emission from photosystems I and II. Biochim Biophys Acta 635: 542-551

Briantais J-M, Vernotte C, Krause GH and Weis E (1986) Chlorophyll fluorescence of higher plants: Chloroplasts and leaves. In: Govindjee, Amesz J, Fork DC. Light emission by plants and photosynthetic bacteria. pp 539-577, New York, Academic Press.

Dietz KJ, Schreiber U and Heber U (1985) The relationship between the redox state of Q_A and photosynthesis in leaves at various CO_2, O_2 and light regimes. Planta 166: 219-226

Heber U, Neimanis S, Setlikova E and Schreiber U (1988) Why is photorespiration a necessity for leaf survival under water stress? Proc Int Congr Plant Physiol, New Delhi (in press)

Horton P and Hague A (1988) Studies on the induction of chlorophyll fluorescence in isolated barley protoplasts. IV Resolution of non-photochemical quenching. Biochim Biophys Acta 932: 107-115

Neubauer C and Schreiber U (1987) The polyphasic rise of chlorophyll fluorescence upon onset of strong continuous illumination: I. Saturation characteristics and partial control by the photosystem II acceptor side. Z Naturforsch 42c: 1246-1254

Quick WP and Horton P(1984) Studies on the induction of chlorophyll fluorescence in barley protoplasts. II Resolution of fluorescence quenching by redox state and the transthylakoid pH gradient. Proc R Soc Lond B 220: 371-382

Radmer RJ and Kok B (1976) Photoreduction of O_2 primes and replaces CO_2 assimilation. Plant Physiol 58: 336-340

Renger G and Schreiber U (1986) Practical applications of fluorometric methods to algae and higher plant research. In: Govindjee, Amesz J, Fork DC. Light emission by plants and bacteria. pp 587-619, New York, Academic Press.

Schreiber U (1986) Detection of rapid induction kinetics with a new type of high-frequency modulated chlorophyll fluorometer. Photosynth Res 9: 261-272

Schreiber U, Bilger W and Schliwa U (1986) Continuous recording of photochemical and non-photochemical chlorophyll fluorescence quenching with a new type of modulation fluorometer. Photosynth Res 10, 51-62

Schreiber U and Bilger W (1987) Rapid assessment of stress effects on plant leaves by chlorophyll fluorescence measurements. In: Tenhunen et al. Plant response to stress. pp 27-53, Berlin, Springer Verlag.

Schreiber U and Neubauer C (1987) The polyphasic rise of chlorophyll fluorescence upon onset of strong continuous illumination: II. Partial control by the photosystem II donor side and possible ways of interpretation. Z Naturforsch 42c: 1255-1264

Schreiber U, Klughammer C and Neubauer C (1988) Measuring P700 absorbance changes around 830 nm with a new type of pulse modulation system. Z Naturforsch, in press.

Walker DA, Sivak MN, Prinsley RT and Cheesbrough JK (1983) Simultaneous measurement of oscillations in O_2 evolution and Chl a fluorescence in leaf pieces. Plant Physiol 73: 542-549

Weis E and Berry J (1987) Quantum efficiency of photosystem II in relation to "energy"-dependent quenching of chlorophyll fluorescence. Biochim Biophys Acta 849: 198-208

STEADY STATE PHOTOSYNTHESIS IN INTACT PLANTS AS ANALYZED BY CHLOROPHYLL FLUORESCENCE AND FAR-RED SPECTROSCOPY

Engelbert Weis and Dorothea Lechtenberg

Institut für Ökologische Pflanzenphysiologie, Universität Düsseldorf, Geb. 26.13, D-4000 Düsseldorf (F.R.G.).

Key words: far-red spectroscopy; *in vivo* chlorophyll fluorescence; intact plants; P700; photosystem II; photosynthesis; quantum yield.

ABSTRACT

We relate photochemical and non-photochemical ('high-energy') quenching of *in vivo* fluorescence and the redox state of P700 (as analyzed by far-red spectroscopy) to electron transport during steady state assimilation. From the data we derived expressions which provide a basis of conducting measurements of the apparent quantum yield and the rate of photosynthesis non-invasively and remotely, using optical approaches which do not require the enclosure of plants in measuring devices.

INTRODUCTION

Depending on light intensity and intercellular CO_2-concentration assimilation in higher plants is thought to be limited either by reactions of the carbon metabolism such as the carboxylation of ribulose-1.5-bisphosphate or by membrane bound primary reactions (Woodrow and Berry, 1988). During light limited assimilation the quantum yield of CO_2fixation is close to its theoretical value (Björkman and Demmig, 1987), suggesting that excitation energy is efficiently trapped by the reaction centers of photosystem (PS) I and II and de-excited by photochemical conversion. When the flux of absorbed energy exceeds that required by biochemical reactions, 'excess' excitation energy not converted by photochemical reactions is emitted from the antennae as fluorescence or dissipated by non-radiative 'thermal' conversion (see Krause and Weis, this volume). Apparently, the distribution of excitation energy between different pathways for photochemical and non-photochemical de-excitation is controlled by the energetic balance of the leaf. When light is in excess, a high proton gradient is built up, which feeds back to both photoreactions. 'Energy'-dependent quenching of excitation energy at PS II as reflected by 'high energy quenching' of fluorescence (Krause et al., 1983) has been shown to regulate the potential rate at which electrons enter the electron transport system at PS II (Weis et al., 1987; Weis and Berry, 1987; Horton et al., 1988; Krause et al., 1988). It may also protect the pigment assembly against excess excitation and related photooxidative side reactions (Krause and Laasch, 1987; Krause et al., 1988). A 'quenching' mechanism may also occur at PS I: when the proton gradient is high and 'photosynthetic control' comes into effect, PS I photochemistry is controlled by a shortage of electrons at the donor side, rather than by an excess of electrons at its acceptor side. Due to this flux control, the photochemically inactive (far-red absorbing) oxidized form of the PS I center, $P700^+$, tends to accumulate in the steady state (Weis et al., 1987; see also Harbinson and Woodward,

H. K. Lichtenthaler (ed.) Applications of Chlorophyll Fluorescence, 71–76.

1987). The radical $P700^+$ efficiently traps excitation energy and converts it radiationless to heat (Nuijs et al., 1986). Both, high energy quenching and formation of oxidized PS I-centers are control mechanisms which serve to match the potential rate of photochemistry to the rate at which their products (ATP and NADPH) can be consumed by biochemical reactions.

A quantitative comparison between the photosynthetic flux in a leaf (as determined by photosynthetic CO_2 or O_2 exchange) and coefficients for fluorescence quenching provides information about the distribution of the energy absorbed by PS II among the different pathways for energy conversion. It is a major advantage of the recently developed light-doubling (Bradbury and Baker, 1981) and saturating pulse approach (Schreiber et al., 1986), that 'photochemical' and 'non-photochemical' quenching of fluorescence can be separated (see also Schreiber, this volume). Recently, a new optical approach (far-red spectroscopy at 820nm) has been developed to monitor the steady state level of the oxidized form of P700 (which absorbs in the near far-red; Ke, 1972), simultaneously with fluorescence and CO_2-uptake in leaves (Weis et al., 1987; Harbinson and Hedley, this volume).

Here, we will show, how optical approaches such as fluorescence and far-red spectroscopy can be used to measure the efficiency of photochemical conversion of excitation energy and the rate of photosynthetic fluxes under steady state condition *in situ*.

MATERIALS AND METHODS

Attached leaves were mounted in a leaf chamber equipped to measure gas exchange. White actinic light was provided for photosynthesis. A fiberoptic, mounted at the upper window of the chamber (3 cm apart from the leaf), was connected to a pulse modulation fluorimeter (PAM-101, Walz. Effeltrich, F.R.G.). In the steady state, a saturating light pulse was given (to obtain the maximal level of steady state fluorescence) and the actinic light was interrupted for a short periode (in presence of a far-red background light, to obtain the steady state level of basal fluorescence). Coefficients for fluorescence quenching were obtained as described recently (Weis and Berry, 1987). A second fiberoptic was connected to a laboratory-build optical system to monitor absorbance changes at 820nm from a modulated (1 KHz), back-scattered measuring light. Steady state illumination was briefly interrupted to allow the spontaneous re-reduction of oxidized P700 (by electron donation from partially reduced plastoquinone). During a saturating far-red pulse (720nm) P700 is supposed to be completely oxidized and a coefficient, a_{820} $(0<a<1)$ is derived to indicate the proportion of P700 which is oxidized during steady state assimilation (Figure 2). Yet, we may not exclude the possibility that other far-red absorbing radical forms of photosynthetic pigments contribute to the 820nm signal.

RESULTS AND DISCUSSION

The balance between 'light-reactions' and electron consuming reactions was varied in attached leaves by changing the incident light flux, I. Experiments were conducted in 300 ppm CO_2 and 2% O_2 (to minimize corrections for O_2-reducing reactions). Net electron transport, J_e, and the apparent quantum yield $\varphi_s = J_e/I$ (on the basis of incident light) have been calculated from photosynthetic gas exchange and quantum flux. Normalized values for φ_s (defined her as Φ_s $(0<\Phi<1)$ may be regarded as fractional coefficients

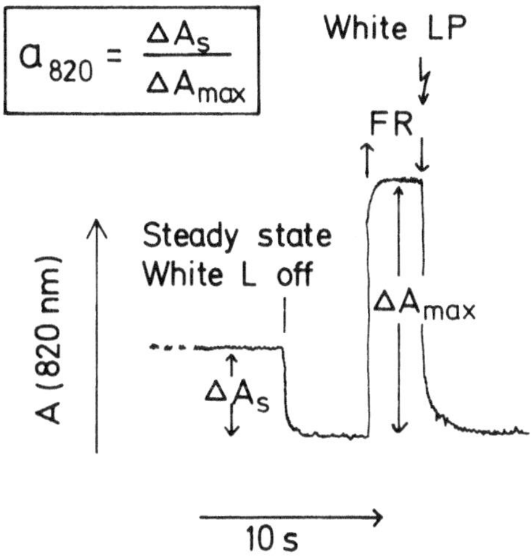

FIGURE 1
Absorbance change at 820nm in an attached leaf. The top of the leaf is irrad ated by a modulated 820nm light, and back-scattered light is collected and analyzed for absorbance changes. From transmitted radiation we obtained almost identical values for a_{820}.

which indicate the proportion of absorbed enerby consumed by net electron transport. Since

$$\Phi_S = \Phi_P \cdot q_Q \qquad \text{equation 1}$$

(where Φ_P is the normalized potential quantum yield of PS II photochemistry with all centers in the 'open' state, and q_Q ($0<q<1$) is a coefficient for photochemical quenching of fluorescence), we can calculate the potential quantum yield from $\Phi_P = \Phi_S/q_Q$. The data show that Φ_P is not constant, but varies in dependence of the energetic balance of the leaf. Figure 2 shows an almost parallel decline in Φ_S and Φ_P.

It suggest that PS II activity is mainly controlled by a decline in its 'intrinsic' photochemical efficiency and that 'closure' of centers by reductions of the acceptor Q_A plays a minor role, i.e., the acceptor Q_A stays in a fairly oxidized state. With spinach (Figure 3) and a large number of other species of C3-plants an inverse, yet not a linear correlation was found between Φ_P and 'high energy quenching' of fluorescence, q_E. It is assumed that a photochemically efficient form, PS II_α, can be converted to an inactive form either by closure of centres (by reduced Q_A) or by forming a 'quenching center' (PS II_ε) in which trapped excitation energy is radiationlessly converted to heat. Changes in the quantum yield of fluorescence and photochemistry are assumed to reflect changes in the relative concentration of these forms (Weis and Berry, 1987). q_E-quenching may actually include several mechanisms for 'radiationless' energy dissipation. Coefficients Φ ($0<\Phi<1$) for radiationless (Φ_d) and 'radiative' (Φ_r) dissipation of excitation energy can be derived as follows: $\Phi_d = 1-\Phi_P$, and $\Phi_r = \Phi_P - \Phi_S$. It was found, that under most steady state conditions, Φ_r stayed fairly low (≤ 0.1), even in high light.

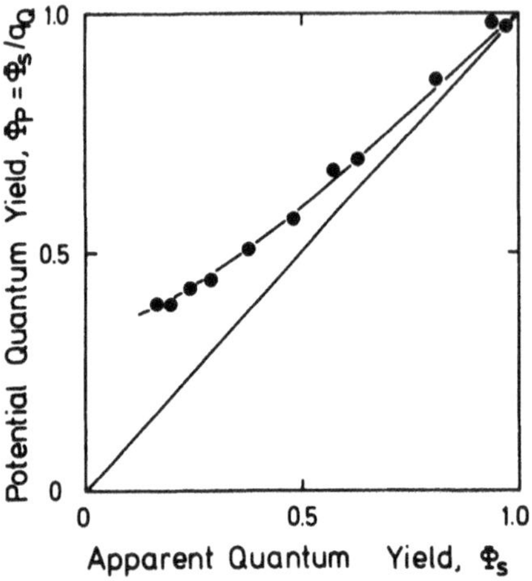

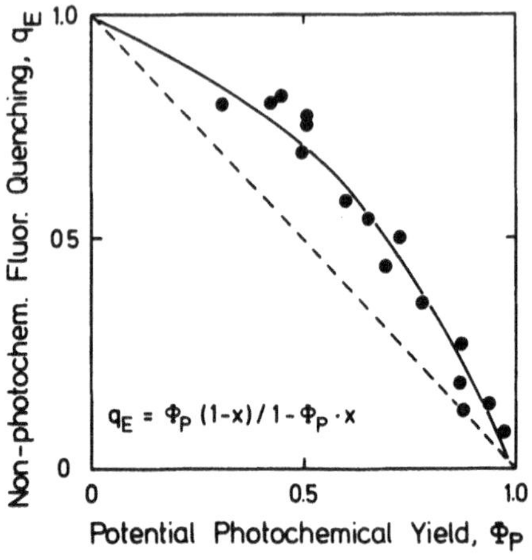

FIGURE 2 (left). The relationship between the apparent quantum yield of net electron transport in the steady state ($\Phi_S = J/I$) and the potential quantum yield, $\Phi_P = \Phi_S/q_Q$. Values obtained from spinach.

(FIGURE 3 (right). The relationship between the potential quantum yield (as in fig. 2) and 'high-energy' quenching, qE.

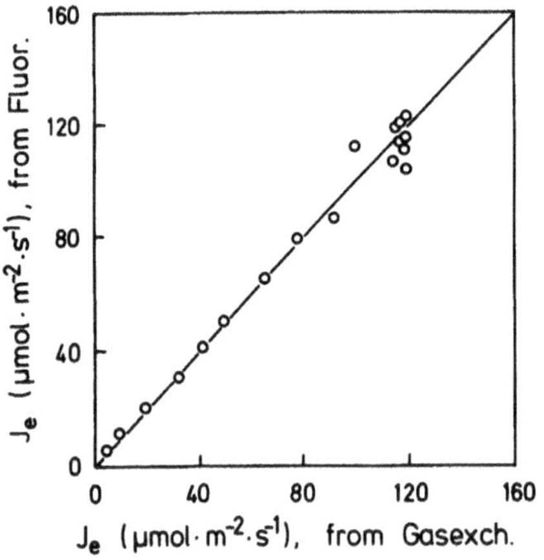

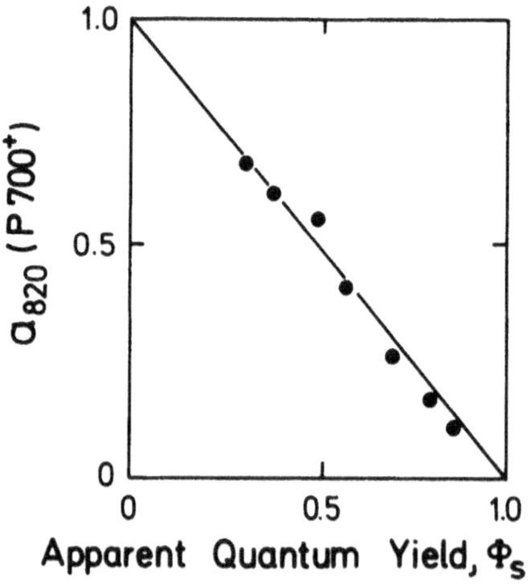

FIGURE 4 (left). Photosynthetic electron transport in spinach leaves as calculated from gas exchange or from fluorescence (using eq. 2).

FIGURE 5 (right). Relationship between the apparent quantum yield, Φ_S and a_{820} (as in figure 1).

From the empirical relationship shown in Fig. 3 (where $x \simeq 0.5$) and eq. 1, an expression is obtained that relates the apparent quantum yield of electron transport to fluorescence quenching:

$$\Phi_S = q_Q \cdot [q_E (1-x)/1-q_E \cdot x] \qquad \text{equation 2}$$

It is similar (though not identical) to an expression published recently (Weis and Berry, 1987) and provides a basis to estimate the photochemical yield and the net rate of electron transport (according to $J_e \approx I \cdot \varphi_P$), from measurements of quantum flux and fluorescence. As shown in figure 4 values for the photosynthetic electron transport in a leaf as calculated from fluorescence are in fairly good agreement with those obtained by conventional gas exchange measurements.

Under most steady state conditions, we observed a fairly linear relationship between the apparent quantum yield of net electron transport and the absorbance signal at 820 nm ($\Phi_S = 1-a_{820}$; Figure 5), suggesting that PS I activity is controlled by electron donation at the donor side (Weis et al., 1987). Only during the induction periode of photosynthesis or during CO_2-saturated photosynthesis a_{820} did not completely match the decline in Φ_S, suggesting that under such conditions accumulation of reduced PS I acceptors contributes to the control of PS I-photochemistry (not shown). But for most steady state conditions, a_{820} may be taken as a measure for the apparent quantum yield of the whole chain electron transport. Together with fluorescence, it provides an excellent tool to analyze photosynthetic control of light-reactions *in vivo*.

CONCLUSIONS

We derived empirical expressions which describe the relationship between fluorescence quenching, 820 nm absorbance change and photosynthetic activity in intact leaves. We explain these expressions in terms of control mechanisms, which serve to match the potential rate of electron transport to the electron demand by biochemical reactions. Since optical measurements as described in this paper can be carried out non-invasively and remotely during steady state illumination, they could provide a means of conducting measurements of photosynthetic activity *in situ*. Usually, the rate of assimilation in plants is measured by photosynthetic gas exchange, but this requires the removal of plants, or parts of a plant from their natural position and enclosure in a chamber. Thus, optical approaches such as described here could help to overcome a major limitation of conventional methods to conduct measurements of assimilation in the field.

REFERENCES

Bradbury M and Baker NR, 1981. Analysis of the slow phases of the *in vivo* chlorophyll fluorescence induction curve. Changes in the redox state of photosystem II electron acceptors and fluorescence emission from photosystem I and II. Biochim. Biophys. Acta 635: 542-551.

Björkman O and Demmig B, 1987. Photon yield of O_2-evolution and chlorophyll fluorescence characteristics at 77K among vascular plants of diverse origins. Planta 170, 489-504.

Harbinson J and Woodward I, 1987, The use of light-induced absorbance changes at 820 nm to monitor the oxidation state of P700 in plants. Plant, Cell & Environm. 10, 131-140.

Horton P, Oxborough K, Rees D and Scholes JD, 1988. Regulation of the photochemical efficiency of photosystem 2; consequences for the light response of field photosynthesis. Plant Physiol. Biochem., in press

Ke B, 1972. The rise time of photoreduction, difference spectrum and oxidation reduction potential of P430. Arch. Biochem. Biphys. 82, 183-186.

Krause GH and Laasch H, 187. Photoinhibition of photosynthesis. Studies on mechanisms of damage and protection in chloroplasts. In: Progress in Photosynthesis Research Vol. 4, p 19-26, Biggins ed., Martinus Mijhoff, Dortrecht.

Krause GH, Briantais J-M and Vernotte C, 1983. Characterization of chlorophyll fluorescence quenching in chloroplasts by fluorescence spectroscopy at 77K. I. ΔpH-dependent quenching. Biochim. Biophys. Acta 723, 169-175.

Krause GH, Laasch H and Weis E, 1988. Regulation of thermal energy dissipation of absorbed light energy in chloroplasts indicated by energy-dependent fluorescence quenching. Plant Physiol. Biochem. 26, in press.

Nuijs AM, Shuvalov VA, van Gorkom HJ, Plijter JJ and Duysens LNM, 1986. Picosecond absorbance difference spectroscopy on the primary reactions and the antenna-excited states in photosystem I particles. Biochim. Biophys. Acta 850, 310-318.

Schreiber U, Bilger W and Schliwa U, 1986. Continuous recording of photochemical and non-photochemical quenching with a new type of modulation fluorometer. Photosynth. Res. 10, 51-62.

Weis E and Berry JA, 1987. Quantum efficiency of photosystem II in relation to 'energy'-dependent quenching of chlorophyll fluorescence. Biochim. Biophys. Acta 894, 198-208.

Weis E, Ball JT and Berry JA, 1987. Photosynthetic control of electron transport in leaves of Phaseolis vulgaris: evidence for regulation of photosystem 2 by the proton gradient. In: Progress in Photosynthesis Research Vol 2, p 553-556, Biggins J ed., Martinus Nijhoff, Dortrecht.

Woodrow IE and Berry JA, 1988. Enzymatic regulation of photosynthetic CO_2-fixation in C3-plants. Ann. Rev. Plant Physiol. 39, 533-594.

COMPLETE FLUORESCENCE EMISSION SPECTRA DETERMINED DURING THE INDUCTION KINETIC USING A DIODE-ARRAY DETECTOR

Claus BUSCHMANN and Hartmut K. LICHTENTHALER

Botanisches Institut II (Plant Physiology and Plant Biochemistry), University of Karlsruhe, Kaiserstr. 12, D-7500 Karlsruhe, F.R.G.

KEY WORDS: Chlorophyll fluorescence, diode-array detector, forest decline, Kautsky effect, fluorometer, photosynthesis, ratio F690/F735, Rfd-values.

ABSTRACT

By using a diode-array detector it has become possible for the first time to measure with one leaf sample complete fluorescence emission spectra during the fast rise and the slow decline of the Kautsky induction kinetic. Thus fast spectral changes in the millisecond range can be detected. Changes in pigment-protein complexes during the first seconds of illumination of an etiolated leaf can be measured at room temperature. During the Kautsky induction kinetic of photosynthetically active green leaves, the maxima of the emission spectrum does not change. However, compared to the long wavelength signal at the maximum at about 735 nm, the short wavelength signal at the maximum at about 690 nm in general increases faster during the fast rise and decreases faster during the slow decline. This is demonstrated for greening and fully green maize leaves. The spectral changes can be expressed by the ratio between the fluorescence signals at the two emission maxima (F690/F735) calculated from the stored spectra taken during the fast rise and the slow decrease of the induction kinetic. The general validity of this ratio change during the Kautsky kinetic is demonstrated for leaves of a variety of different plant species each with a high and a low chlorophyll content.

INTRODUCTION

The Kautsky induction kinetic of the chlorophyll fluorescence is normally measured at one or two fixed wavelengths. The fluorescence signal is detected through a filter, the transmission of which has a fixed wavelength range. Most kinetics are given for the fluorescence in the low wavelength maximum at about 690 nm, only few kinetics are measured at two wavelengths simultaneously (e.g. Schreiber et al. 1977, Kocsányi et al. 1988). Conventional fluorescence spectrometers need about one or two minutes to measure the complete fluorescence emission spectrum. Thus only spectra under steady state conditions can be determined, otherwise the decrease of the fluorescence signal during the induction kinetic would interfere with the spectral measurement. Recently diode-array detectors have become available, with which it becomes possible to measure one complete spectrum in the millisecond-range. These detectors replace the photomultiplier at the exit of the monochromator. Because the exit slit is removed, the whole spectral range

H. K. Lichtenthaler (ed.) Applications of Chlorophyll Fluorescence, 77–84.

falls upon a line of about 1000 diodes. The complete spectral range is measured simultaneously and the signals are transferred via an electronic device (OMA = optical multichannel analyzer) to the data output (oscilloscope) or to a computer which stores the data. With diode-array detectors it has become possible for the first time to measure with one leaf sample complete fluorescence emission spectra during the fast rise and the slow decline of the Kautsky induction kinetics.

MATERIALS AND METHODS

Radish seedlings (Raphanus sativus L.) were grown for 6 days in the dark (etiolated leaves) or under continuous white light (green leaves) on tap water. Maize seedlings (Zea mays L.) were grown on peat. Greening primary leaves of maize were taken from plants cultivated for 10 days in the dark and then transferred for 4 or 5 hours into white light (1.3 µg chlorophyll a and b per cm^2). Fully green tertiary leaves of maize (41.2 µg chlorophyll a and b per cm^2) were taken from plants grown for 3 weeks in white light (14 h per day). Bean plants (Phaseolus vulgaris L.) and tabacco plants (Nicotiana tabacum L.) were grown for several weeks on peat in the greenhouse of the Botanical Garden of the University of Karlsruhe. The other plants (Populus nigra, Abies alba, Picea abies) were taken from fully developed trees growing on the University campus.

Leaves or needles of the plants were illuminated with a 5 mW He/Ne-laser. The fluorescence light was guided via a monochromator on to a diode-array detector (OMA I: SIT-vidicon, OMA II: ISIT-vidicon, OMA III: PAR 1420 with channelplate intensifier, EG&G, D-8000 München). The data gathered by the optical multichannel analyzer were either transferred to an oscilloscope and filmed with a video camera (OMA I) or stored in a computer (OMA II and III). The spectra were redrawn from the video pictures on a TV screen (OMA I) or plotted from the stored digitized data (OMA II and III). The calculations of the difference spectra, of the second derivative of the spectra and of the ratio F690/F735 was carried out with the stored data using the computer software of OMA II and III. Further details can be found in the publications of: Buschmann and Schrey 1981 (OMA I), Buschmann and Sironval 1984 (OMA II) and Lichtenthaler and Buschmann 1987 (OMA III).

RESULTS AND DISCUSSION

Induction kinetic of etiolated leaves

An etiolated leaf contains protochlorophyllide as precursor of chlorophyll. In the light protochlorophyllide accumulated in the dark is transformed into chlorophyllide within microseconds (Sironval and Brouers 1984). During the subsequent 30 seconds of first illumination of the etiolated leaf different chlorophyllide-protein complexes (chlorophyllide forms) undergo changes in the pigment protein binding, which have long been determined as changes of the fluorescence spectrum after freezing samples in liquid nitrogen at different times after the onset of illumination (for radish leaves: Buschmann and Sironval 1979). At room temperature the 690 nm fluorescence of the photosynthetically inactive etiolated leaf shows an induction kinetic similar to that of a green plant (e.g. Buschmann 1981). This kinetic has been explained by the transformation of a chlorophyllide form with high fluorescence yield to another chlorophyllide form with low fluorescence yield

TABLE 1. Maximum position of the fluorescence emission spectra at room temperature taken from an etiolated radish cotyledon during the first 30 seconds of illumination (taken from Buschmann and Sironval 1984).

time after onset of illumination	emission maximum	changes compared to the previous spectrum (maxima and minima of the difference spectra)	chlorophyllide forms (minima of the 2nd derivative)
1 second	685 nm	-	660, 678, <u>685</u>, 690 nm
3 seconds	685 nm	increase: 673 and 695 nm decrease: 685 nm	660, 678, <u>685</u>, 690 nm
30 seconds	685 nm	increase: 675 and 703 nm decrease: 685 nm	<u>675</u>, <u>681</u>, <u>690</u> nm

(Jouy and Sironval 1979). The first complete room temperature fluorescence spectra taken with the OMA I from etiolated leaves and imaged with a video camera from an oscilloscope screen showed one maximum, the position of which remained constant during the initial illumination period (Buschmann and Schrey 1981). Later (Buschmann and Sironval 1984), when the data could be stored in a computer (OMA II), it was demonstrated that changes in the chlorophyllide forms show up as a broadening of the fluorescence peak. The calculation of the difference spectra and the second derivative indicated that this broadening is due to the appearance of long and short wavelength chlorophyllide forms and to the dissappearance of the 685 nm chlorophyllide form present at the onset of the illumination (Table 1).

Induction kinetic of green leaves

During the Kautsky induction kinetic the fluorescence emission maximum of green leaves does not change. This was shown for the first time by measuring complete room temperature spectra with the OMA I system (Buschmann and Schrey 1981). The emission spectra taken during the slow decrease of the induction kinetic show a faster decrease of the short-wavelength maximum than of the long-wavelength maximum (Fig. 1). The decrease in the ratio F690/F735 confirmed earlier data determined by measuring the fluorescence simultaneously at the two maxima positions using filters with maximal transmittance at these two wavelengths (Schreiber et al. 1977). A greening primary leaf blade of maize (10 d darkness + 4 h light), which exhibits a very low chlorophyll content, shows chlorophyll fluorescence emission spectra with a maximum in the 690 nm region (exact position: 683 nm) and a shoulder near 735 nm (Fig. 2 and 4, left part). Compared to the young greening leaf the fully green tertiary leaf blade, which exhibits a much higher chlorophyll content, is characterized by a lower fluorescence yield and also by a different shape of the chlorophyll fluorescence spectra (Fig. 2 and 4, left part). The main fluorescence maximum of the fully green leaf lies in the long-wavelength region at 740 nm. In contrast to the greening leaf there is only a small shoulder in the 690 nm region. Consequently, the ratio of the fluorescence in the 690 and 735 nm region (ratio F690/F735) of the fully green leaf is much lower than in the greening leaf (Table 2).

Spectra during the fast fluorescence rise. During the fast rise of the fluorescence induction kinetic, the wavelength positions of the maxima and shoulders do not change. For both leaf types the rise to the kinetic maximum

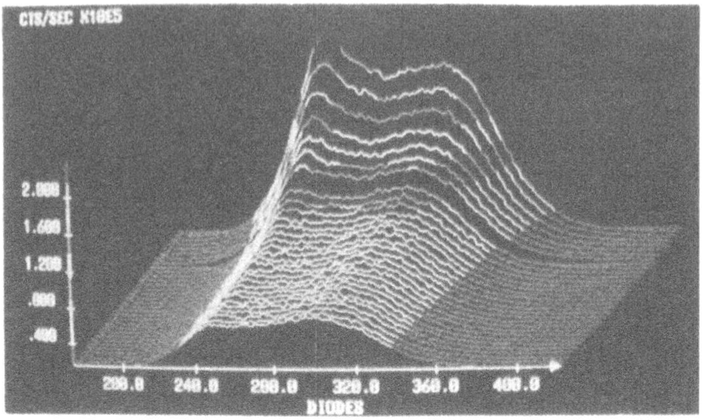

Figure 1. (see colour plate No 10 at the end of the book) Fluorescence emission spectra of a bean leaf (Phaseolus vulgaris L.) measured every 200 ms during the slow decline of the induction kinetic. For a better visibility the sequence of the spectra were plotted inversely (1st spectrum at the end of the z-axis). The spectra were taken with a detector system equivalent to the OMA II (set up and software from Instruments SA, D-8025 Unterhaching).

is accomplished about 200 ms after the onset of the illumination of the predarkened leaf (i.e. within 10 measuring periods of 20 ms). The kinetic recorded in the 690 nm region is shown in Fig. 2 for the greening leaf (left part) and the fully green leaf (right part). For the greening leaf the fluorescence reached during the 1st measuring period amounts to 70% and during the 2nd period to 88% of the maximum after 10 periods (Fig. 3). The fully green leaf (Fig. 2, right part) shows a lower fluorescence yield than the greening leaf. The fluorescence rise proceeds somewhat more slowly than in the greening leaf (Fig. 3). For the fully green leaf during the first measuring period 59% and during the 2nd period 70% of the maximum fluorescence is reached (Fig. 3). In both leaf types the fluorescence increase is slightly higher in the 690 nm than in the 735 nm region. Therefore the ratio F690/F735 increases during the kinetic rise. In the greening leaf the ratio F690/F735 increases from 3.8 (1st measuring period of 20 ms) to 4.8 (10 periods) and declines again to a value of 4.3 (50 periods, 1s after the onset of illumination). For the fully green leaf these values rise from 0.4 (1st period) to 0.5 (10th period) and decrease again to 0.45 (50 periods) as is shown in Table 2.

Spectra during the slow fluorescence decline. With two other leaf samples of maize plants we studied the chlorophyll fluorescence emission spectra during the fluorescence decline (slow component of the Kautsky effect), which proceeds in the range of minutes. The fluorescence spectra of continuously illuminated predarkened samples were recorded in measuring periods of 200 ms with intervals of 4.8 s (= 24 ignore periods of 200 ms). During the slow fluorescence decline, the intensity of the fluorescence becomes lower in the whole spectral range both for the greening and for the fully green leaf (Fig. 4). Again in both cases no changes of the wavelength position of the maxima and/or shoulders could be observed. The different shape and intensity of the fluorescence spectra measured in the first 200 ms measuring period is contrasted for both leaf types at the same amplification (Fig.4, left part).

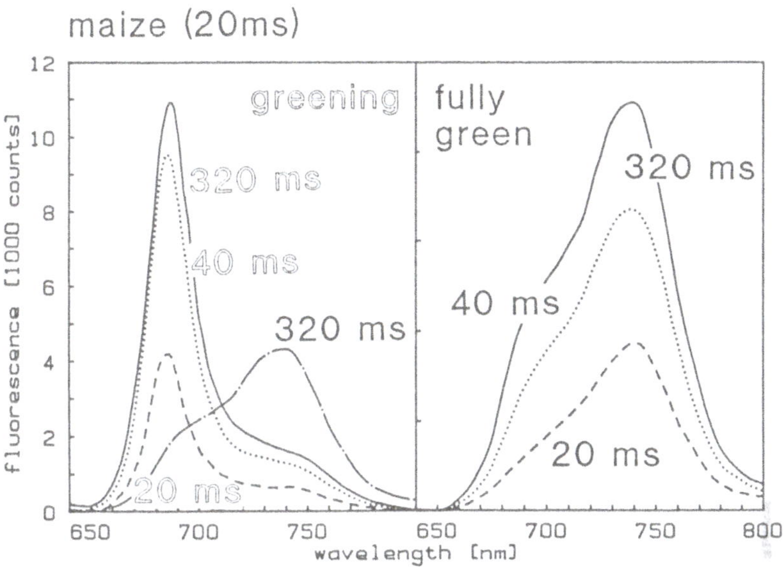

Figure 2. Chlorophyll fluorescence emission spectra of maize leaves measured with the OMA III in 20 ms periods during the fast rise of the induction kinetic. The measurement was completed at the time indicated. Left part: Greening leaf (open letters) and for comparison the 320 ms spectrum of the fully green leaf (filled letters) plotted with the same amplification, right part: fully green leaf.

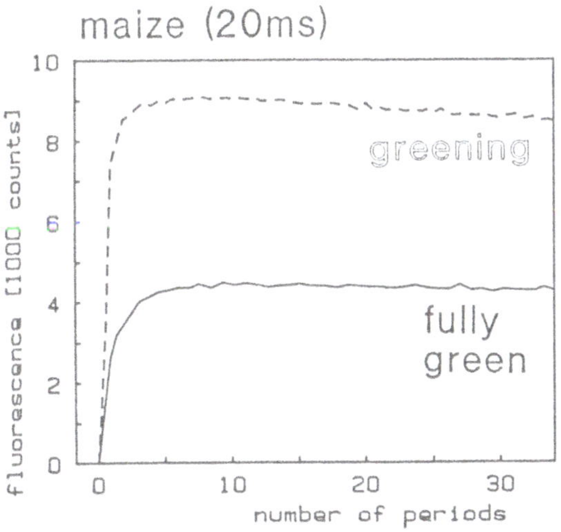

Figure 3. Induction kinetic of the chlorophyll fluorescence of a greening and a fully green maize leaf measured at 690±3.6 nm during the fast rise to the kinetic maximum. Data plotted from the stored digitized data for each of the 34 spectra taken every 20 ms (confer Fig. 2).

The fluorescence decline, as shown in the kinetics at the 690 nm region, proceeds faster for the younger greening leaf than for the fully green leaf (Fig. 5). The ratio of the fluorescence decrease (Rfd-value) was calculated from the kinetics for the 690 and the 735 nm regions, although the steady state was not yet reached after the measuring time of 170 s. In both leaf types the Rfd-values were higher in the 690 nm region (greening leaf: 1.9, fully green leaf: 2.1) than in the 735 nm region (1.8 for both leaves).

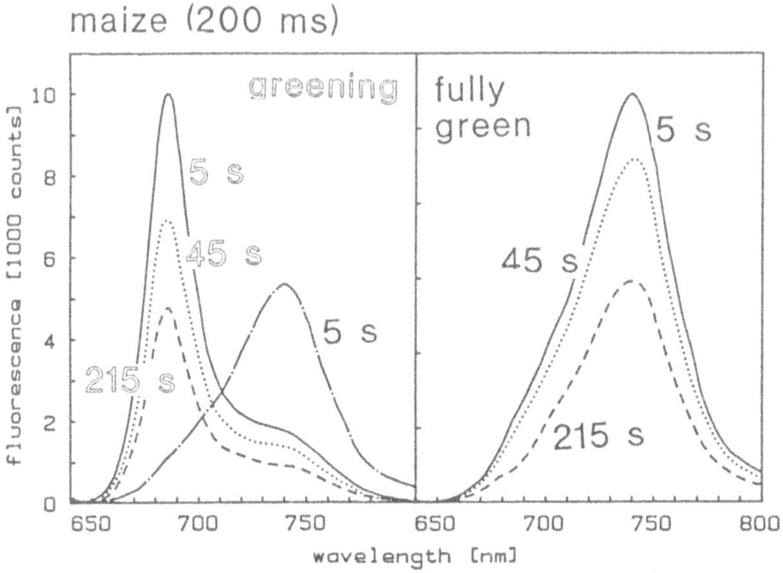

Figure 4. Chlorophyll fluorescence emission spectra of maize leaves measured with the OMA III in 200 ms periods during the slow decrease of the induction kinetic. The measurement was completed at the time indicated. Left part: Greening leaf (open letters) and for comparison the 5 s spectrum of the fully green leaf (filled letters) plotted with the same amplification, right part: fully green leaf.

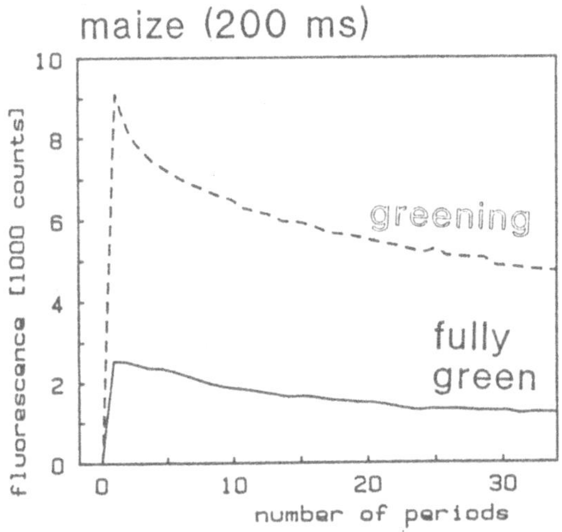

Figure 5. Induction kinetic of the chlorophyll fluorescence of a greening and a fully green maize leaf measured at 690+3.6 nm during the slow decrease from the kinetic maximum to the steady state. Data plotted from the stored digitized data for each of the 34 spectra taken within 200 ms every 5 s (confer Fig. 4).

For both leaves the fluorescence in the 690 nm region decreases to a larger degree than in the 735 nm region. Therefore the values for the ratio F690/F735 decrease between 5 and 215 s (i.e. between the 1st and the 43rd measuring period after onset of illumination) from 3.5 to 3.0 in the young greening leaf and from 0.32 to 0.29 in the fully green leaf (Table 2). The ratio F690/F735 during the fast rise and the slow decrease of the induction kinetic is summarized in Table 2 for leaves of a variety of different plants

Table 2. Ratio of the fluorescence signal at 690+3.6 nm to the signal at 735+3.6 nm for different plants with (a) low and (b) high chlorophyll content. The ratios were determined from the stored emission spectra measured with the OMA III during the fast rise (left column) and the slow decrease of the induction kientic at the time indicated in brackets below the ratio.

	fluorescence ratio F690/F735			
	fast rise		slow decrease	
Phaseolus vulgaris				
a) greening leaf	**1.29** (20 ms)	**1.73** (1.2 s)	**0.75** (5 s)	**0.69** (200 s)
b) green leaf	**0.4** (20 ms)	**0.5** (1.2 s)	**0.53** (5 s)	**0.44** (200 s)
Nicotiana tabacum				
a) aurea form Su/su	**1.27** (20 ms)	**1.41** (300 ms)	**1.40** (5 s)	**1.13** (5 min)
b) green form su/su	**0.96** (20 ms)	**1.06** (300 ms)	**1.07** (5 s)	**1.05** (5 min)
Populus nigra				
a) senescent green leaf	**0.68** (20 ms)	**0.78** (300 ms)		
b) fully active green leaf	**0.72** (20 ms)	**0.88** (300 ms)		
Zea mays				
a) greening primary leaf	**3.8** (20 ms)	**4.8** (300 ms)	**3.5** (5 s)	**3.0** (200 s)
b) fully green tertiary leaf	**0.4** (20 ms)	**0.5** (300 ms)	**0.32** (5 s)	**0.29** (200 s)
Abies alba				
a) 1986, youngest needle year, light-green)	**1.45** (4 ms)	**1.61** (280 ms)	**1.10** (5 s)	**0.99** (200 s)
b) S-needles 1985 (rise: green, decrease: dark-green)	**0.92** (4 ms)	**0.72** (280 ms)	**0.28** (5 s)	**0.43** (200 s)
Picea abies				
a) S-needles 1985 (1-year-old, yellowish-green)	**0.64** (4 ms)	**0.65** (280 ms)		
b) S-needles 1985 (1-year-old, dark-green)	**0.33** (4 ms)	**0.18** (280 ms)		

each with a high and a low chlorophyll content. In general the ratio F690/F735 rises during the first second (except for dark-green needles) and declines during the subsequent minutes. However the changes in the ratio are small compared to the differences between leaves with high and low chlorophyll content (Table 2) or the differences due to stress and damage of the plants (Rinderle and Lichtenthaler 1988).

CONCLUSIONS

By means of diode-array detectors fast spectral changes can be determined with a spectral resolution comparable to that of conventional spectrometers. Complete fluorescence emission spectra measured within milliseconds can provide further fast information on the functioning of photosynthesis and can be applied to great advantage also in remote sensing.

REFERENCES

BUSCHMANN C, 1981. The characterization of the developing photosynthetic apparatus in greening barley leaves by means of (slow) fluorescence kinetic measurements. In: Photosynthesis, Vol 5, Akoyunoglou G ed, pp 417-426. Balaban International Science Services, Philadelphia.

BUSCHMANN C and SCHREY H, 1981. Fluorescence induction kinetics of green and etiolated leaves by recording the complete in-vivo emission spectra. Photosynthesis Research 1: 233-241.

BUSCHMANN C and SIRONVAL C, 1979. Influence of kinetin on protochlorophyll(ide) accumulation and on the Shibata shift in Raphanus seedlings. Planta 139: 127-132.

BUSCHMANN C and SIRONVAL C, 1984. Fluorescence emission spectra of etiolated leaves measured at 296 and 77 K during the first seconds of continuous illumination. In: Protochlorophyllide reduction and greening, Sironval C and Brouers M, eds, pp 139-148.

JOUY M and SIRONVAL C, 1979. Quenching of the fluorescence emitted by P695-682 at room temperature in etiolated illuminated leaves. Planta 147: 127-133.

KOCSANYI L, HAITZ M and LICHTENTHALER HK, 1988. Measurement of laser-induced chlorophyll fluorescence kinetics using a fast acoustooptic device. In: Applications of chlorophyll fluorescence. Lichtenthaler HK ed, pp 92-100, Kluwer Academic Press, Dordrecht.

LICHTENTHALER HK and BUSCHMANN C, 1987. Chlorophyll fluorescence spectra of green bean leaves. J Plant Physiol 129: 137-147.

LICHTENTHALER HK and BUSCHMANN C, 1988. Changes in the chlorophyll fluorescence spectra during the Kautsky induction kinetics. In: Proceedings of the 4th International Colloquium on Spectral Signatures of Objects in Remote Sensing, Aussois, pp. 245-250, ESA Publications Division, Noordwijk.

LICHTENTHALER HK, BUSCHMANN C, RINDERLE U and SCHMUCK G, 1986. Application of chlorophyll fluorescence in ecophysiology. Radiation and Environmental Biophysics 25: 297-308.

RINDERLE U and LICHTENTHALER HK, 1988. The chlorophyll fluorescence ratio F690/F735 as a possible stress indicator. In: Applications of chlorophyll fluorescence, Lichtenthaler HK ed, pp 176-183, Kluwer Academic Press, Dordrecht.

SCHREIBER U, FINK R and VIDAVER W, 1977. Fluorescence induction in whole leaves: Differentiation between the two leaf sides and adaptation to different light regimes. Planta 133: 121-129.

SIRONVAL C and BROUERS M, 1984. Protochlorophyllide reduction and greening. Martinus Nijhoff/Dr W Junk Publishers, The Hague.

CHANGES IN CHLOROPLAST FLUORESCENCE DURING LEAF DEVELOPMENT

Z. ŠESTÁK* and P. ŠIFFEL**

Institute of Experimental Botany, Czechoslovak Academy of Sciences,
Department of Physiology of Photosynthesis,
* CS-160 00 Praha 6, Na Karlovce 1a, Czechoslovakia
** CS-370 05 České Budějovice, Branišovská 31, Czechoslovakia

Key words: chlorophyll-protein complexes; chlorophyll in vivo; chlorophyll synthesis; chlorophyll degradation; fluorescence emission spectra; fluorescence excitation spectra; fluorescence kinetics; fluorescence polarization; leaf ontogeny

ABSTRACT

In the course of chloroplast and leaf development the formation and degradation of photosynthetic apparatus is reflected in changes in the amounts of individual pigment-protein complexes that have various functions in the photochemistry of photosynthesis. The ontogenic differences in composition of the set of pigment-protein complexes are reflected in spectral properties of chloroplasts and leaves. Changes in shapes of fluorescence emission and excitation spectra, patterns of induction fluorescence and fluorescence polarization occurring during leaf development are reviewed.

INTRODUCTION

During the development of a leaf many of its photosynthetic characteristics, such as anatomical parameters, size, number and ultrastructure of chloroplasts, contents of pigments and pigment-protein complexes, composition and activities of photosystems (PS) and photophosphorylation mechanisms, activities of enzymes of the carbon fixation pathways, rates of gas exchange and biomass production are changed (see Šesták 1985).

When a leaf develops in a normal environment, chlorophyll (chl) accumulation starts with leaf unfolding. Throughout the leaf life span both synthesis and degradation of chl proceed, but their ratio changes, with synthetic processes prevailing in the first phase of leaf development, a relative balance in the phase of the so-called "leaf photosynthetic maturity", and a prevalence of degradative processes at the end of leaf life. Therefore protochlorophyllides may be detected in the major part of leaf life.

Chl *a* synthesis precedes that of chl *b*. The synthesized chls enter chl-protein complexes participating in PS 1 and PS 2; these complexes appear in thylakoids of young as well as adult leaves. After some 6 to 10 h further phases of formation of the pigment apparatus are connected with (1) incorporation of carotenoids into the pigment antenna; (2) prevailing accumulation of light-harvesting complexes (LHC), and (3) appearance of

H. K. Lichtenthaler (ed.) Applications of Chlorophyll Fluorescence, 85–91.

long-wavelength chl a bands. The changes in contents of individual pigment-protein complexes with specific functions in PS 1 and PS 2 take place during the whole ontogeny of the photosynthetic apparatus (chloroplast, leaf, plant), and are - together with ultrastructural changes in thylakoids - reflected in its spectral properties. Both the absorption and fluorescence spectra of leaves and isolated chloroplasts are interrelated (see Šesták 1985). They have a very complicated structure and in spite of a great effort in this field during the last three or four decades they have not yet been fully understood. A high overlapping of the bands in the chloroplast spectra also restrains the use of visible spectroscopy for precise analysis of components of the pigment apparatus in vivo. Nevertheless, major pigment-protein complexes have specific spectral bands which can be used for a rough but fast and nondestructive estimation of their contents and affiliation as LHC, antenna complexes or reaction centres (RC) of PS 1 and PS 2. Also some minor components (protochlorophyllide, short-wavelength chl a and b emission - Lebedev et al. 1985, Šiffel et al. 1987) are detectable in vivo only using the fluorescence spectroscopy.

In addition to excitation and emission fluorescence spectra that detect the synthesis, presence and degradation of chl in vivo complexes, fluorescence kinetic (fluorescence yield, induction fluorescence, Kautsky effect) is often determined as a characteristic of photosynthetic activity and its changes induced by various stresses, of energy transfer within the photosystems, of size of the photosynthetic unit, etc. (see Walker 1987).

The goal of this paper is to review the changes in fluorescence emission and excitation spectra, induction fluorescence and fluorescence polarization connected with leaf development in natural conditions. All the spectral changes are better distinguished when measured at liquid nitrogen temperature (cf. Šiffel and Šesták 1988).

FLUORESCENCE EXCITATION SPECTRA

In Cucurbita maxima cotyledons no difference was found in the shape of fluorescence excitation spectra in the expansion phase and in advanced age (Harnischfeger 1973). Nevertheless, measurements of isolated chloroplasts placed as a monolayer on the surface of a membrane filter showed that at 77 K the chloroplasts from younger (7 or 9 d) primary leaves of Phaseolus vulgaris had a much higher ratio of excitation peaks E436/E472 than chloroplasts isolated from the 13 to 37 d leaves (Šiffel et al. 1985, 1986). Prolongation of leaf life, e.g. by the continuous removal of newly formed buds, restores chl synthesis, and thus the spectral characteristics of younger leaves are temporarily restored (Šiffel et al. 1985, 1986). The ratio E436/E472 expresses relative changes in the concentrations of chlorophyll-protein complexes belonging to PS 1 + PS 2 and the LHC (high ratio = enhanced rate of chl accumulation and high concentration of PS 1 + PS 2 complexes, and vice versa).

FLUORESCENCE EMISSION SPECTRA

Emission spectra of green leaves have usually two to three main peaks belonging to chl a : one (at room temperature) or two (at 77 K) between/at 685 and 695 nm, and one at 734-740 nm. These peaks are usually connected with antenna and RC of PS 2, and LHC of PS 1, respectively. The LHC of PS 2 may be detected by curve fitting and derivative spectroscopy as F680. Such peaks may be found in special cases also in dark-grown cotyledons, e.g. in 21-d cotyledons of Pinus jeffreyi, when they replace the emission bands of protochlorophyllide forms characteristic for young cotyledons (Michel-Wolwertz 1977). The position of the long-wavelength peak may shift by about 3 nm to longer wavelength during leaf ontogeny. The ratio of the emission peaks (F695/F735 or similar) varies during leaf ontogeny (in the range of 0.4 to 2.0). It either increases with leaf age (Sinapis alba - Horváth et al. 1973; P. vulgaris - Raafat et al. 1969, Jenkins et al. 1981b; Hordeum vulgare - Mader et al. 1981; Picea pungens - Khodasevich and Lis 1980) or declines (Pisum sativum - Paromenskaya et al. 1975; Prunus laurocerassus - Lichtenthaler 1986; Theobroma cacao - Baker and Miranda 1981). An increase followed by a decline was found at room temperature in the course of development of the fourth leaf of various cultivars of Solanum tuberosum (Vacek et al. 1979).

Measurements on whole leaves may be erroneous due to spectra flattening, distortion, self-absorption, etc. caused by high pigment concentration, differences in leaf thickness, non-uniform leaf surface, variance in optical properties of cells, etc. These factors certainly cause the differences between fluorescence spectra measured on the upper or lower side of a leaf (Acer platanoides - Lichtenthaler 1986). In addition to this, differences in various segments of long leaves reflect also the different age of individual tissues: thus in 7-d Zea mays leaves the ratio of emission peaks declines from leaf base to tip (Baker and Miranda 1981). Lowest ratios of peaks in the middle of leaf blade of H. vulgare may be partially explained by a low area mass, high chl a and b contents and low chl a / b ratio in these segments (Nauš et al. 1985).

The inaccuracies caused by measuring fluorescence spectra of whole leaves may be overcome by using isolated chloroplasts: nevertheless, the results obtained may be very different. Thus while the emission peaks ratio increased with age (22 to 50 d) of the 3rd leaf of H. vulgare, an opposite trend was observed in chloroplast suspensions from the same leaves (Mader et al. 1981). Also the leaf age dependent shift in the positions of emission peaks (mainly the long-wavelength one) was more expressed in chloroplast suspensions than in whole leaves (Mader et al. 1981). In isolated chloroplasts suspended on a surface of membrane filter a decline followed by an increase in the F685/F735 ratio was observed during development of the P. vulgaris primary leaf (Lebedev et al. 1986, Šiffel et al. 1985, 1986).

A detailed analysis of emission spectra detected also emission bands of chl a and b with maxima at about 673 and 653 nm, respectively. These pigments can be separated from thylakoids only by ultracentrifugation. They do not transfer excitation energy to the reaction centres, and their amounts correspond with the rate of chl accumulation. The emissions corresponding to these pigments were high in greening leaves, and the removal of leaf buds induced a rapid increase of both emission bands in older leaves (Šiffel et al. 1987).

Fluorescence spectroscopy enables to detect also the degradative changes in senescing leaves until a very low chl concentration (ca. 0.1 mg chl m-2) remains. In P. vulgaris fluorescence emission belonging to PS 1 and PS 2 declined, while F681 characteristic for LHC was most stable (this was confirmed also by comparison of the excitation spectrum with excitation spectra of LHC complexes isolated by electrophoresis). Similar changes were found in Beta vulgaris, P. sativum, Vicia faba and Petunia hybrida, but not in Spinacia oleracea or Hibiscus rosa-sinensis. In the last two species the bands of LHC and PS 1 were degraded most rapidly (Šiffel, unpublished).

In addition to the above changes observed in leaves developing in light, the age of etiolated leaves at first irradiation affects also the processes of chl formation and the related balance of protochlorophyll(ide) and chlorophyll(ide) forms (for review see Šesták 1984) and their reflex in fluorescence spectra. Thus, e.g. the ratio of protochlorophyll(ide) emissions at 628-632/650-657 nm declines and again increases with leaf age in plants grown under long far-red and then red radiation (Quercus robur - Axelsson et al. 1981), the chl complex fluorescing at 695 nm needs not to be formed in very young leaves (Goedheer and van der Cammen 1981).

FLUORESCENCE POLARIZATION

Fluorescence polarization degree at room temperature changed with age of the 4th leaf of S. tuberosum in a sinusoid curve, different for each of four cultivars (Vacek et al. 1979). In P. vulgaris an increase in polarization degree accompanied leaf development (Raafat et al. 1968).

FLUORESCENCE KINETICS

Fluorescence kinetics changes during leaf development have been studied in many details. In the well-known OIDPSMT pattern all parameters of both the rapid and slow phases may change with leaf age. The height of the first peak (FP) generally declines (P. vulgaris - Raafat et al. 1969, Jenkins et al. 1981a; Z. mays - Kozel et al. 1983). The M peak may not appear after the first irradiation in very young (6 d) leaves of P. vulgaris (Jouy 1982), its level (FM) similarly to the level of steady fluorescence (FT) declines with leaf age (P. vulgaris - Jenkins et al. 1981a).

Nesterenko and Sid'ko (1985) distinguish three types of induction patterns appearing during leaf development: (1) one-peak curves with a rapid decline to stationary level, without the SM phase (typical for 2-4 d leaves of Triticum aestivum); (2) double-peak curves with an expressed PSMT part, with P = 1-2 s, M = 8-10 s (mature 15-25 d leaves); (3) one-peak curves with a slow decline to stationary level (senescing leaves, > 25 d). Similar, but less expressed were these types in a leaf age gradient in Cucumis sativus (Nesterenko and Sid'ko 1986).

With leaf age also the times to reach P (τP) (P. vulgaris - Jenkins et al. 1981a; P. sativum - Somersalo and Aro 1987) or M (τM) (P. sativum - Somersalo and Aro 1987; Nicotiana tabacum - Šiffel et al. 1988) decline. Among the various intensity ratios, the FP/FT ratio usually increases and then declines (T. aestivum - Nesterenko and Sid'ko 1980; N. tabacum - Šiffel et al. 1988; C. sativus - Nesterenko and Sid'ko 1986). The (FP-FT)/FT(= Rfd) ratio declines with leaf age of needles (Picea abies - Lichtenthaler and Buschmann 1984, Lichtenthaler et al. 1985a,b, 1986) or cotyledons (Raphanus sativus - Bach and Lichtenthaler 1983). The ratio (FP-FO)/FO(= Fv/FO) that estimates the maximum yield of primary photochemistry of PS 2 either continuously declines with leaf age (P. sativum - Somersalo and Aro 1987; N. tabacum - Chia et al. 1986; P. abies - Schmuck and Lichtenthaler 1986) or increases and then declines in cotyledons (R. sativus - Bach and Lichtenthaler 1983) or leaves (T. cacao - Baker and Miranda 1981). The steepest slope of the decline after P increases and then declines in the course of leaf development (P. sativum - Somersalo and Aro 1987).

By comparing the kinetics of fluorescence measured at various wavelengths, a similarity of leaf ontogeny related courses of the ratio of (FP-FO)/FP measured at 695/735 nm to that of the characteristic (FP-FO)/FO measured at 735 nm was evident (increase followed by a decline), while the (FP-FO)/FP ratio measured at 695 nm increased to a plateau (T. cacao - Baker and Miranda 1981).

The relative level of DCMU-induced fluorescence changes declined from young to senescent leaves of Ipomoea pentaphylla (Kulandaivelu and Daniell 1980).

The changes in shape of fluorescence kinetics and various calculated parameters for leaf blade segments from base to tip in monocotyledons (T. aestivum - Nesterenko and Sid'ko 1985) or dicotyledons (T. cacao - Baker and Miranda 1981) often recall those induced by leaf ageing. The lowest fluorescence yield found in some cases in the middle part of a leaf (H. vulgare - Nauš et al. 1985) or the lowest (FP-FO)/FP ratio for similar segments (T. aestivum - Bredenkamp et al. 1986) may be connected with a low leaf thickness and high chl content. In flat leaves the zone of most homogeneous fluorescence is in the middle of the leaf blade lacking large conducting bundles (C. sativus - Nesterenko and Sid'ko 1986).

REFERENCES

(for other references see the book of Šesták 1985)

Bach TJ and Lichtenthaler HK, 1983. Inhibition by mevinolin of plant growth, sterol formation and pigment accumulation. Physiologia Plantarum 59: 50-60.

Bredenkamp GJ, Percival MP, Webber AN and Baker NR, 1986. Organization of the light-harvesting apparatus during chloroplast biogenesis in wheat. In: Akoyunoglou G and Senger H. Regulation of chloroplast differentiation. Pp. 259-265. A.R. Liss, New York.

Chia CP, Duesing JH and Arntzen CJ, 1986. Developmental loss of photosystem II activity and structure in a chloroplast-encoded tobacco mutant, lutescens-1 . Plant Physiology 82: 19-27.

Kozel U, Buschmann C and Lichtenthaler HK, 1983. Influence of nitrogen deficiency on chlorophyll accumulation and photosynthesis of wheat seedlings grown at high and low light-quanta fluence rates. In: Metzner H. Photosynthesis and plant productivity. Pp. 122-126. Wissenschaftliche Verlagsgesellschaft, Stuttgart.

Lebedev NN, Šiffel P and Krasnovskiĭ AA, 1985. Detection of protochlorophyllide forms in irradiated green leaves and chloroplasts by difference fluorescence spectroscopy at 77 K. Photosynthetica 19: 183-187.

Lebedev NN, Šiffel P and Šesták Z, 1986. Composition of chlorophyll-protein complexes and fluorescence spectrum changes during ontogeny of primary leaves of Phaseolus vulgaris L. Photosynthetica 20: 146-152.

Lichtenthaler HK, 1986. Laser-induced chlorophyll fluorescence of living plants. Proceedings of IGARSS'86 Symposium ESA SP-254: 1571-1579.

Lichtenthaler HK and Buschmann C, 1984. Photooxidative changes in pigment composition and photosynthetic activity of air-polluted spruce needles (Picea abies L.). In: Sybesma C. Advances in photosynthesis research. Vol.

IV, pp. 245-250. Martinus Nijhoff/Dr W.Junk, The Hague - Boston - Lancaster.

Lichtenthaler HK, Schmuck G and Döll M, 1985a. Photosyntheseaktivität bei Nadeln gesunder und geschädigter Koniferen. LIS-Berichte 57: 87-105.

Lichtenthaler HK, Schmuck G, Döll M and Buschmann C, 1985b. Untersuchungen über die Funktionsfähigkeit des Photosyntheseapparates bei Nadeln gesunder und geschädigter Koniferen. Projekt Europäisches Forschungszentrum für Massnahmen zur Luftreinhaltung, 1. Statuskolloquium : 81-105.

Lichtenthaler HK, Schmuck G, Nagel E and Buschmann C, 1986. Messung der Photosynthese gesunder und geschädigter Bäume: Suche nach Parametern zur Früherkennung von Baumschäden. Projekt Europäisches Forschungszentrum für Massnahmen zur Luftreinhaltung, 2. Statuskolloquium : 97-114.

Michel-Wolwertz MR, 1977. Chlorophyll formation in cotyledons of Pinus jeffreyi during germination in the dark. Occasional accumulation of protochlorophyll(-ide) forms. Plant Science Letters 8: 125-134.

Nauš J, Dvořák L and Kupka Z, 1985. Heterogeneity of chlorophyll-a fluorescence parameters of green leaves. Acta Universitatis Palackianae Olomucensis Facultas Rerum Naturalium 82 (Physica 24): 25-34.

Nesterenko TV and Sid'ko FY, 1985. Vozrastnye izmeneniya medlennoĭ induktsii fluorestsentsii khlorofilla list'ev pshenitsy. Fiziologiya Rasteniĭ 32: 440-448.

Nesterenko TV and Sid'ko FY, 1986. Medlennaya induktsiya fluorestsentsii khlorofilla v ontogeneze list'ev ogurtsa. Fiziologiya Rasteniĭ 33: 672-683.

Raafat A, Gausz J, Szalay L and Horváth I, 1968. Chlorophyll photobiology of ageing bean leaves in vivo. Acta Biochimica et Biophysica Academiae Scientiarum Hungaricae 3: 453.

Schmuck G and Lichtenthaler HK, 1986. Application of laser-induced chlorophyll-a fluorescence in the forest decline research. Proceedings of IGARSS'86 Symposium ESA SP-254: 1587-1590.

Šesták Z, 1984. Effects of leaf age on protochlorophyllide and chlorophyllide formation (a review). In: Sironval C and Brouers M. Protochlorophyllide reduction and greening. Pp. 365-375. Martinus Nijhoff/Dr W Junk, The Hague - Boston - Lancaster.

Šesták Z, 1985. Photosynthesis during leaf development. Academia, Praha, Dr W. Junk, Dordrecht - Boston - Lancaster.

Šiffel P, Durchan M, Květoň J, Ondřej M and Šesták Z, 1988. Photosynthsis in regenerants of tobacco transformed by plasmids of Agrobacterium 2. Fluorescence emission spectra and age induced changes in slow fluorescence induction. Photosynthetica 22:214-220.

Šiffel P, Lebedev NN and Krasnovskiĭ AA, 1987. Detection of short-wavelength chlorophyll a emission in green leaves. Photosynthetica 21: 23-28.

Šiffel P, Lebedev NN and Šesták Z, 1985. Changes in chloroplast absorption and fluorescence spectra during ontogeny of primary bean leaves. Photosynthetica 19: 127-137.

Šiffel P, Lebedev NN and Šesták Z, 1986. Changes in shape of chloroplast spectra during ontogeny of primary leaves of French bean correlate with rate of chlorophyll accumulation. In: Akoyunoglou G and Senger H. Regulation of chloroplast differentiation. Pp. 691-696. A.R.Liss, New York.

Šiffel P and Šesták Z, 1988. Low temperature fluorescence spectra of chloroplasts: methodical aspects and possible applications. This volume.

Somersalo S and Aro E-M, 1987. Fluorescence induction in pea leaves of different ages. Photosynthetica 21: 29-35.

Vacek K, Nauš J, Barthová V and Zrůst J, 1979. Fluorescence, absorption and photosynthetic rate measurements of potato leaves (Solanum tuberosum L.) during their vegetation. Acta Universitatis Carolinae, Mathematica et Physica 20(2): 3-9.

Walker D, 1987. The use of the oxygen electrode and fluorescence probes in simple measurements of photosynthesis. Research Institute of Photosynthesis, University of Sheffield.

MEASUREMENT AND ANALYSIS OF CHLOROPHYLL FLUORESCENCE IN PLANTS USING MODULATED ACTINIC ILLUMINATION

Jan F.H. Snel

Agricultural university, Dept. of Plant Physiological Research,
Gen. Foulkesweg 72, 6703 BW Wageningen, The Netherlands.

KEYWORDS: photosynthesis, light emitting diodes, frequency domain, fluorescence quenching.

ABSTRACT

This communication reports on the measurement and analysis of variable fluorescence in the frequency domain. Leaves of plants were illuminated with light modulated in intensity using a voltage controlled lightsource. The actinic light was sinusoidally modulated in the frequency range from 0.02 to 100 Hz. The observed chlorophyll fluorescence changes are a more or less linear function of the actinic light intensity at modulation frequencies above 1 Hz. Based on measurements of the phase shift between the modulated actinic light and the fluorescence changes, we suggest that this high-frequency component is mainly attributable to photochemical quenching. At lower frequencies a strong deviation of linearity is observed.

ABBREVIATIONS: LED: light emitting diode, LHCP: light harvesting chlorophyll protein, VCLS: voltage controlled lightsource.

INTRODUCTION

Measurement of chlorophyll fluorescence has been shown to be a valuable tool in plant physiology. Chlorophyll fluorescence induction curves contain information about the redox state of the acceptor(s) of photosystem 2 and about the state of "energization" of the thylakoidmembrane (for reviews see Schreiber, 1983; Krause and Weis, 1984; Vredenberg, 1986). Chlorophyll fluorescence in plants originates mainly from photosystem 2. Several mechanisms of fluorescence quenching have been described. The major mechanisms are photochemical quenching, related to competing photochemistry in photosystem 2, and non-photochemical quenching which is mainly caused by enhanced radiationless deexcitation related to energization of the thylakoid membrane (Krause and Weis, 1984; Bilger and Schreiber, 1986).

The kinetics of chlorophyll fluorescence induction have been analysed mainly in the time domain. Usually a pulse of light is applied to a more or less dark-adapted plant, leaf or chloroplast suspension. The resulting fluorescence induction pattern often shows complex kinetics with frequently a dampened oscillation due to incomplete regulation of ATP and NADPH levels

H. K. Lichtenthaler (ed.) Applications of Chlorophyll Fluorescence, 93–98.

in the stroma during the onset of carbon fixation. Analysis of chlorophyll fluorescence (Kolbowski, 1986) and oxygen evolution (Lam and Bungay, 1986) in the frequency domain below 1 Hz, has revealed several time constants. The various time constants were attributed to biochemical reactions, e.g. LHCP-phosphorylation (Kolbowski, 1986), Calvin cycle activity (Kolbowski, 1986, Lam and Bungay, 1986) or mass flow (Lam and Bungay, 1986).
In this report chlorophyll fluorescence is analyzed in the frequency domain from 0.02 to 100 Hz in order to obtain information about the photochemical steps as well. It is shown that at frequencies above 1 Hz the fluorescence changes are caused by photochemical quenching.

MATERIALS AND METHODS

The experimental setup is schematically depicted in figure 1. It consists basically of the following components: a pulsed chlorophyll fluorometer, a function generator, a voltage controlled actinic lightsource, a four-armed

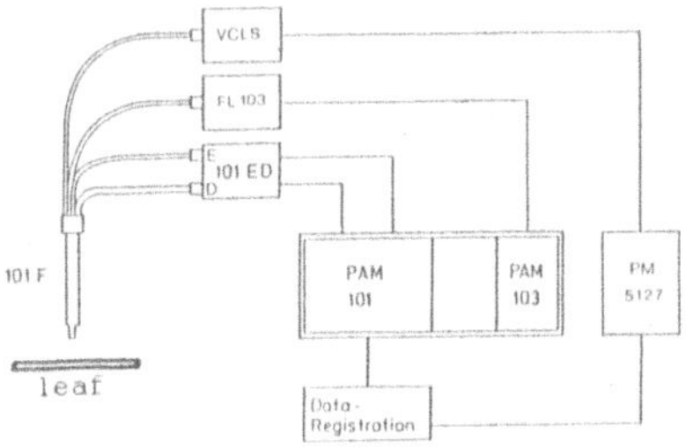

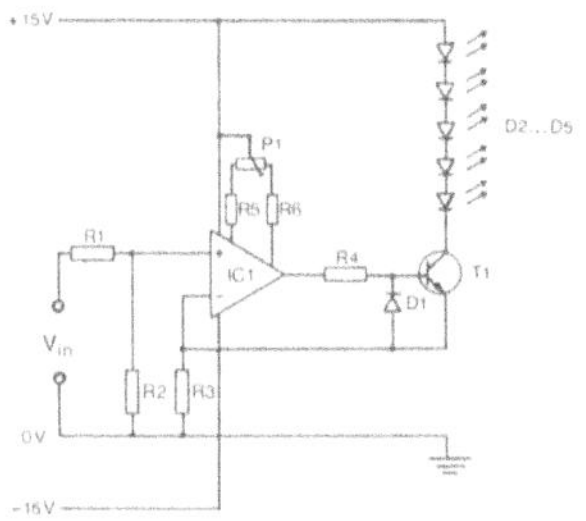

Figure 1 (left): Schematic representation of the experimental setup. Abreviations: 101 F: polyfurcated optical fiber; PAM 101: pulse-amplitude-modulation fluorometer (basic unit); 101 ED: emitter/detector unit; FL 103: high-intensity halogen lightsource; PAM 103: timer module for FL 103; VCLS: Voltage Controlled Light Source; PM 5127: function generator.

Figure 2 (right): Schematic diagram of the voltage-to-current converter used in the VCLS. IC1= LF 356; T1= BC 140; D1= 1N4148; D2..D6: H-2k; R1=4k7; R2=R4=1k; R3=0.1k; R5=R6=10k; P1=20k.

optical fiber for the measuring pulse, the fluorescence signal and the actinic light and finally data registration equipment (transient recorder and XY-plotter). Chlorophyll a fluorescence was measured using a PAM fluorometer (Heinz Walz, Mess- und Regeltechnik, Effeltrich, BRD). The configuration consisted of the basic module PAM-101, the emittor/detector unit ED-101, the high-intensity lightsource FL 103 with its control module PAM 103, and the optical fiber assembly 101 F. The apparatus and the measuring principles have been described (Schreiber, 1986; Schreiber et al, 1986).

Modulated actinic light was derived from a voltage controlled lightsource. This laboratory built lightsource uses five current-driven LED's (Stanley Electric Co., Ltd type H-2k). The emitted red light is centered around

660 nm with a spectral half-bandwith of approx. 25 nm. The main reasons for using LED's are their stability, their high electrical bandwith which extends from DC to more than 1 MHz and the linear relation beween LED current and intensity of the emitted light. The driver for the LED's is a simple voltage to current converter (see fig. 2). The light intensity of the VCLS is nearly linear with the input voltage over more than two decades (fig. 3). The linearity could be improved by applying negative feedback from the (light)output to the input via a photodiode.

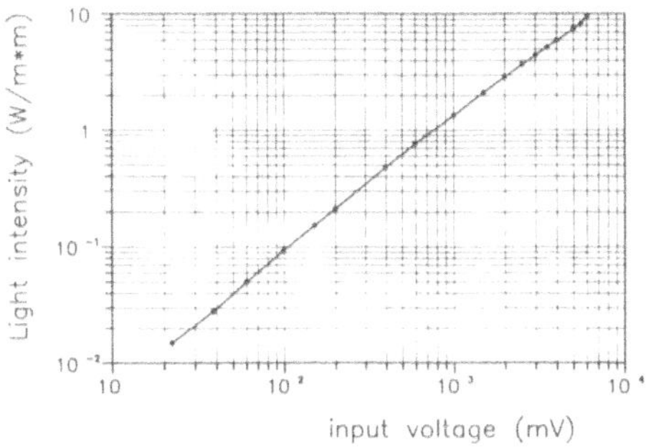

Figure 3: Intensity of the light output of the VCLS as a function of the input voltage. The light intensity was measured at the leaf surface using a United Detector Technology 80X Optometer.

The output signal of the fluorometer was recorded on a Nicolet 2090 digital storage scope in the single channel mode and plotted on a BBC XY-recorder. The input voltage for the VCLS was provided by a function generator (Philips, type PM-5127) which also delivered the triggering pulse for the Nicolet 2090 storage scope.

RESULTS

Figure 4 shows the response of a detached spinach leaf on modulated illumination, measured as chlorophyll fluorescence yield. The leaf was illuminated with sinusoidally modulated light at frequencies ranging from 0.02 Hz to 100 Hz and the response was registrated in the stationary state. Figure 4 shows two striking features of the signals: 1) the signals are extremely small, approx. 14% of Fo at 1 Hz, and 2) the signals are more or less sinusoidal at frequencies above 0.5-1 Hz and strongly distorted at lower frequencies.
Based on the reported time constants for flash-induced fluorescence decay ($\tau < 1$ ms, e.g. Schreiber, 1986), caused by changes in photochemical quenching and the time constant for non-photochemical ($\tau >> 1$ s, Schreiber et al., 1986) quenching, we have suggested that the fluorescence changes above 0.5-1 Hz are mainly due to changes in photochemical quenching. At lower frequencies both photochemical and non-photochemical quenching were suggested to be involved in the fluorescence changes (Snel et al., 1987). This hypothesis can be tested by determining the phase shift between input (actinic light) and output (fluorescence yield). Photochemical quenching is maximal in open reaction centers and minimal in closed reaction centers and therefore the phase shift between light input and fluorescence output is less than 90°. In contrast non-photochemical quenching is low in non-energized thylakoids (dark state) and increases with increasing light

intensity and therefore we expect that changes in energization induced by modulated illumination result in a phase shift of at least 180° between input and output.

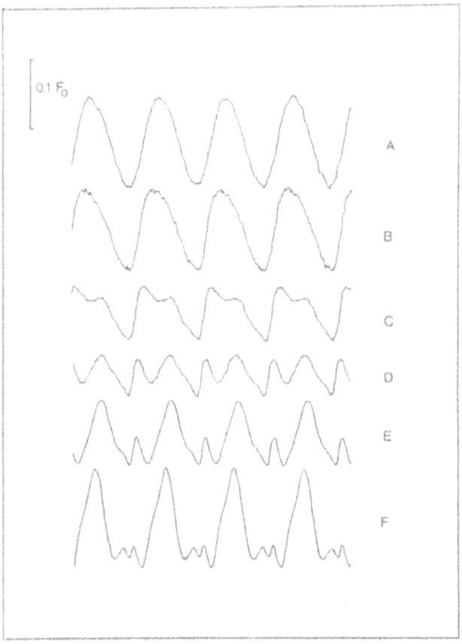

Figure 4: Chlorophyll fluorescence response of a detached spinach leaf to sinusoidally modulated actinic light. Input voltage: $V(t) = 4.965+4.9*\sin(2\pi ft)$ with f = 1 (A), 0.5 (B), 0.2 (C), 0.1 (D), 0.05 (E) and 0.02 (F) Hz. The measuring light intensity was maximal, the pulse repetition rate 1.6 kHz and the low-pass filter was adjusted to the sampling rate to minimize noise.

Figure 5 shows the amplitude and phase plots of the fluorescence signals from a maize leaf measured at a higher measuring light intensity. Fig 5A shows the amplitude of the fluorescence signals as a function of the modulation frequency. There is a relative maximum at 0.6 Hz, a frequency at which the fluorescence signal is relatively undistorted and a relative minimum at 0.1 Hz, corresponding with a high degree of distortion.

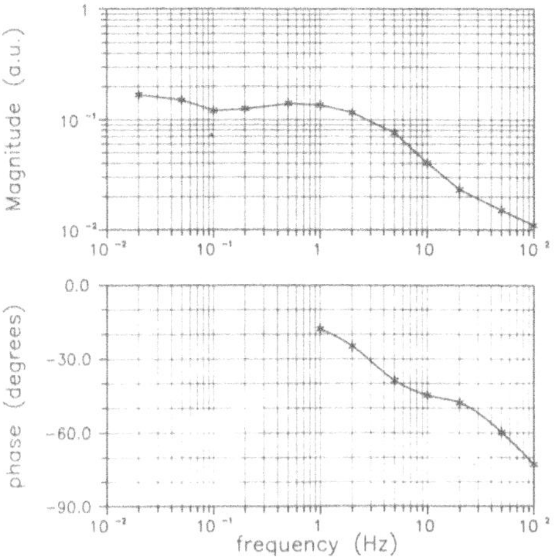

Figure 5: Bode plot of the chlorophyll fluorescence response of a Maize leaf to modulated actinic light. Input voltage: $V(t) = 4.965 + 4.9*\sin(2\pi ft)$. The measuring light intensity was maximal, the pulse repetition rate 1.6 kHz and the low-pass filter was adjusted to the sampling rate to minimize noise.

Figure 5 clearly shows that the phase shift is less than 90°, even at the highest frequency. This implies that the changes in fluorescence yield observed in the frequency range from 1 to 100 Hz are due to changes in photochemical quenching.

Further evidence comes from experiments (data not shown) in which non-photochemical quenching was determined at various instants during modulation using the saturating pulse method described before (Schreiber et al., 1986). These data show that at frequencies below 0.1 Hz non-photochemical quenching contains a modulated component, while no modulation occurs at higher frequencies.

DISCUSSION

Analysis of photosynthesis in the frequency domain using modulated actinic light has several advantages. First the measurements are made in a stationairy state, which means that averaging of signals is relatively easy and fast, improving the sensitivity of the measurements. Moreover one can apply linear analysis techniques to determine the time constant, delay and (relative) magnitude of the various processes which are involved in the regulation of photosynthesis (e.g. Kolbowski, 1988).

The phase response shown in Fig.5 indicates that chlorophyll fluorescence yield changes in the frequency range above 1 Hz are the result of changes in photochemical quenching. This is in agreement with data based on extrapolation of low-frequency measurements (Kolbowski, 1988). Moreover Fig.4 shows that, at frequencies above 0.5 Hz, chlorophyll fluorescence yield contains no higher harmonics, which means that chlorophyll fluorescence is a linear indicator of photochemical quenching. This property of the photosynthetic apparatus makes chlorophyll fluorescence suitable to determine time-constants of electron flow in vivo using linear analysis.

ACKNOWLEDGEMENTS: The author wishes to thank Drs. Wim J. Vredenberg, Olaf van Kooten and Jörg Kolbowski for stimulating discussions.

REFERENCES

Kolbowski J, 1986. Steady-state kinetics of chlorophyll fluorescence and light scattering. In: Progress in Photosynthesis Research (J Biggins, ed.), Vol. II, Martinus Nijhoff/Dr W Junk, The Hague.

Kolbowski J, 1988. Lineare analyse der signale aus dem photosynthetischen apparat grüner pflanzen mit besonderer berücksichtigung der wechselwirkung zwischen photosynthese und plasmalemma-transport. Diss., Christian-Albrecht-Universität, Kiel.

Krause G and Weis E, 1984. Chlorophyll fluorescence as a tool in plant physiology. II. Interpretation of fluorescence signals. Photosynth. Res. 5: 139-157.

Lam HLY and Bungay HR, 1986. Frequency response analysis of oxygen evolution by algae. J. Biotechnol. 4: 125-142.

Lam HLY, Bungay HR and Culotta LG, 1986. An engineer looks at photosynthesis. Appl. Biochem. Biotechnol. 13: 37-73.

Schreiber U, 1983. Chlorophyll fluorescence yield changes as a tool in plant physiology. I: The measuring system. Photosynth. Res. 4: 361-373.

Schreiber U, 1986. Detection of rapid induction kinetics with a new type of high-frequency modulated chlorophyll fluorometer. Photosynth. Res. 9: 261-272.

Schreiber U, Schliwa U and Bilger W, 1986. Continuous recording of photochemical and non-photochemical chlorophyll fluorescence quenching with a new type of modulated fluorometer. Photosynth. Res. 10: 51-62.

Snel JFH, van Kooten O and Vredenberg WJ, 1987. Plant Physiol. 83: 161.

Vredenberg WJ, 1986. Fluorescence and absorbance measurements in leaves: sensors of photosynthetic performance. In: Advanced Agricultural Instrumentation (WG Gensler, ed.), Nato ASI series. Series E, No. 111, Martinus Nijhoff Publishers, Dordrecht.

MEASUREMENT OF THE LASER-INDUCED CHLOROPHYLL FLUORESCENCE KINETICS USING A FAST ACOUSTOOPTIC DEVICE

László KOCSÁNYI[1), Michael HAITZ[2) and Hartmut K. LICHTENTHALER[2)

1) Department of Atomic Physics, TU Budapest,
Budafoki ut 8, H-1111 Budapest, Hungary

2) Botanisches Institut II (Plant Physiology and Plant Biochemistry)
University of Karlsruhe, Kaiserstr. 12,
D 7500 Karlsruhe, F.R.G.

KEY WORDS: acoustooptic device, Bragg cell, chlorophyll fluorescence fluorescence induction kinetics, fluorescence ratio F690/F730, Kautsky effect.

ABSTRACT

A laser induced computer-aided fluorometer system (LICAF) is presented which permits monitoring of the chlorophyll-fluorescence induction kinetics at two different wavelengths (690 and 730 nm regions) excited by a HeNe-laser. The exciting laser beam is controlled by a fast acoustooptic device. The rising time of the illumination is shorter than 200 ns. The data acquisition rate of the LICAF is at maximum 2000 data/sec, at each of the two wavelength regions.

The newly developed system permits measurement in 5 µs periods, every 500 µs of the complete chlorophyll fluorescence induction kinetics (fast and slow components of the Kautsky effect) of predarkened leaves. The ratio of the fluorescence in the 690 and 730 nm region (F690/F730) is calculated and plotted together with the induction kinetics. Examples of the application of the LICAF in the stress detection in plants e.g. water stress and herbicide (diuron) treatment is demonstrated.

INTRODUCTION

Chlorophyll fluorescence induction kinetics (fast and slow components of the Kautsky effect) reveal valuable information on the state of the photosynthetic apparatus in plants (Krause and Weis, 1984; Larcher and Cernuska, 1985; Lichtenthaler 1984, 1986; Lichtenthaler et al. 1986; Neubauer and Schreiber, 1986).

There are several types of fluorometers proposed to investigate the properties of the fluorescence induction kinetics. However, most of these methods do not give any information about the spectral characteristics of the induced chlorophyll fluorescence. With a new portable field fluorometer we can sense the fluorescence induction kinetics simultaneously in the 690 and 730 nm regions, which correspond to the fluorescence maxima of leaves (Lichtenthaler, 1986; Lichtenthaler and Rinderle 1988). From the corresponding values of the fluorescence decrease ratio (Rfd 690 and Rfd 730) as a vitality index one can calculate the stress adaptation index Ap, which has

H. K. Lichtenthaler (ed.) Applications of Chlorophyll Fluorescence, 99–107.

additional information on the photosynthetic apparatus (Strasser et al. 1987; Lichtenthaler and Rinderle, 1988). Using an optical multichannel analyser it was shown that during the fluorescence induction kinetics spectral changes of the chlorophyll fluorescence can occur (Lichtenthaler and Buschmann, 1987). On the other hand the fluorescence ratio F690/F730 changes mainly depending on the chlorophyll content of the leaf and to a lower degree also on the photosynthetic activity (Lichtenthaler,1987; Lichtenthaler and Rinderle, 1988). It is therefore desirable to register the ratio of the two fluorescences F690/F730 together with the induction kinetics simultaneously.

On the basis of our laser-equipped portable field fluorometer, which permits separate fluorescence registration at 690 and 730 nm, we tried to establish a system which is suitable to indicate the ratio F690/F730 also. In this article we introduce a laser-induced, computer-aided fluorometer (LICAF-system) to monitor the fluorescence intensity at different wavelength (690 nm and 730 nm) as well as the fluorescence ratio F690/F730 during the complete fluorescence induction period or selected parts of it.

MATERIALS AND METHODS

Leaves of different physiological condition of laurel cherry (*Prunus laurocerasus*) and bean (*Phaseolus vulgaris*) were kept 15 min in darkness prior to the measurement. After different times of water stress (hours after abscission) the fluorescence measurements were repeated. Measurements were made in air at room temperature. The herbicide diuron was added in one experiment in a 10^{-3} M solution.

The chlorophyll fluorescence induction kinetics at the two wavelength regions (690 and 730 nm) as well as the fluorescence ratio F690/F730 were determined during the whole induction period, or in particular parts of it by using the newly developed LICAF-system described in detail in the following chapter. This new system also permits resolution of the major part of the fast fluorescence rise.

THE LICAF MEASURING SYSTEM

The main parts of the newly developed laser-induced and computer-aided fluorometer (LICAF) are (Fig. 1):

- a triggerable laser light source
- a detector unit
- a data-acquisition and control unit.

Triggerable laser light source. To excite the fluorescence a 10 mW HeNe-laser (Spectra Physics) is applied. The advantages of a HeNe-laser compared to a LED source are:

- a good collimated, high-power beam and
- a narrow spectral radiation band, below the excited fluorescence.

To trigger the laser light, the beam is guided through a /4 circular polarizer and an acoustooptic modulator (also called Bragg cell). Driving the Bragg cell by a radiofrequency electric signal (fo=48 MHz) part of the laser beam will be diffracted (see Fig. 2). The diffracted light of the Bragg cell illuminates the leaf via the input arm of a fiberoptic system.

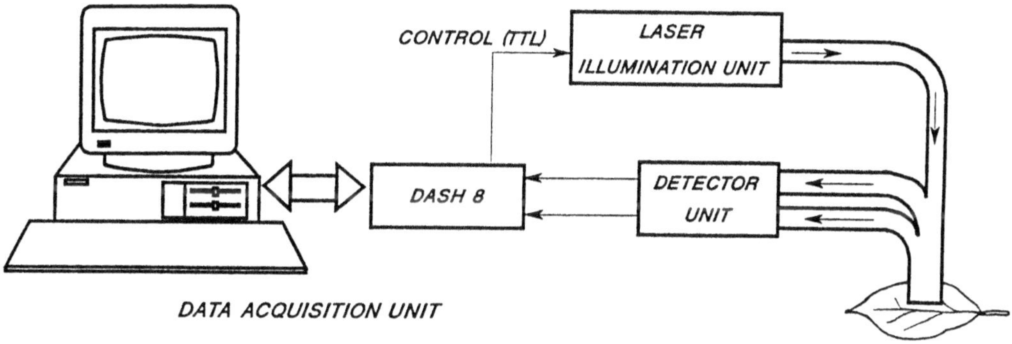

Figure 1. Scheme of the newly developed laser-induced, computer-aided fluorescence system for the registration of the chlorophyll fluorescence induction kinetics.

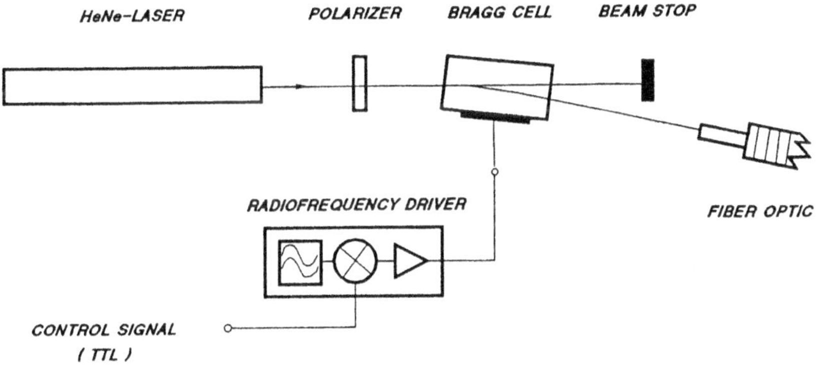

Figure 2. Scheme of the electronically triggerable laser source using an acoustooptic Bragg cell.

The latter is also used to guide the fluorescence signals to the detector unit. The radiofrequency driving electronic includes a 48 MHz signal generator, a mixer and an amplifier. When the control input of the mixer is logical "1" the mixer couples the 48 MHz signal of the generator to the amplifier, and when the control signal is logical "0" it damps the signal of the generator, and no radiofrequency signal arrives the amplifier.

The Bragg cell and the driving electronic applied in the LICAF was developed at the TU Budapest, Department of Atomic Physics, and has the following main specifications:

diffraction efficiency ($\eta = I_{diff}/I_o$)	
- for circularly polarised light	90 %
- for random or linear polarised light	45 %
driving frequency	48 MHz
rising time	200 ns
interacting material	TeO_2
ultrasonic transducer	piezoelectric, $LiNbO_3$.

The advantage of this acoustooptic system is the very fast switching of the laser-beam (rising time 200 ns).

Detector unit. To detect the signals two photodiodes (Type: SD-444-41-11-261, Silicon Detector Corp.) in photoconductive operating mode are used. In front of the diodes optical band-pass filters are placed ($_0$= 690 ± 18 nm, T_0=49 % and $_0$=730 ± 18 nm T_0=48 %). Detectors and filters are mounted in a dark housing to reduce the dark current.

Data acquisition and control-hardware. For data acquisition and control an IBM compatible PC (in our case a HP-Vectra ES) with an A/D converter and timer board (Metrabyte, DAS 8) is applied. The board includes an 8 channel, 12 bit (resolution: 2,44 mV at ± 5 V) successive approximation, high-speed analog/digital converter with sample/hold and a 8254 programmable timer/counter unit. To control the acoustooptic modulator and to generate interrupts two timers are applied. To measure both fluorescence intensities, only two input channels of the multiplexer must be monitored.

Data acquisition and control-software. The whole induction measurement takes 5 minutes. The A/D conversion rate of both fluorescence signals in the first 5 seconds is 2000 samples/sec, whereas in the rest of the 5 minutes 25 sample/sec. The programme is written by us in BASIC and it uses machine-code subroutines to enhance communication with the A/D board. The main program functions during the measurement are the following:

- prescription of measuring circumstances for the A/D converter and the counters
- starting measurements
- handling of interrupts
- saving data

whereas after the measurement:

- calculating quotient (F690/F730)
- printing of optional parts of the induction kinetics of fluorescence intensity and quotient
- calculating and printing of different characteristic values of the kinetic curves (for example Rfd values, time delay of maxima).

APPLICATION OF THE LICAF SYSTEM TO LEAVES

The possibility to register and determine the chlorophyll-fluorescence induction kinetics in the 690 and 730 nm region as well as the ratio F690/F730 is shown in Fig. 3a, b for a green and a light-green aurea leaf of cherry laurel (*Prunus laurocerasus L.*). One registers the full fluorescence induction kinetics (rise and decrease of the fluorescence) until the steady state is reached after ca. 4 to 5 min. By selection of optional time intervals one can study then the fast fluorescence rise in the first 250 ms (ms-range) determine the maximum fluorescence and beginning fluorescence decline by choosing the range of seconds and analyze the slow fluorescence decline (min-range) from which the Rfd-values will be calculated. At the same time one can follow the changes in the fluorescence ratio F690/F730 during the different phases of the induction kinetics.

The cherry laurel leaf (Fig. 3b) with its lower chlorophyll content has a higher chlorophyll-fluorescence yield than the normal green leaf (Fig. 3a) and also a higher intensity of the 690 nm fluorescence (solid line) with relation to the 730 nm fluorescence (dotted line). As as consequence the ratio F690/F730 exhibits in the aurea significantly higher values than in the normal green leaf. A further difference is seen in the development of the ratio F690/F730 during the fluorescence-induction kinetics. In the aurea leaf there is a slight and reproducible increase of the ratio F690/F730 up to about 1 second which then decreases from values of ca. 1.0 to ca. 0.83,

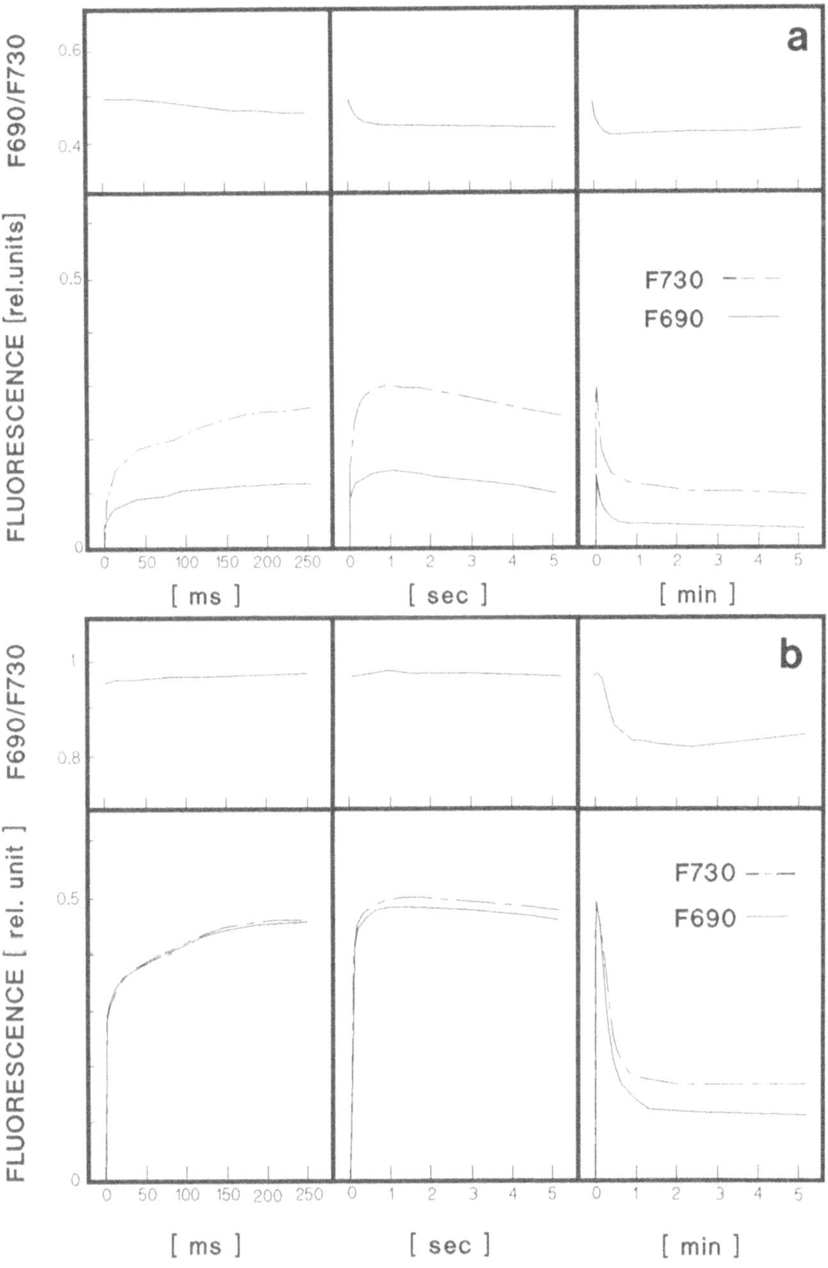

Figure 3. Redrawing of the laser-induced chlorophyll fluorescence induction kinetics in the 690 and 730nm region and fluorescence ratio F690/F730 in optional time intervals measured with the LICAF-system in cherry laurel leaves (Prunus laurocerasus) of different chlorophyll content. **a**) normal green leaf (ca 35 µg chlorophyll·cm^{-2}) and **b**) aurea leaf (ca. 12 µg chlorophyll·cm^{-2}).

whereas in the green leaf the values of F690/F730 decline from the beginning of the induction kinetic (Fig. 3a, b).The original tracings with one point per each 5 µs measuring period is shown in Fig. 4.

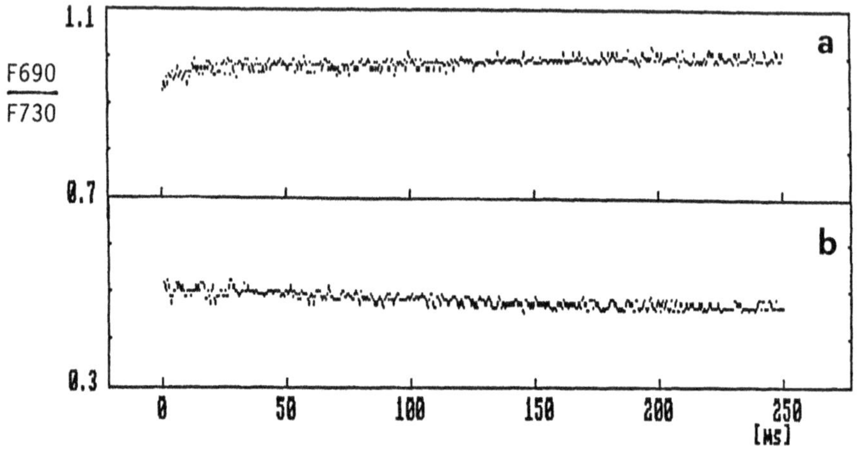

Figure 4. Changes in the values of the chlorophyll fluorescence ratio F690/F730 as measured with the LICAF-System during the first 250 ms of the laser-induced chlorophyll-fluorescence rise kinetic (fast component of the Kautsky effect) in cherry laurel leaves (Prunus laurocerasus) of different chlorophyll content. **a**) aurea leaf (ca. 12 µg chlorophyll·cm^{-2}) and **b**) normal green leaf (ca. 35 µg chlorophyll·cm^{-2}). Each point corresponds to a sensing period of 5 µs.

Effect of water stress on the induction kinetics. The effect of water stress on the chlorophyll-fluorescence induction kinetics and the ratio F690/F730 was investigated in abscissed light-green Prunus leaves (Fig. 5). After 4 h water stress the fluorescence intensity and yield decreases over the whole range of the induction kinetics with the slow desiccation of the tissue in both wavelength regions as compared to the starting kinetic (control in Fig. 5). The Rfd-values increase somewhat as given in the legend of Fig. 5 and indicate that the photosynthetic functions is still there. When after 24 h water stress (Fig. 5c) the photosynthetic quantum conversion is almost fully lost (Rfd values below 0.25), the fluorescence intensity at the end of the induction kinetics (after ca. 240-300 seconds) is higher than that of the controls (Fig. 5a). The maximum fluorescence at the initial phase of the fluorescence induction is further decreased. The decline of the fluorescence ratio F690/F730 during the early phase of the induction kinetics of the control, which is still visible after 4 h water stress, can no longer be observed after 24 h of water stress. The changes in the ratio F690/F730 apparently only occur in still photosynthetically active leaf tissue.

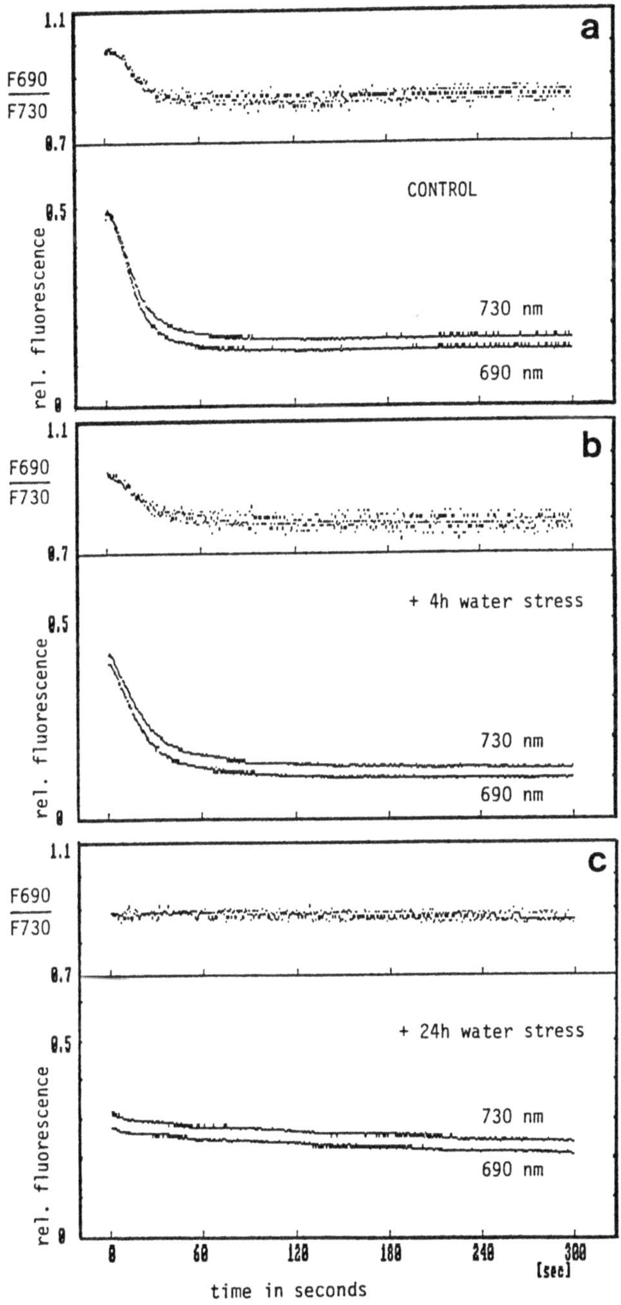

Figure 5. Laser-induced chlorophyll-fluorescence induction kinetics in the 690 and 730 nm region and fluorescence ratio F690/F730 in a light green cherry laurel leaf at different times of water stress induced by leaf abscission. **a**) control **b**) 4 h and **c**) 24 h water stress. The Rfd values somewhat increase from the control to 4 h water stress and then quickly decline for Rfd 690 from 2.5 to 2.7 and 0.25 and Rfd 730 from 1.9 to 2.3 and 0.24. The stress adaptation index Ap declines from 0.17 via 0.11 to 0.01.

Effect of the herbicide diuron. Treatment of the leaves with the herbicide diuron, which is known to block photosynthetic electron transport at the Q_B-binding site (B-protein) between the quencher Q_A and the large plastoquinone pool, decreases the initial maximum fluorescence and increases the steady-state fluorescence in both wavelength regions in a similar way to water stress (Fig. 6). Also in this case one can no longer observe the changes in the ratio F690/F730 at the beginning of the induction kinetics.

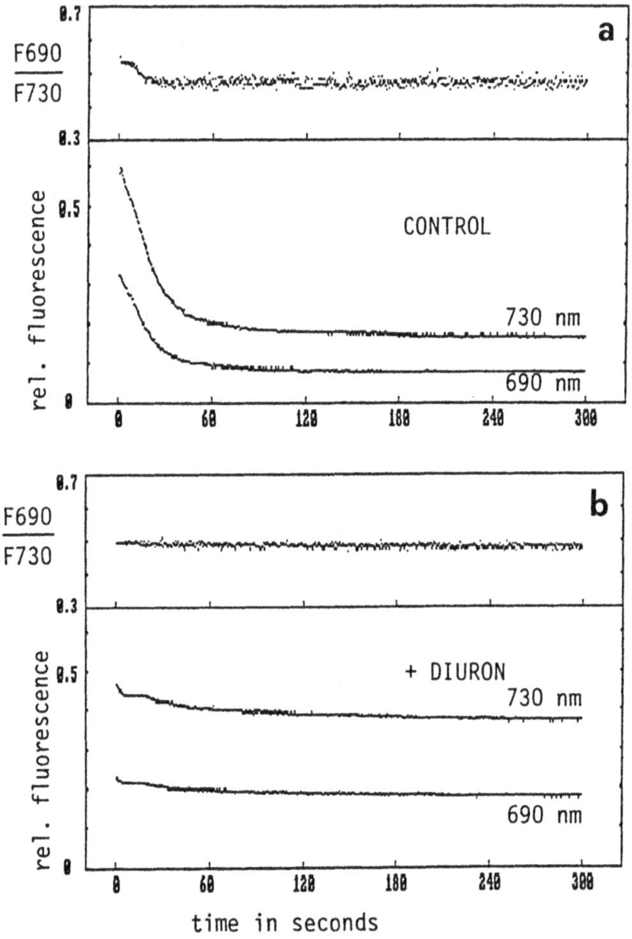

Figure 6. Laser-induced chlorophyll-fluorescence induction kinetics in the 690 and 730 nm region and fluorescence ratio F690/F730 in bean leaves (Phaseolus vulgaris) **a**) control, **b**) leaf treated with 10^{-3}M diuron.

CONCLUSION: The newly developed LICAF system opens new possibilities to detect follow and screen stress conditions in plants and possible regenerations via the laser-induced in vivo chlorophyll fluorescence. With its computer-aided data aquisition and processing it not only permits the measuring and study of different optional parts of the fluorescence induction kinetics, but also to determine the changes in the fluorescence ratio F690/F730, the Rfd690 and Rfd730-values, the stress adaptation index

Ap as well as the changes in the fluorescence intensity (in the maximum and the steady-state) with increasing stress duration and strength.

ACKNOWLEDGEMENTS

We wish to thank Doz. P. Richter and Mr. L. Jakab, Budapest, for help in the construction of the acoustooptic modulator and Mr. B. Volk, Karlsruhe, for valuable technical assistance.

REFERENCES

JAKAB L, KOCSANYI L, GIBER J, 1985. Bandwith limiting factors of the acoustic transducers of TeO_2 acoustooptic devices. In: Proceed. Symposium on Electronics Technology Budapest, Vol. II, pp. 198-201.

KRAUSE GH and WEIS E, 1984. Chlorophyll fluorescence as a tool in plant physiology II. Interpretation of fluorescence signals. Photosynth. Res. 5: 139-157.

LARCHER W and CERNUSCA A, 1985. Microcomputer-gesteuerte mobile Anlage zum fluorometrischen Nachweis von Photosynthesestörungen. Sitzungsber. Österr. Akad. Wissensch., Math.-naturw. Kl. Abt. I 194, pp. 45-64, Springer Verlag, Wien.

LICHTENTHALER HK, 1984. Influence of environmental factors on composition and function of the photosynthetic apparatus. In: Advances in Photosynthesis Research, Sybesma C ed., Vol. IV, pp. 241-244, Martinus Nijhoff/Dr. W. Junk Publisher, The Hague.

LICHTENTHALER HK, 1986. Laser-induced chlorophyll fluorescence of living plants. In: Proceed. Remote Sensing Symp. IGARSS Zürich Vol. III, pp. 1587-1590, ESA Publications Division, Noordwijk.

LICHTENTHALER HK, 1987. Chlorophyll fluorescence signatures of leaves during the autumnal chlorophyll breakdown. J. Plant Physiol. 131: 101-110.

LICHTENTHALER HK and BUSCHMANN C, 1987. Chlorophyll fluorescence spectra of green bean leaves. J. Plant Physiol. 129: 137-147.

LICHTENTHALER HK and RINDERLE U, 1988. Role of chlorophyll fluorescence in the detection of stress conditions in plants. CRC Critical Reviews in Analytical Chemistry 19, Suppl. I, S29-S85.

LICHTENTHALER HK, BUSCHMANN C, RINDERLE U und SCHMUCK G, 1986. Application of chlorophyll fluorescence in ecophysiology. Radiation Environmental Biophysics 25: 297-308.

NEUBAUER C and SCHREIBER U, 1986. The polyphasic rise of chlorophyll fluorescence upon onset of strong continuous illumination. I. Saturation characteristics and partial control by the photosystem II acceptor side. Z. Naturforsch. 42c: 1246-1254.

STRASSER R, SCHWARZ B and BUCHER J, 1987. Simultane Messung der Chlorophyll Fluoreszenz-Kinetik bei verschiedenen Wellenlängen als rasches Verfahren zur Frühdiagnose von Immissionsbelastungen an Waldbäumen. Ozoneinwirkung auf Buchen und Pappeln. Europ. J. Forest Pathology 17: 149-157.

FLUORESCENCE PROPERTIES OF PARAQUAT RESISTANT CONYZA LEAVES

Z. Szigeti[1], E. Pölös[2], E. Lehoczki[3] 1/ Department of Plant Physiology, Eötvös University, Budapest PO Box 33o., H-1445, 2/Research Institute of Viticulture and Enology, Kecskemét, H-6ool, 3/ Department of Biophysics, József Attila University, Szeged, H-67ol, Hungary

Key words: Conyza canadensis, fluorescence induction, Gramoxone, paraquat resistance

ABSTRACT

Fluorescence parameters /Fv, Fs, Fo, Rfd/ derived from fast and slow induction curves of paraquat treated susceptible and resistant Conyza plants were investigated by a newly developed microcomputerized apparatus. Resistant plants sprayed with Gramoxone /4 µM paraquat/ showed a transient inhibition which means that paraquat can penetrate into the chloroplasts. The recovery of the fluorescence parameters began 2 h after spraying, and it was completed in more than 48 h. pI_{5o} values of induction parameters were determined in Conyza leaves floated on paraquat solution of different concentration for 24 h. Fv and Rfd values were the most sensitive parameters giving a resistance factor of about 3oo and more than 45o, respectively.

INTRODUCTION

Resistance to paraquat has been demonstrated in a series of weed species. It may be the result of different alterations in the resistant plants: 1. reduced penetration of paraquat to the active site in the chloroplasts /Fuerst et al. 1985/, 2. increase of the activities of the enzymes involved in the catabolism of active oxygen species /Shaaltiel and Gressel 1986/, 3. alteration in the redox potential of the primary electron acceptor of PS I.

The effect of paraquat on the photosynthetic electron transport can also be measured by the chlorophyll fluorescence induction method, because bipyridinium compounds quench fluorescence by accepting electrons from PS I, thereby keeping the plastoquinone pool oxidized /Fuerst et al. 1985/. Recently it was demonstrated that paraquat caused quenching of fluorescence in the susceptible Conyza biotype but not in the resistant ones /Pölös et al. 1987/.

The aim of the present work was to characterize the photosynthetic responses of paraquat susceptible and resistant horseweed /Conyza canadensis/ plants by fluorescence induction measurements under different conditions using a newly developed microcomputerized apparatus.

H. K. Lichtenthaler (ed.) Applications of Chlorophyll Fluorescence, 109–114.

MATERIAL AND METHODS

Plant material. Seeds of susceptible and resistant Conyza plants were collected from vineyards near Kecskemét. Plants were grown in soil under laboratory conditions /illumination 13o $\mu E/m^2s$ PAR, 16 h light/8 h dark period, 22-25 ^oC/ for 3-4 months.Then the plants in soil containers were transferred to field conditions. About 6-month-old plants having 12-15 leaves in rosetta stage were used in the experiments.

Paraquat treatment. Intact Conyza plants were sprayed with commercial formulated Gramoxone /25 % paraquat/ solution. Paraquat concentration of the spraying solution was about 4 μM. The treatment was carried out on moderate light. After spraying samples /leaf discs of lo mm in diameter/ were taken in different time to obtain time curve of paraquat effect. In other experiments the leaves were floated on paraquat solution of 1 μM- lmM concentration range at 13o $\mu E/m^2s$ for 24 h.

Fluorescence induction measurements. 12 leaf discs from different sprayed plants were collected and placed for dark adaptation into the sample holder for 3o minutes. Fluorescence induction curves were measured by a newly developed apparatus. The induction kinetics and characteristic parameters derived from the curves are means of 12 independent determinations.

The new apparatus /Chlorophyll fluorometer, Micro and Experimental GMK, Szeged/ consists of 3 parts: 1. measuring head containing LED-s, photodiode, trifurcating fiber optics and adequate filters, 2. power and control unit with amplifier, 3. data recording and controlling microcomputer system.

The light sources are two LED-s /TLRA 15oC, Toshiba/ with a continuous current of 2o mA, emitting light peaking at 66o nm resulting 15o $\mu E/m^2s$ PAR on the surface of the leaf discs. The photodiode /BPX 6o, Siemens/ operates in photovoltaic mode providing high speed and output current which is linear to the signal intensity. The third arm of the trifurcating fiber optics was used for guiding the emitted fluorescence to the photodiode. Photodiode detector was protected by a far red cut-off filter /RG 9, Schott/.

The measuring head is connected to the power and control unit by a five-lead cable. This unit is attached to a microcomputer system /ZX Spectrum/ for data storage, data processing and controlling of measurements. The controlling system is capable of capturing signal changes in 1 ms and 2oo ms, respectively, which makes possible simultaneous measurement of fast and slow fluorescence transients. Acquisition and storage of 2o points from zero to 2 ms ensures clear resolution of Fo. Data processing programme is able to average characteristic parameters of max. 12 fast and 12 slow induction curves.

RESULTS AND DISCUSSION

The fast and slow fluorescence induction curves of susceptible Conyza plants measured in different time after spraying with Gramoxone are shown in Fig.1.A,B.

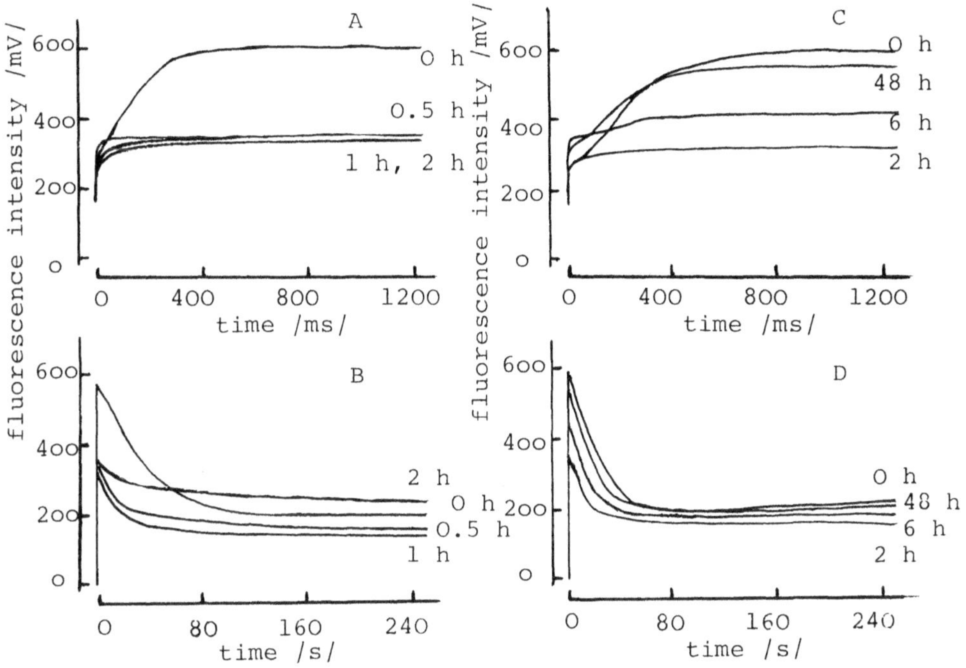

FIGURE 1. A./ Fast fluorescence induction curves of paraquat susceptible Conyza canadensis sprayed with Gramoxone /about 4 μM paraquat/. B./ Slow fluorescence decay of the same samples. C./ Fast fluorescence induction curves of paraquat resistant Conyza canadensis sprayed with Gramoxone /about 4 μM paraquat/. D./ Slow fluorescence decay of the same samples.

The development of inhibition was very fast. According to the changes of Fs /steady-state fluorescence/ and Fv /variable fluorescence/ a strong effect was detected as early as 3o minutes after spraying /Fig. 2./. After 2-4 h the plants were irreversibly damaged: the leaves were almost completely wilted, dehydrated. This tendency was reflected in the monoton decrease of Rfd value, the ratio of fluorescence decrease /Fmax - Fs/ to the steady-state fluorescence /Fs/ /Fig.3./. Rfd is considered to be an indicator for the potential photosynthetic activity of a leaf /Strasser et al. 1987/ and is extensively used in the ecophysiology /Lichtenthaler et al. 1986/.

In the first period of treatment /o.5 - 2 h/ resistant Conyza plants showed a marked inhibition of fluorescence induction, similarly to the susceptible ones /Fig. 1.C,D/. This means that paraquat can penetrate into the chloroplasts of

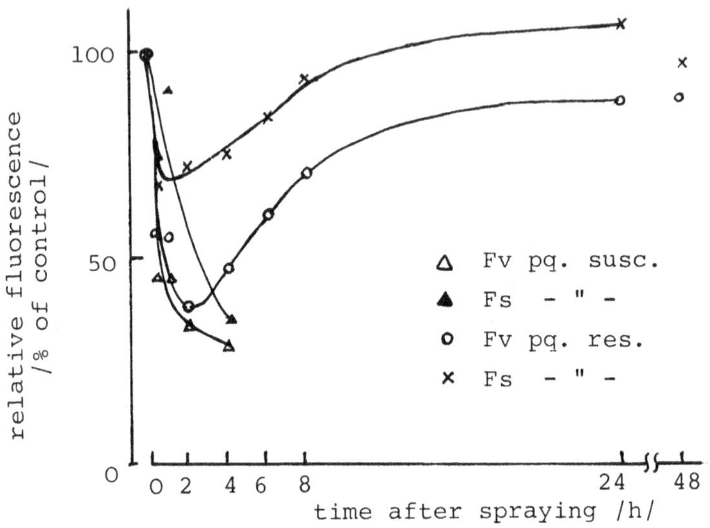

FIGURE 2. Time dependence of different fluorescence parameters /Fv and Fs/ of intact susceptible and resistant Conyza plants sprayed with Gramoxone /about 4 μM paraquat/

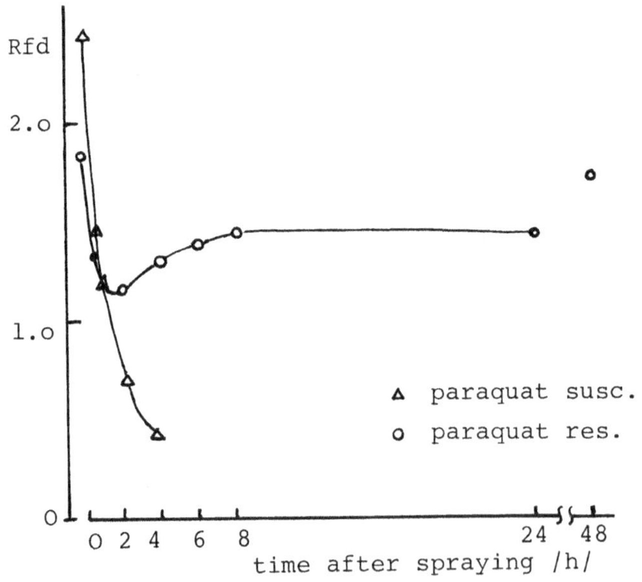

FIGURE 3. Time dependence of Rfd [(Fmax - Fs) / Fs] values of intact susceptible and resistant Conyza plants sprayed with Gramoxone /about 4 μM paraquat/.

resistant plants and is able to accept electrons and to exert its phytotoxic effect. After this 2 h period fluorescence induction curves step-by-step became similar to the untreated control curves. This was clearly reflected in the changes of parameters Fv and Fs /Fig. 2./. 48 h after spraying these parameters were near to those of untreated control. However, the reestablishment of Rfd was slow and did not reach the initial value /Fig. 3./. A similar transient reduction of CO_2 fixation was found in sprayed paraquat resistant Conyza by Shaaltiel and Gressel /1987/. Our fluorescence induction measurements are in good agreement with this observation, but the rate of restoration of these sensitive fluorescence parameters was slower in our case.

The lower Rfd values of the untreated resistant plants /Rfd <2/ reflects a lower vitality as it was proved earlier by in vivo CO_2 fixation experiments /Pölös et al. 1988/. It may be due to the simultaneous atrazine resistance of these Conyza plants.

The transient reduction of photosynthetic activity was not clearly seen on resistant leaves floated on paraquat solution /Pölös et al. 1988/. Therefore this experiment was repeated and the time dependence of different parameters of fluorescence induction curves were investigated. This transient inhibition was really not present in resistant Conyza leaves floated on lo µM paraquat solution, only some insignificant changes of Fv, Fs and Rfd values were obtained.

Compairing the sensitive and resistant Conyza biotypes we determined the concentration dependence of fluorescence induction parameters of floated leaves to gain pI_{50} values for the determination of the resistance factor /Table 1./.

fluorescence induction parameter	pI_{50} susceptible biotype	pI_{50} resistant biotype	resistance factor
Fv	5.85	3.33	331
Fs	4.54	2.9o	44
Fs -Fo	4.7o	3.83	7.4
Rfd	5.66	<3.0	>457

TABLE 1. Resistance factors calculated on the basis of pI_{50} values of different parameters of fluorescence induction curves in susceptible and resistant Conyza leaves floated on paraquat solution for 24 h in light.

Resistance factors calculated for Fv and Rfd were the highest, the most sensitive ones. These factors are higher than those obtained in our earlier experiments, where this factor was only about 17o /Pölös et al. 1988/. The difference is due to the higher sensitivity and better reproducibility of the new measuring system described here.

According to our results the experiments using floated, detached leaves are suitable for demonstration of paraquat resistance. These experiments, however, are not the most suitable for investigation of mode or mechanism of resistance to paraquat. All transient events /CO_2 fixation, fluorescence induction/ were detectable only in intact sprayed resistant plants.

The temporary inhibition of photosynthesis in these plants showed that paraquat can penetrate into the chloroplasts of resistant plants, too. The recovery of fluorescence induction curves indicates that in resistant chloroplasts must exist a mechanism which is responsible for gradual elimination of paraquat effect.

REFERENCES

Fuerst EP, Nakatani HY, Dodge AD, Penner D and Arntzen CJ, 1985. Paraquat resistance in Conyza. Plant Physiol. 77:984-989.

Lichtenthaler HK, Buschmann C, Rinderle U and Schmuck G, 1986. Application of chlorophyll fluorescence in ecophysiology. Radiat. Environ. Biophys. 25:297-3o8.

Pölös E, Laskay G, Szigeti Z, Pataki Sz and Lehoczki E, 1987. Photosynthetic properties and cross-resistance to some urea herbicides of triazine resistant Conyza canadensis Cronq /L./. Z. Naturforsch. 42c:783-793.

Pölös E, Mikulás J, Szigeti Z, Matkovics B, Do Qui Hai, Párducz Á and Lehoczki E, 1988. Paraquat and atrazine coresistance in Conyza canadensis /L./ Cronq. Pestic. Biochem. Physiol. 3o: 142-154.

Shaaltiel Y and Gressel J, 1986. Multienzyme oxygen radical detoxifying system correlated with paraquat resistance in Conyza bonariensis. Pestic. Biochem. Physiol. 26:22-28.

Shaaltiel Y and Gressel J, 1987. Kinetic analysis of resistance to paraquat in Conyza. Evidence that paraquat transiently inhibits leaf chloroplast reactions in resistant plants. Plant Physiol. 85:869-871.

Strasser RJ, Schwarz B and Bucher JB, 1987. Simultane Messung der Clorophyll Fluoreszenz-Kinetik bei verschiedenen Wellenlängen als rasches Verfahren zur Frühdiagnose von Immissionsbelastungen an Waldbäumen: Ozoneinwirkungen auf Buchen und Pappeln. Eur. Zeitschr. Forstpath. 17:149-157.

CHARACTERIZATION OF PARAQUAT-RESISTANT CONYZA LEAVES THROUGH DELAYED FLUORESCENCE

E. LEHOCZKI AND Z. SZIGETI

Department of Biophysics, Jozsef Attila University, Szeged, Hungary and Department of Plant Physiology, Eotvos Lorand University, Budapest, Hungary

Key words: delayed fluorescence, paraquat resistance, Conyza canadensis

ABSTRACT

The kinetics of chlorophyll delayed fluorescence (DF) induction was used as a tool for the detection of paraquat action in paraquat resistant (R) and susceptible (S) biotypes of Conyza canadensis. In both biotypes, a physiologically active level of paraquat rapidly inhibits the DF within 30 min after the spraying of whole plants. The inhibition of the DF in the S biotype is irreversible, while in the R plants the DF is slowly recovered with time. This transient inhibition of photosynthesis in intact leaves is very important as concerns the explanation of the mechanism of paraquat resistance. In the present study a solid-state device for measurement of the DF is described, which is suitable for studies of photosynthesis and stress physiology under field conditions.

INTRODUCTION

A rapid and transient inhibition of prompt fluorescence induction during the first hour was earlier observed when whole plants of a paraquat-resistant (R) biotype of Conyza canadensis were sprayed with formulated paraquat. The inhibitory effect of paraquat was strong within 30 min after spraying, and the inhibition effect subsequently decreased with time (Szigeti et al. 1988). A similar transient reduction of CO_2 fixation was found in paraquat R Conyza by Gressel's group (Shaaltiel and Gressel, 1987) and by Polos et al. (Polos et al., 1988). Since these inhibitions of photosynthesis in intact leaves are very important for an understanding of the paraquat-resistance mechanism, another sensitive in vivo method (Havaux and Lannoye, 1985) the technique of induction of delayed fluorescence (DF), was used to characterize the mode of action of paraquat in susceptible (S) and R Conyza leaves. DF is produced by plants after their illumination has ceased, and induction transients can be observed when intermittent excitation is applied for a longer time. DF is a complex phenomenon not yet fully understood. It is closely related with the primary processes of photosynthesis, and a current theory suggests that it involves a dark reaction

H. K. Lichtenthaler (ed.) Applications of Chlorophyll Fluorescence, 115–120.

between Q^- and Z^+. The high-energy state of phosphorylation, the pH gradient a cross the thylakoid membrane and the membrane potential are known to stimulate this process (Itoh and Murata, 1973; Lavorel, 1975). In order to provide relevant data, a solid-state device for the monitoring of DF has been constructed.

MATERIALS AND METHODS

Plant material. Seeds of paraquat R Conyza were collected from vineyards near Kecskemét (Hungary). Paraquat resistance was confirmed (Polos et al., 1988). The growth conditions and the conditions of paraquat treatment are described elsewhere (Szigeti and Lehoczki, 1988). The experiments were performed at 25 ± 0,5 °C on leaf discs 10 mm in diameter freshly cut from herbicide-treated or control leaves. For the measurement of DF, the discs were transferred to a special, thermostated sample holder for a 30-min dark adaptation. The DF measuring system consists of low-power solid-state components with fiber optics. The core of the system is an emitter-detector unit (A) connected to a special trifurcated fiber optics head (B). The head constitutes the optical link between the emitter-detector unit and the plant sample (Figure 1). The DF was excited by two flashing high-intensity LEDs (type TLRA 150-C, Toshiba) and measured 0.8 - 5.0 ms after cessation with 0.5 ms flashes. The LEDs and their circuitry are key components in the operation of the apparatus. The TLRA 150-C LED has a peak emission at 660 nm and is slightly overdriven (operated at 100 mA). The intensity of the exciting light pulses at the surface of the discs is about 300 $\mu Em^{-2}s^{-1}$ (FAR). The delayed light emitted from the front surface of the leaves was detected via one third of the trifurcated fiber optics by a photodiode (BPX-60, Siemens) in photocurrent mode and protected by a long-pass filter (RG 9, Schott). LEDs and detector are enclosed in a compact housing (A in Figure 1), which contains circuitry for a preamplification of the pulse signal. A five-lead cable connects the emitter-detector unit with the power and control unit, which contains the circuitry for driving the LEDs and for processing the pulse signal, and also a BNC output for recording of the DF signal. A special gating circuit coupled with the LED trigger unit prevents saturation of the amplifier. This measuring system displays a high degree of compactness, and therefore an appropriate connection of the trifurcated fiberoptics head to the sample holder (C).

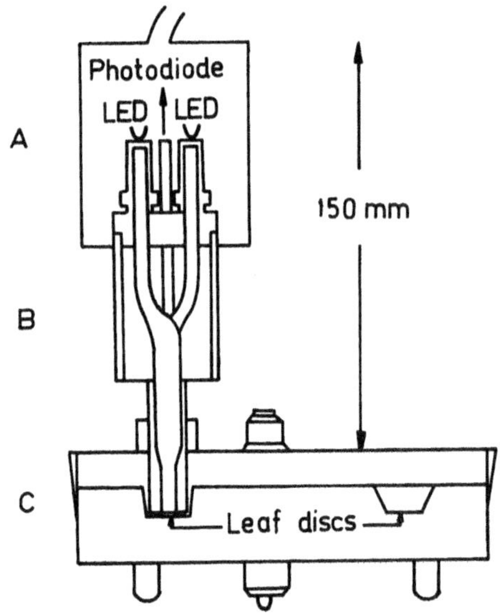

Figure 1. Experimental setup for the measurement of delayed fluorescence induction curves in leaves discs.

This virtually prevents contributions of stray light to the DF. This is essential for a reliable determination of the DF, which requires a high amplifier gain. Therefore, our fluorometer can easily be transported, and measurements under field conditions can readily be performed.

RESULTS

After the switching of the light pulses, the intensity of the delayed light rapidly increases from zero to an initial rise level (I), and then to the maximum level (P) within 1 s, and DF decreases from P to the steady-state level (S) in 3-6 min (Figure 2). Typical patterns of the transients in the DF appear for untreated leaves of the S and R biotypes. Qualitatively, both biotypes have the same type of induction curves, but with minor differences. The level of the I level was significantly higher in the R biotype, but the P level was higher in the S biotype; the S level was the same for the two. Paraquat rapidly caused about the same decrease (quenching) of the DF intensity within 30 min in the two biotypes at the same paraquat dose. The decrease in the DF indicates a rapid penetration of paraquat from the epidermis into the chloroplasts, thereby causing an inhibition of photosynthesis. In paraquat S plants, the intensity of DF at the P level continuously decreased after a 30-min paraquat treatment; the plants were desiccated within 3-4 h and became practically dead. In contrast, the DF from the R biotype recovered after an initial inhibition period (Figure 3).

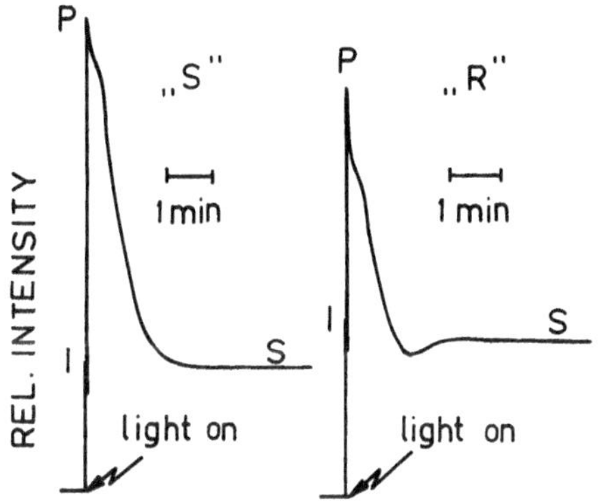

Figure 2. Induction of delayed fluorescence of chlorophyll in paraquat sensitive (S) and resistant (R) biotypes. I, S and P represent the initial rise, the steady-state and maximum levels, respectively.

After the spraying of the plants with paraquat, the characteristic parameters of the DF transients (I, P and S levels) in the S and R biotypes changed differently. In the S plants the I and S level remained constant and changes were found only in the P level, which decreased rapidly (Figure 3). However, in the R biotype the decrease in intensity of the P level was smaller, and after 30-min of treatment it continuously increased and slowly recovered. The recovery was not completed within 72 h after spraying. At the same time, the I and S levels of the DF became higher after spraying and decreased only slowly in a reverse manner as compared to the P level (Figure 3).

To obtain more information about the processes leading to these changes, we estimated the proportions of the I, P and S levels of the DF. Figure 4 shows the time response curves of the ratios (P-I)/I and (P-S)/S. From the data in Figure 4, we con-

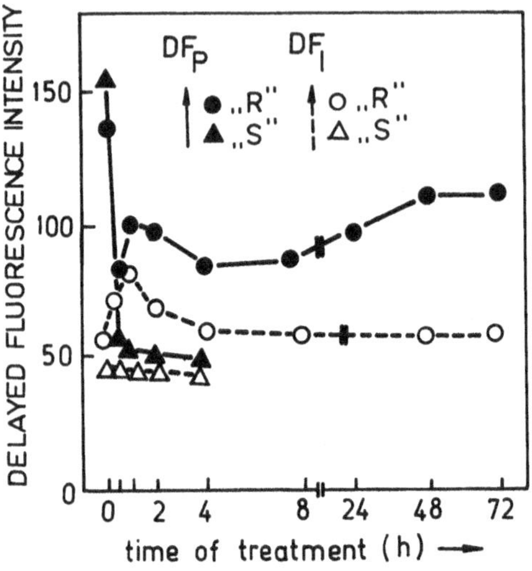

Figure 3. Time response dependence of delayed fluorescence at the initial (I) and the maximum (P) levels in paraquat S and R Conyza.

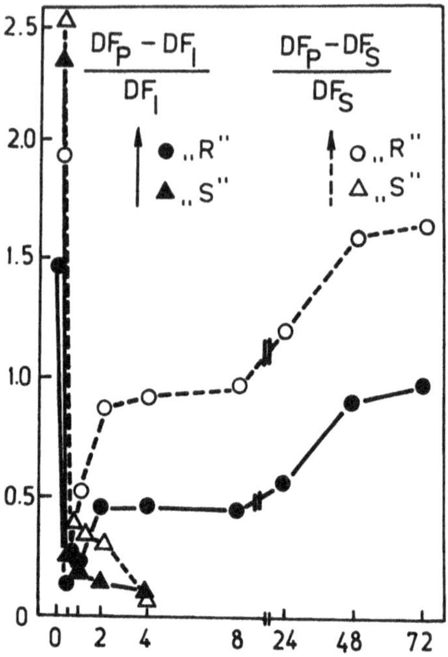

Figure 4. Time response curves of the ratios (P-I)/I and (P-S)/S of delayed fluorescence.

cluded that the accumulation of the high-energy state of the thylakoid membranes is inhibited in the S biotype in the presence of paraquat (see the ratio (P-S)/S). Partial inhibition of the accumulation of the high-energy state also occurs in the R biotype (see the ratio (P-S)/S), but the primary processes of PS II in the R type are stimulated in the presence of paraquat (see the high I level (Figure 3) and the small ratio (P-I)/I). Figure 4 also shows that recovery of the accumulation of the high-energy state occurs more quickly than stimulation of the primary processes in PS II.

DISCUSSION

In numerous studies it was suggested that the primary mode of resistance is due to the different mechanism of the sequestration of paraquat occuring before paraquat reaches the chloroplasts (Harvey et al., 1978; Fuerst et al., 1985; Vaughn et al., 1985; Powles and Cornic, 1987). The other explanation of paraquat resistance is that there is an elevation of the enzyme level leading to the detoxification superoxide anion radical (Shaaltiel and Gressel, 1986). A rapid and transient inhibition of CO_2 fixation (Shaaltiel and Gressel, 1986) and, on the other hand, the appearance of paraquat quenched fluorescence in intact leaves during the first hour of treatment in both biotypes (Polos et al., 1988) are very important facts for the explanation of the primary mode of the mechanism of paraquat resistance. As it is known, paraquat interacts with photosystem I by efficiently accepting electrons and as activated paraquat it reacts with O_2 forming toxic superoxide (Summers, 1980). Our measurements of the time response of in vivo inhibition of photosynthesis by paraquat can be utilized to decide between the two possibilities mentioned above.

In our experiments paraquat rapidly caused the same decrease of the DF in both biotypes at the same paraquat dose. This decrease indicates a rapid penetration of paraquat from the epidermis of leaves into the chloroplasts. After the penetration paraquat it serves as an electron sink serving oxygen as terminal acceptor. This allows an oxidation of the primary acceptor Q trough linear electron transport. According to this explanation the changes we found in DF indicate how PS I and PS II reactions are inbalanced via electron transport chain between them in the presence of paraquat. The decrease of the DF in the S biotype is irreversible, while there is a slow recovery in the R plants. This observation suggests that paraquat can reach its site of action and exerts its phytotoxicity already at the beginning of treatment. This means that in R biotype paraquat (or its metabolites) are being sequestered by some protective processes present only in resistant plants.

REFERENCES

Fuerst EP, Nakatani HY, Dodge AD, Penner D, Arntzen CJ, 1985. Paraquat-resistance in Conyza. Plant Physiology 77: 984-989.

Harvey BMR, Muldoon J and Harper DB, 1978. Mechanism of paraquat tolerance in perennial ryegrass. I. Uptake, metabolism and translocation of paraquat. Plant Cell Environ. 1: 203-209.

Havaux M and Lannoye R, 1985. In vivo chlorophyll fluorescence and delayed light emission as rapid screening techniques for stress tolerance in crop plants. Z. Pflanzenzüchtg. 95: 1-13.

Itoh S and Murata N, 1973. Correlation between delayed light emission and fluorescence of chlorophyll a in system II particles derived from spinach chloroplasts. Photochemistry and Photobiology 18: 209-218.

Lavorel J, 1975. Luminescence. In: Govindjee ed. Bioenergetics of Photosynthesis. Academic Press, New York. pp. 223-317.

Polos E, Mikulas J, Szigeti Z, Matkovics B, DoQuy H, Parducz A and Lehoczki E, 1988. Paraquat and atrazine co-resistance in Conyza canadensis. Pesticide Biochemistry and Physiology 30: 142-154.

Powles SB and Cornic G, 1987. Mechanism of paraquat resistance in Hordeum glaucum. Studies with isolated organelles and enzymes. Aust. J. Plant Physiology 14: 81-89.

Shaaltiel Y and Gressel J, 1986. Kinetic analysis of resistance to paraquat in Conyza. Plant Physiology 85: 869-871.

Summers LA, 1980. The Bipyridilium Herbicides. Academic Press, London.

Szigeti Z, Polos E and Lehoczki E 1988 (In this proceeding).

Vaughn KC and Fuerst EP, 1985. Structural and physiological studies of paraquat-resistant Conyza. Pesticide Biochemistry and Physiology 24: 86-94.

TEMPERATURE DEPENDENCE OF IN VIVO CHLOROPHYLL FLUORESCENCE AS AFFECTED BY DCMU

P.R. van Hasselt, University of Groningen, Department of Plant Physiology, Biological Centre, PO Box 14, 9750 AA Haren (Gn), the Netherlands

ABSTRACT

Temperature dependent changes in maximal variable chlorophyll fluorescence were determined between 30°C and 0°C in cucumber and tomato leaf discs. In both species two abrupt changes occurred in the maximal variable chlorophyll fluorescence plotted against temperature. Cucumber showed breaks around 20°C and 9°C and tomato around 24°C and 15°C. In the presence of DCMU, which blocks electron transport, the low temperature breakpoint disappeared; the maximal variable fluorescence decreased at temperatures below the high temperature breakpoint in tomato leaf discs but not in cucumber leaf discs. The causes of the temperature dependent abrupt changes in variable fluorescence and practical application of the breakpoint temperature as a screening method for the ability of plants to adapt to suboptimal temperature are discussed.

INTRODUCTION

The fast rising energy costs in the seventies prompted plant breeders in the Netherlands to start programmes for the breeding of glasshouse crops like cucumbers and tomatoes with a low-energy requirement for growth and production. Conventional selection programmes in glasshouses, kept at suboptimal temperature conditions during the winter, were started. These programmes, however, are laborious and take several years. For a faster selection of young plants new selection criteria had to be developed. At the Dept. of Plant Physiology of the University of Groningen a research project has been started in 1985 to study the possibility of using chlorophyll fluorescence as a screening method for better growth capacity at suboptimal temperature. The great advantage of using chlorophyll fluorescence as a screening method is that it can easily and fastly be measured in intact living plants or plant parts. After measurement the plants can grow and develop further and can be used for breeding purposes. The difficulty of the method is that the interpretation of the fluorescence signal is complicated due to the fact that the chlorophyll fluorescence is the resultant of different and sometimes counteracting reactions in the chloroplast (Krause and Weis 1984). To select appropriate fluorescence parameters which correlate well with adaptive capacity to a certain stress, more background information is needed about the effects of the stress on the physiological processes underlying the fluorescence output. Recently an automatical chlorophyll fluorescence measuring system was completed at our laboratory. This system enables us to study the effects of temperature stress on chlorophyll fluorescence of up to 100 leaf discs and will be of great help in studying correlations between low temperature resistance and fluorescence

H. K. Lichtenthaler (ed.) Applications of Chlorophyll Fluorescence, 121–126.

parameters.

It was observed earlier in cucumber leaf discs that abrupt changes occurred in the maximal induced fluorescence Fv(P) plotted against temperature (Van Hasselt, Woltjes and De Jong 1983; Van Hasselt and De Jong 1984). Similar breaks were observed in maize (Havaux and Lannoye 1984). The temperature of one of the breaks shifted to a lower value when plants were grown at suboptimal temperature. This shift may be a measure for the ability of the plant to adapt to suboptimal temperature. Little is known about the underlying mechanisms. The aim of the present study was to get more insight in the physiological backgrounds of the abrupt changes in chlorophyll fluorescence induced by low temperature above the freezing point.

MATERIALS AND METHODS

Plant material. Cucumber plants (*Cucumis sativus* L.) cv. Groene Scherpe were grown in Conviron growth cabinets on 30 l. tanks with half strength aerated Hoagland nutrient solution as described before (Janssen and Van Hasselt 1988). Tomato plants (*Lycopersicon esculentum* Mill.) line F from the Institute of Horticultural Plant Breeding, Wageningen, were grown in a similar way except that growth temperature was kept constant at 22°C.

Leaf discs (diameter 7 mm) were punched from nearly expanded leaves of 3 to 4 week-old plants. Prior to the measurements the discs were kept floating on tap water at room temperature for 30 min. Two discs were used for fluorescence measurements. The discs were placed on wet filterpaper on an aluminium water temperated cuvet connected with a cooling bath. The cuvet was covered with a lid with two quartz windows over the leaf discs. A flow of humidified air over the discs ensured a constant gas phase during the experiments. Leaf discs were treated with DCMU (-3(3,5-dichlorophenyl)-1,1-dimethylurea) by incubation during 2 h on a 50 μM solution as described before (Janssen and Van Hasselt 1988).

Measurement of temperature dependent chlorophyll fluorescence. Slow fluorescence induction of the leaf discs has been measured at decreasing temperatures from 30° to 0°C, the temperature decreased in steps of 2° or 3°C at a rate of 18°C/h. Next there was a dark incubation of 20 min at the measuring temperature, followed by 10 min of fluorescence measurement. Then a new cycle started. Fluorescence was induced by irradiation of the leaf discs through the quartz windows with 10 μE m^{-2} s^{-1} red light of a light emitting diode and measured as described before (Janssen and Van Hasselt 1988). In some experiments blue-green light from a light source with a Schott BG39 filter was used. The light was guided by one arm of bifurcated fiberoptics to the leaf disc. The fluorescence has been measured by a photomultiplier shielded with a 687 nm Balzers filter connected to the other arm. The slow induction kinetics were registered on a recorder. The slow induction kinetics were registered on a recorder. The fluorescence parameters qQ and qE were measured with a pulse amplitude modulation fluorometer (PAM 101, H. Walz, Effeltrich, Germany).

RESULTS AND DISCUSSION

When the temperature of cucumber leaf discs was decreased from 30°C to o°C, two breakpoints in the maximal variable fluorescence, Fv(P) plotted against temperature were evident in most experiments. Fv(P) increased from 30° to 18°C. This increase was followed by a decrease of Fv(P) to 11°C.

Below 11°C a fast increase was evident (Fig. 1a). The use of breakpoint temperatures as a selection criterion is complicated by the fact that breakpoints cannot be measured in all leaf samples. The histogram (Fig. 1b) shows that the high temperature breakpoint was measured in 59% and the low breakpoint in 70% of the experiments. The reason for the absence of breaks is yet unknown. At present the effect of pretreatments, e.g. light quality, on the frequency of the breaks are studied.

The maximal variable fluorescence, used as a measure in this study, reflects the fluorescence yield when the electron acceptor of photosystem II (PSII), Qa, is highly reduced and the thylakoid proton gradient is small. At the low light intensity applied in this study, Fv(P) will reflect a transient equilibrium between the electron flow from PSII to Qa and the flow from Qa to photosystem I (PSI). Sudden changes in Fv(P), therefore, can be caused by an abrupt alteration of the reducing capacity of PSII or a sudden change of the electron transport capacity beyond Qa. Seeking an explanation for the occurrence of the breaks, DCMU an inhibitor of electron transport beyond Qa was used. The results are shown in Fig. 1c,d.

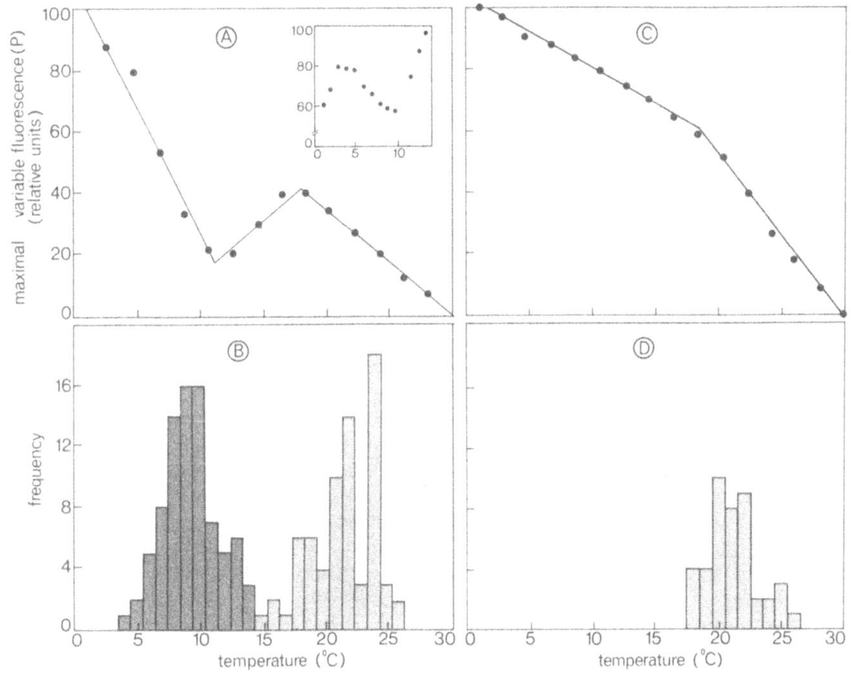

FIGURE 1. a) Temperature dependent changes in maximal variable chlorophyll fluorescence, Fv(P), of cucumber leaf discs. b) Frequency of breakpoint temperatures of Fv(P) of 119 cucumber leaf discs (mean values 9.2 ± 2.2°C and 20.8 ± 4.4°C). c) Temperature dependent changes in Fv(P) of DCMU-treated cucumber leaf discs. d) Frequency of breakpoint temperatures of maximal variable fluorescence of 62 DCMU.treated cucumber leaf discs (mean value 20.6 ± 2.1°C).

The maximal fluorescence of DCMU-treated leaf discs also increased when the temperature was lowered from 30° to 0°C but only the high temperature breakpoint at 20°C was present. The absence of the low temperature breakpoint, when electron transport beyond Qa was blocked, indicates that an abrupt decrease of the electron transport capacity beyond Qa causes the fast increase of Fv(P) below 8°C and therefore underlies the low temperature breakpoint. An abrupt decrease of electron transport capacity induced by low temperature may be caused by a phase transition (solidification) of certain polar lipids in the thylakoid (Raison and Orr 1986). Such a phase transition could decrease the rate of lateral diffusion of the plastoquinon electron carrier in the membrane. When the leaf discs were kept for a longer period at temperatures below 5°C, chilling damage occurred resulting in a decrease of Fv(P) (Fig. 1a, insert). An important reason for the decrease of Fv(P) under chilling conditions is supposed to be a decrease of the reducing capacity of PSII caused by inhibition of the watersplitting system (Smillie and Nott 1979). The results of Fig. 2 sustain this explanation. Below 6°C the time to reach Fv(P), a measure for the reducing capacity of PSII, decreased while Fv(P) is still increasing. This result indicates that under these conditions the decrease of reducing capacity of PSII is less than the inhibition of electron transport beyond Qa.

The fysiological backgrounds of the high temperature breakpoint, which remains in the presence of DCMU, are currently studied. Recent results with tomato leaf discs show that in the presence of DCMU a decrease of Fv(P) occurred at temperatures below the breakpoint. This fluorescence quenching in the presence of DCMU was enhanced when the light intensity was increased (Fig. 3). Its nature is still unknown. The quenching may represent an increase of nonradiative decay at lower temperature, when less light energy can be used for photosynthesis. A second possibility is that photoinhibition of the PSII reaction centre occurs in the presence of DCMU. Also the fact that temperature effects on the distribution of light over the photosystems (Weis 1985) may contribute to the sudden increase of fluorescence quenching at the high temperature breakpoint. An explanation of the processes underlying the abrupt change of Fv(P) at the high temperature breakpoint is of special interest for the practical application of breakpoint temperatures of Fv(P) in plant breeding since it was found that this breakpoint shifted to a lower temperature when plants were grown at suboptimal temperature (Van Hasselt and De Jong 1983). Fig. 4 shows the effect of temperature on the chlorophyll fluorescence quenching by reduced Qa (qQ) and by pH (qE) during the first 50 seconds of induction; both quencing components have a high value at high and a low value at low temperature. The rapid increase of qE at 30°C may explain the relative low value of Fv(P) at this temperature. The origin of the high temperature breakpoint appears to be the result of several temperature dependent processes interacting in the chloroplast membrane.

ACKNOWLEDGEMENTS

The author thanks Luuk Janssen and Geert Krol who skilfully performed most of the experiments for this study. This work was supported by the Committee for Energy Consumption of the Ministry of Agriculture and Fisheries of the Netherlands.

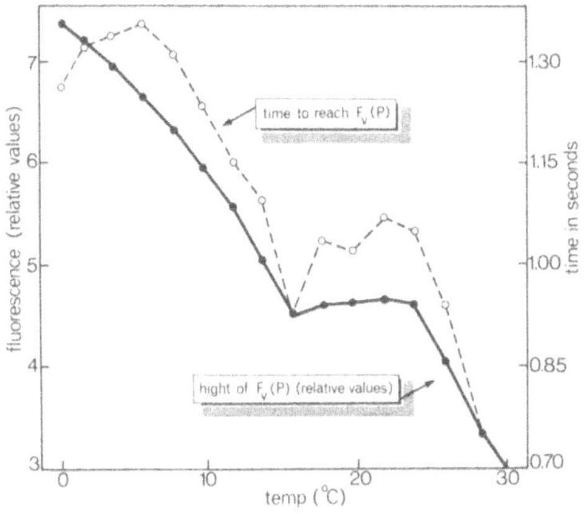

FIGURE 2. Temperature dependent changes in maximal variable chlorophyll fluorescence, Fv(P), and in the time to reach Fv(P) of tomato leaf discs.

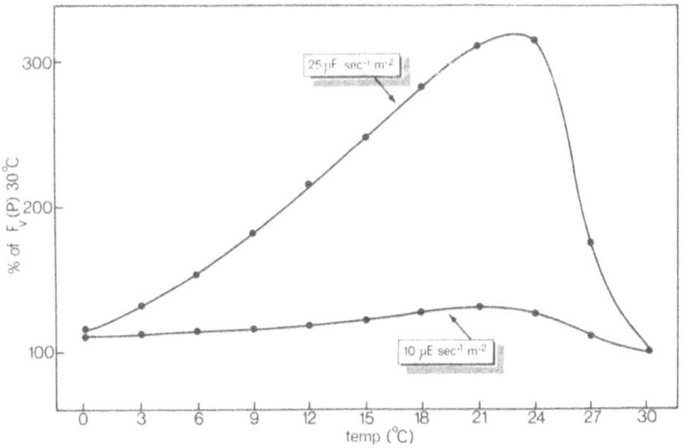

FIGURE 3. The effect of DCMU on temperature dependent changes in maximal variable chlorophyll fluorescence, Fv(P), of tomato leaf discs measured at the represented light intensities.

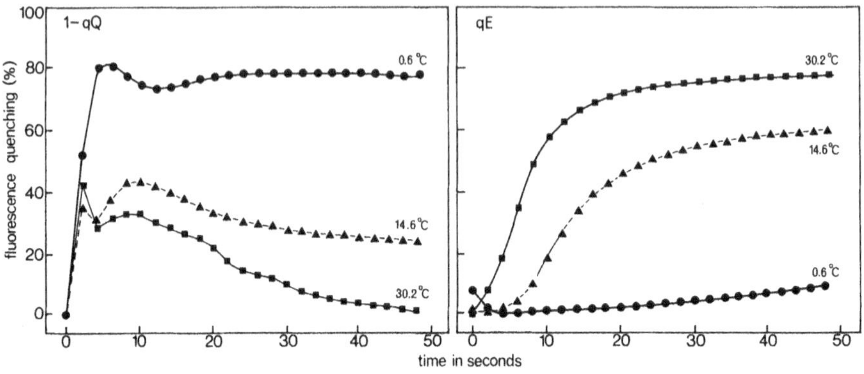

FIGURE 4. The effect of temperature on qQ (photochemical quenching) and qE (energy dependent quenching) of the chlorophyll fluorescence of tomato leaf discs.

REFERENCES

Havaux M and Lannoye R, 1984. Effects of chilling temperatures on prompt and delayed chlorophyll fluorescence in maize and barley leaves. Photosynthetica 18(1): 117-127.

Krause GH and Weis E, 1984. Chlorophyll fluorescence as a tool in plant physiology II: interpretation of the fluorescence signal. Photosynthesis Research 5: 139-157.

Janssen LHJ and Van Hasselt PR, 1988. Temperature-induced alterations of in vivo chlorophyll-a fluorescence induction in cucumber as affected by DCMU. Photosynthesis Research 15: 153-162.

Raison JK and Orr GR, 1986. Phase transitions in thylakoid polar lipids of chilling sensitive plants. Plant Physiology 80: 638-645.

Smillie RM and Nott R, 1979. Assay of chilling injury in wild and domestic tomatoes based on photosystem activity of the chilled leaves. Plant Physiology 63: 795-801.

Van Hasselt PR, Woltjes J and De Jong F, 1982. The effect of temperature on chlorophyll fluorescence induction of cucumber lines differing in growth capacity at suboptimal conditions. In: Marcelle R, Clysters H and Poucke M van (eds). Effects of stress on photosynthesis, pp 257-265. Martinus Nijhoff/Dr W Junk Publishers, The Hague.

Van Hasselt PR and De Jong F, 1984. Effects of leaf and growth temperature on temperature dependent chlorophyll fluorescence induction in cucumber leaves. In: Sybesma C (ed.). Advances in Photosynthesis Research, pp 475-478. Martinus Nijhoff/Dr W Junk Publishers, The Hague.

Weis E, 1985. Light and temperature induced changes in the distribution of excitation energy between the photosystem I and the photosystem II in spinach leaves. Biochim Biophys Acta 807: 118-126.

CHAPTER 2

Applications in stress physiology and environmental research

IN VIVO CHLOROPHYLL FLUORESCENCE AS A TOOL FOR STRESS DETECTION IN PLANTS

Hartmut K. LICHTENTHALER

Botanisches Institut II (Plant Physiology and Plant Biochemistry)
University of Karlsruhe, Kaiserstr. 12, D-7500 Karlsruhe 1, F.R.G.

KEY WORDS: chlorophyll fluorescence, Kautsky effect, fluorescence ratio F690/F735, photosynthesis, Rfd-values, plant stressors, stress concepts in plants, stress detection by chlorophyll fluorescence, vitality index.

ABSTRACT

A definition of stress to plants and a general stress concept are presented including a list of possible major natural and anthropogenic stress factors. It is shown that the white light or laser-induced chlorophyll fluorescence of predarkened leaves is a very suitable tool for early stress detection in plants as well as for regeneration studies. The different chlorophyll fluorescence parameters to be measured are: variable fluorescence (ratio fmax/fo), the vitality index (fluorescence decrease ratio, Rfd-values), the stress adaption index Ap, the fluorescence spectra (ratio F690/F735) and the photochemical Q- and non-photochemical E-quenching (qQ and qE). Some examples of the measurement of these parameters under stress conditions are given.

INTRODUCTION

For plants there exist optimum conditions for growth and development as well as for individual metabolic pathways such as the process of photosynthesis (CO_2 assimilation). The optimum conditions for plant growth and cell metabolism are guaranteed and influenced by many environmental and endogenous factors, and they will differ considerably from plant to plant depending on habitat and genetic determination. As an example plants growing at a high northern latitude possess much lower temperature optima for photosynthesis and respiration than plants growing in temperate or tropical climate zones. Within certain limits plants can also adapt their morphology and habitus as well as particular organelles and metabolic pathways to a particular environmental condition. Thus plants grown in full sunlight or artificial high-light conditions will adapt their leaf morphology to high-light intensity and their chloroplasts for high rates of photosynthetic quantum conversion (e.g. a high number of electron transport chains and reactions centers per total chlorophyll and low grana stacks). In contrast the much thinner and larger shade leaves of shade and low-light plants primarily invest in the build-up of a large pigment antenna (e.g. high and broad grana stacks and a much higher proportion of light-harvesting chlorophyll proteins, LHCPs, than sun chloroplasts) to compensate for the light-deficiency situation of their habitat (Lichtenthaler et al. 1981, 1982, 1983). Physiologically viewed such adaptations make sense since they provide the plants, under a particular environmental condition of the habitat, with a better biomass production and growth.

H. K. Lichtenthaler (ed.) Applications of Chlorophyll Fluorescence, 129–142.

Under the particular environmental and climatic conditions of a habitat, which can be a matter of short-term or long-term changes, it rarely or never occurs that all factors e.g. temperature, water supply, light intensity, mineral content of the soil etc. are in the optimum range for plant growth. Therefore plants seldom grow under full optimum conditions. This does not necessarily mean that plants are under stress. A reduced rate of photosynthesis can induce a reduced rate of growth without inducing damage to the plant. Stress, in turn, is generally regarded as an event which can cause damage. Since the transitions between the situation of a reduced metabolic activity and an actual damaging stress may be fluent, the term stress needs a definition, which is given below.

Once plant stress and stress effects are defined, one has to ask how stress to plants can be detected. The photosynthetic quantum conversion and CO_2-assimilation is the central metabolic process which governs and regulates plant growth. Many environmental and stress factors will either directly or indirectly effect the rate of photosynthesis. It is well established that the rate of photosynthetic quantum conversion and the health state of the photosynthetic apparatus can be measured and monitored by the non-destructive in vivo chlorophyll fluorescence (Krause and Weis, 1984; Lichtenthaler, 1984, 1986; Lichtenthaler et al. 1986; Lichtenthaler and Rinderle, 1988). This method therefore not only permits an early detection of reduced rates of photosynthesis but also of stress conditions in green plants (Lichtenthaler et al. 1986, Lichtenthaler and Rinderle, 1988). In this paper the present stress concept of plants and the various possibilities for stress detection in plants via the in vivo chlorophyll fluorescence are presented.

A. DEFINITION OF STRESS

Among stress effects on living organisms and living plants one has to differentiate between mild and strong stresses. A mild stress may activate cell metabolism, increase the physiological activity of a plant, and does not cause any damage effects even at long duration of the stress factors. Such mild stimulating stress is favourable for the plant. It may therefore be termed eu-stress following a proposal of W. Larcher (personal communication). Examples of this are many inhibitors of plant growth and cell metabolism, which at very low concentrations exhibit promotion effects which are positive and stimulating for the organism. At higher concentration such inhibitors (e.g. herbicides) cause damage to plant metabolism e.g. photosynthesis and to growth processes, which will decrease the physiological activity of the plant and may induce early senescence and death. These inhibiting and damaging stress conditions can then be regarded as a distress. The transitions between eu-stress and distress may be fluent.

J. Levitt's (1980) definition of stress is:
"**any environmental factor potentially unfavourable to living organisms**".

A very detailed definition of stress in plants is given by W. Larcher, 1987:

"**Every organism experiences stress, although the way in which it is expressed differs according to its level of organization. From the botanist's point of view, stress can be described as a state in which increasing demands made upon a plant lead to an initial destabilization of functions, followed by normalization and improved resistance. If the limits of tolerance are exceeded and the adaptive capacity is overtaxed,**

permanent damage or even death may result. Stress thus contains both destructive and constructive elements: it is a selection factor as well as a driving force for improved resistance and adaptive evolution".

One can also apply the well developed stress concept in physics to plants. Correspondingly, the terms stress, strain and damage would be defined:

stress: a state of the plant under the condition of a force applied.

strain: is the response to the stress and the force applied to the plant (i.e. the expression of stress before damage occurs)

damage: is the result of too high a stress, which can no longer be compensated for.

Among the plants' responses to stress one has to differentiate according to Stocker 3 stages (see Larcher, 1987):

1. RESPONSE PHASE: (begin of stress)

alarm reaction
- deviation of the functional norm
- decline of vitality
- catabolic processes exceed anabolism

2. RESTITUTION PHASE: (continuing stress)

stage of resistance
- adaptation processes
- reparation processes
- hardening (reactivation)

3. END PHASE: (long-term stress)

stage of exhaustion
- stress intensity too high
- overcharge of the adaptation capacity
- chronic desease or death

To my opinion a fourth stage of plants' responses should be added, when the stress ceases and the damage, set by the stress, was not too high:

4. REGENERATION PHASE:
- partial or full regeneration of the physiological function, when the stressor is removed.

Stress factors acting on plants. Among the various stress factors one has to differentiate between natural and anthropogenic ones. The major stress factors acting on the plants are:

1.natural stress factors:
high light (photoinhibition, photooxidation),
heat (increased temperature),
water shortage (desiccation problems),
natural mineral deficiency (e.g. nitrogen shortage),
longer rain periods,
low chilling temperatures,
sudden and late frost.

2. **anthropogenic stress factors:**
herbicides, pesticides, fungicides,
air pollutants e.g. SO_2, NO, NO_2, NOx
ozone (O_3) and photochemical smog,
photooxidants (e.g. peroxyacylnitrates),
acid rain, acid fog, acid morning dew,
acid pH of soil and water,
mineral deficiency of the soil induced by acid rain
(shortage of the basic cations K, Mg, Ca, often Mn and sometimes Zn),
over-supply of nitrogen (dry and wet NO_3-deposits),
heavy metal load (lead, cadmium, etc.)
overproduction of NH_4^+ in breeding stations (uncoupling of electron transport),
increased UV-B radiation.

In general, stress conditions are not caused by a single stress factor alone, but by several stress factors which come together and act jointly. Thus high-light stress, which may cause photoinhibition of photosynthesis and photooxidation of the leaf pigments, is always associated with heat stress and with water shortage and desiccation problems of the plant and the cytoplasm. High-light stress, including heat and water stress, is particularly damaging when the air contains pollutants such as NO and NO_2 as well as small organic and hydrocarbon compounds, which by photochemical reactions are transformed into strongly oxidizing agents such as ozone and peroxyacylnitrates. When different stress factors are acting simultaneously, they often exert an enhanced influence which has been shown for the combined action of SO_2 and NOx. Thus the combined effect of the gases SO_2 + NOx is much higher than the additive effects of SO_2 and NOx given alone, since SO_2 destroys the nitrate reductase which detoxificates the NOx (Whitmore and Freer-Smith, 1982; Wellburn, 1982). Forest decline as a whole (Waldsterben, dying of conifer and broadleaf trees), which has arisen in the last 10 years, is also a combined action of many natural and anthropogenic stress factors, which set limitations for photosynthesis, plant growth and plant vitality (e.g. Lichtenthaler and Buschmann, 1984 a,b; Lichtenthaler et al., 1985; Rock et al., 1986; Schmuck and Lichtenthaler, 1986; Strasser et al., 1987). When the natural climatic stress is more favourable for plant growth and leaf/needle development, the plants can endure a higher amount of anthropogenic stress conditions. In contrast, in summers with continuous sunshine and practically no rain for 6 months such as in Central Europe 1983, natural stress and anthropogenic stress factors will enhance each other and not only reduce the plants' vitality but cause such severe damage that e.g. many forest trees for example will die off.

B. STRESS DETECTION BY IN VIVO CHLOROPHYLL FLUORESCENCE

Light absorbed by the photosynthetic pigments (chlorophylls or carotenoids) will be used for photosynthesis (photosynthetic quantum conversion) or dissipated as heat or as red chlorophyll fluorescence. Chlorophyll fluorescence is inversely related to the rate of photosynthesis (Krause and Weis, 1984; Lichtenthaler and Pfister, 1978; Lichtenthaler, 1986). Under optimum conditions of photosynthesis the dissipation of absorbed light energy via chlorophyll fluorescence is low. The light induced red chlorophyll fluorescence is particularly strong when the photosynthetic quantum conversion is blocked by the herbicide diuron (see colour plate No 1 at the end of this book). Chlorophyll fluorescence is not only shown by chlorophyll a but also by the isolated chlorophyll b and phaeophytin (see colour plate

No 2 at the end of this book). In vivo only chlorophyll a fluorescence is found, since chlorophyll b transfer its excited states to chlorophyll a. When the overall process of photosynthesis is disturbed in either the light or the dark reactions, the fluorescence emission increases and several chlorophyll fluorescence parameters change their characteristics. This opens the possibility for stress detection in green plants via the non-destructive in vivo chlorophyll fluorescence, which can be applied to intact leaves (see the review of Lichtenthaler and Rinderle, 1988).

There are several aspects and parameters of the chlorophyll fluorescence, which can be used for stress detection in intact plants. These are:

1. The light-induced fluorescence induction kinetics (Kautsky effect) in predarkened samples with
   - a fast fluorescence rise signal (ms-range) and
   - a slow fluorescence decrease (min-range) from the maximum to the steady state fluorescence

2. Determination of particular fluorescence ratios (calculated from the induction kinetics)
   - the variable fluorescence ratio: fmax/fo (= ratio of maximum fluorescence fmax to ground fluorescence fo)
   - the fluorescence decrease ratio Rfd as vitality index (Rfd = fd/fs = fluorescence decrease/steady-state fluorescence)

3. Simultaneous registration of the induction kinetics in the 690 and 730 nm region with the laser-equipped portable field fluorometer:
   - determination of Rfd 690 and Rfd 730
   - determination of the stress-adaptation index Ap

$$Ap = 1 - \frac{Rfd\ 730 + 1}{Rfd\ 690 + 1}$$

4. Intensity and shape of the chlorophyll fluorescence emission spectra:
   - position of the fluorescence maxima/shoulder in the 690 and 735 nm region
   - ratio of the intensity of the two fluorescence maxima: ratio F690/F735

5. Application of the PAM-fluorometer with repetative saturation pulses, which permits differentiation between photochemical and non-photochemical fluorescence quenching.
   - determination of the quenching coefficients qQ and qE
   - height of the saturation pulse-induced fluorescence spikes (distance g-h = capacity of Q_A reoxidation in the photosynthetic electron-transport chain per leaf area unit)

6. Low-temperature chlorophyll fluorescence-emission spectra in whole leaves or ground up leaves:
   - shape of the spectra and position of the maxima
   - ratio of the fluorescence in the emission maxima
   - differentiation between the fluorescence contribution of photosystem I, photosystem II and the light-harvesting antenna.

In general not one but several chlorophyll fluorescence parameters are measured, since some of the diverse fluorescence parameters possess different and complementary information on the physiology of photosynthesis, the amount of stress as well as on the site of damage within the photosynthetic apparatus.

C. EXAMPLES FOR THE MEASUREMENT OF FLUORESCENCE PARAMETERS UNDER STRESS CONDITIONS.

Chlorophyll fluorescence induction kinetics. The registration of the fluorescence induction kinetics (slow component of the Kautsky effect, measuring time ca. 4-5 min) with determination of the **Rfd-values as vitality index** (Fig. 1) provides very much information on the physiological state of photosynthesis and has been applied with great success in forest decline research, ecophysiology and stress detection in plants (Lichtenthaler 1984, 1986, 1987; Lichtenthaler and Buschmann, 1984, 1986; Lichtenthaler et al. 1982; Lichtenthaler and Rinderle, 1988; Schmuck and Lichtenthaler, 1986; Strasser et al. 1987). The height of the Rfd-values (Rfd = fd/fs) indicates the potential photosynthetic activity of a leaf, as was confirmed by parallel measurements with an infra-red gas analyzer (IRGA) system (Lichtenthaler, 1984; Lichtenthaler et al. 1982) and a CO_2 /H_2O porometer (Lichtenthaler et al. 1986; Lichtenthaler and Rinderle, 1988; Schmuck and Lichtenthaler 1986). Under stress conditions (e.g. water stress and desiccation) the fluorescence decrease (fd)-from the maximum (fmax) to the steady state - becomes increasingly lower and the steady state fluorescence fs continuously rises.

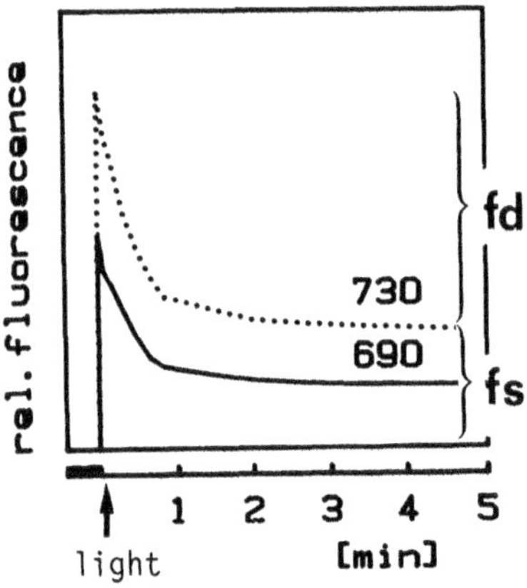

Figure 1. Laser-induced chlorophyll fluorescence induction kinetics of dark adapted bean leaves (Kautsky effect slow component, min- range). The slow fluorescence decline fd from fmax to the steady state fluorescence fs is paralleled by the energetization of the photosynthetic membrane and the onset of photosynthetic oxygen evolution. The fluorescence decrease ratio (Rfd = fd/fs) is a measure of the potential photosynthetic capacity of a leaf. Whether the full potential capacity of a leaf can be used depends upon the water status and the opening conditions of the stomata.

As a consequence the Rfd-values (Rfd=fd/fs) steadily decline with increasing stress from Rfd 690-values of ca. 3-5 (young greening and greenleaf tissue) via values of 1-2 to values far below 1 (e.g. 0.1-0.9) which indicate irreversible damage to the photosynthetic apparatus (s. also Table 1 and 2). In plants with closed stomata and no net CO_2-assimilation e.g. in conifers in springtime before the new needle year is developed - high Rfd-values indicate that the photosynthetic apparatus is fully functional. When applying our portable field fluorometer the Rfd-value method permits a very fast and extensive outdoor screening of the vitality of trees and agricultural plants. It allows the measurement of 40 to 50 leaf samples in one morning and thus provides a 10-fold higher number of leaf sample registrations per time unit than the CO_2/H_2O porometer method. Once lower Rfd-values and a loss of vitality are recognized, other fluorescence methods (see above points 2 to 7) and other ecophysiological methods can be applied to further define and localize the damage to the photosynthetic apparatus. The Rfd-value method has also successfully been used as a superior method of groundtruth control measurements in the remote sensing of vegetation (Rock et al. 1986; Schmuck and Lichtenthaler, 1986; Lichtenthaler et al. 1987).

Table 1. Development of several chlorophyll-fluorescence parameters maximum fluorescence fmax steady-state fluorescence fs and Rfd-values, stress adaptation index Ap) at two different wavelength regions (690 and 730 nm) during a water-stress treatment induced by abscission of a maple leaf (Acer platanoides) from the tree. Mean of 3 determinations, standard deviation 4 % or less.

	time	690 nm region			730 nm region			Ap
		fmax	fs	Rfd 690	fmax	fs	Rfd 730	
control	0 h	167	37	3.5	628	183	2.4	0.24
water stress	1 h	133	31	3.3	610	176	2.5	0.20
	2 h	125	29	3.3	589	172	2.4	0.20
	4 h	91	25	2.6	508	160	2.2	0.12
	6 h	44	20	1.2	226	113	1.0	0.12
	8 h	49	30	0.6	211	140	0.5	0.07
	10 h	58	42	0.38	223	164	0.36	0.02

Table 2. Effects of abscission (indicated as hours water stress) on the water content (% H_2O) and various chlorophyll fluorescence parameters (Rfd 690, Rfd 730, Ap and the ratio F690/F735 as well as the coefficients qQ, qE and the height of the fluorescence spikes g-h after a 10 min saturation pulse kinetic) of spruce needles (Picea omorika). The fluorescence spikes are given in relative units. The chlorophyll content was 76 (1988 needles) and 100 $\mu g \cdot cm^{-2}$ (1987 needles). Mean of 3 determinations, standard deviation 5% or less.

time	% H_2O	Rfd-values		A_p	F690	qQ	qE	g-h
		690	730		F735			
Picea, 1988								
0 h	71	2.9	2.1	0.21	0.8	0.95	0.41	91
1 h	69	2.6	1.9	0.20	0.8	0.95	0.50	71
2 h	70	2.4	1.9	0.15	0.8	0.93	0.55	57
4 h	68	2.4	1.9	0.15	0.8	0.94	0.69	36
6 h	67	2.3	1.8	0.15	0.8	0.94	0.74	25
8 h	63	2.1	1.6	0.16	0.8	0.89	0.75	20
24 h	54	1.9	1.4	0.17	0.9	0.75	0.77	3
Picea, 1987								
0 h	60	4.0	3.0	0.19	0.9	0.94	0.26	92
2 h	59	3.8	2.9	0.19	0.8	0.94	0.27	79
4 h	58	3.5	2.8	0.16	0.8	0.93	0.35	68
6 h	57	3.6	2.8	0.17	0.8	0.94	0.36	60
8 h	56	3.2	2.6	0.14	0.8	0.93	0.38	66
24 h	46	3.0	2.4	0.15	0.9	0.90	0.65	35

The Rfd values measured in the 690 nm region are always higher than those measured in the 730 nm region (Lichtenthaler and Rinderle, 1988). Since both Rfd-values (Rfd 690 and Rfd 730) decline with increasing stress (s. Table 1 and 2), both can be used as a vitality index. Thus it will depend on the filter system of a fluorometer present in a laboratory whether Rfd 690 or Rfd 730 or both Rfd-values can be measured. With the PAM-fluorometer the Rfd 730 values can be measured (Haitz and Lichtenthaler, 1988), whereas our portable field fluorometer (Lichtenthaler and Rinderle, 1988) permits the simultaneous determination of Rfd 690 and Rfd 730 values.

From the Rfd-values measured in the 690 and 730 nm region one can determine the **stress adaptation index Ap** (Strasser et al., 1987; Lichtenthaler and Rinderle, 1988). The Ap-index contains additional information on the photosynthetic apparatus and its degree of damage. Though the absolute values of the index Ap may depend upon the type of fluorometer applied (e.g. bandwidth of filter systems, intensity and wavelength of the excitation light), they decline later than the Rfd-values. Using our portable field fluorometer we determined values of Ap of 0.10 to 0.3 for intact photosynthetically active leaves of healthy plants (s. also Tables 1 and 2). Sun leaves which are exposed to more light and heat stress in general exhibit higher Ap-values (0.25 to 0.3) than shade leaves (0.19 to 0.26). Due to a short-term stress factor the Rfd-values of leaves and needles may decline very much, whereas the Ap-values are little affected (Lichtenthaler and Rinderle, 1988). In such cases the leaves and their photosynthetic activity can be regenerated (visible in increasing Rfd-values) when the stressor is removed. From our present experience it appears that at Ap-values below 0.16 an irreversible damage of the photosynthetic apparatus begins and that senescent leaves with an Ap below 0.1 can no longer recover.

When there is stress on the process of photosynthesis the **variable fluorescence (vF or Fv)**, which is the variable part of the fluorescence above the ground fluorescence (fo) and corresponds to the fluorescence increase from fo to the fluorescence maximum fmax of the induction kinetic, declines. This variable fluorescence is often also expressed in relative units as **ratio fmax/fo**. Stress decreases the variable fluorescence of dark-adapted leaves. The ratio fmax/fo, though different in absolute values, similarly decreases with increasing stress and damage as the Rfd-values. The ground fluorescence fo, which is thought to derive from the pigment antenna, very often increases under stress conditions. This is seen when the photosynthetic electron transport is blocked by the herbicide diuron (Fig. 2) and also when the reaction-center chlorophyll of photosystem II is e.g. photooxidatively destroyed (Lichtenthaler and Buschmann, 1984b; Lichtenthaler and Rinderle, 1988).

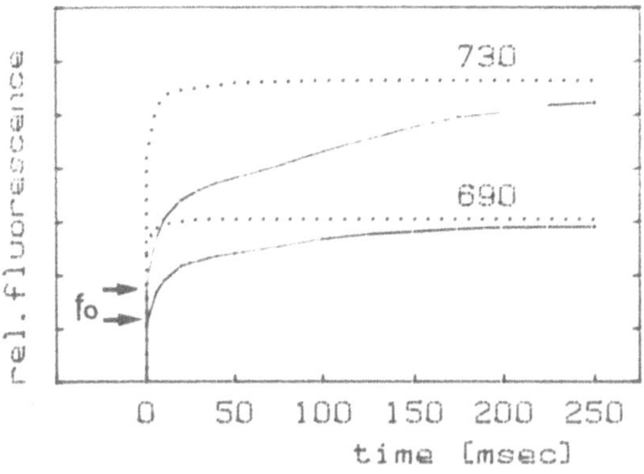

Figure 2. Fast chlorophyll fluorescence rise kinetics (Kautsky effect: fast component) of a dark-adapted tobacco leaf (solid line) with indication of the ground fluorescence fo and the maximum fluorescence fmax. The dotted lines are obtained after a 20 min treatment of the leaf with a 10^{-3}M solution of the herbicide diuron.

The absolute height of the ground fluorescence fo depends very much on the chlorophyll content of the leaf per leaf area unit. Decreasing or lower chlorophyll content will increase fo as well as fmax, because the reabsorption of the emitted fluorescence by the remaining chlorophyll is lower. For this reason the absolute values of fo and fmax, are only then valid parameters, when the chlorophyll content remained the same before and after stress treatment. With the relative variable fluorescence ratio fmax/fo, in turn, one can distinguish, whether a lower chlorophyll content e.g. in damaged forest trees, is associated with a loss of photosynthetic function and a relative increase of fo as compared to fmax. In many laboratories the fluorescence equipment available may not permit the resolution of the fast fluorescence rise and the determination of fo and fmax. This, in fact, is not necessary. Since fo and fmax only cover the first 300 to 400 ms of the fluorescence induction kinetics, which may not be fully representative of the photochemical processes in a leaf, it is better to use for screening of stress effects the whole fluorescence induction kinetics (ca. 4 min) and the determination of the Rfd-values.

The chlorophyll fluorescence ratio F690/F735, as determined from the fluorescence emission spectra of leaves obtained at room temperature, is a very suitable indicator in particular of long-term stress conditions in plants (s. Rinderle and Lichtenthaler, 1988). In normal green leaves a large proportion of the in vivo chlorophyll-fluorescence in the 690 nm region is reabsorbed by the leaf chlorophyll, since the absorption bands of the latter overlap with the wavelength of the emitted fluorescence. The 735 nm region fluorescence, in turn, is little affected. A lower chlorophyll content per leaf area unit is very often found as a result of nitrogen or Fe or other mineral deficiencies. In forest-decline research one has also detected lower chlorophyll content in needles of damaged trees (Lichtenthaler and Buschmann 1984a,b), often associated with low levels of potassium, manganese, calcium and mangan (Lichtenthaler et al. 1985). Consequently the ratio F690/F735 is much increased with decreasing chlorophyll content and also enhanced - though to a lower degree - with a loss of photosynthetic function e.g. by blocking of the electron flow by diuron (s. Rinderle and Lichtenthaler, 1988). Therefore, the ratio F690/F735 is a very suitable stress indicator in green plants and possibly also in remote sensing.

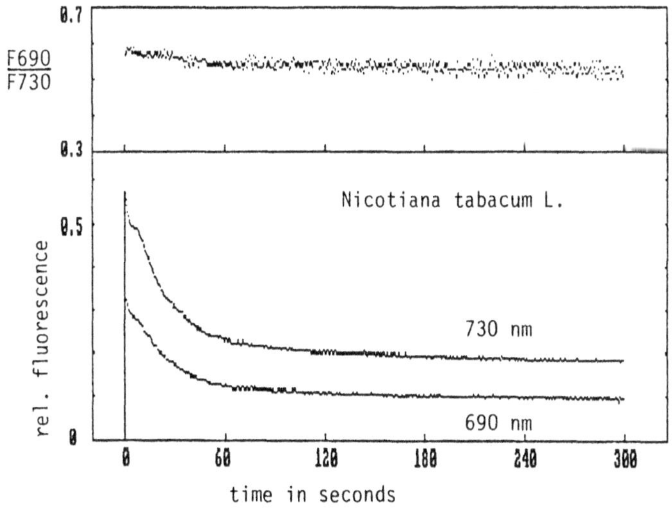

Figure 3. Measurement of the laser-induced chlorophyll fluorescence-induction kinetics in tobacco leaves (Nicotiana tabacum) in the 690 and 730 nm region using the new laser-induced, computer-aided LICAF system (Kocsanyi et al. 1988). The LICAF system also permits registration of the induction kinetics, as well as the changes in the fluorescence ratio F690/F730 during the induction kinetics.

The newly developed LICAF system of laser-induced computer-aided fluorescence registration , which is based on our He-Ne laser-equipped portable field fluorometer, not only permits measurement of the chlorophyll fluorescence-induction kinetics at two wavelengths but also the ratio of the fluorescence in the 690 and 730 nm region (F690/F730) (s. Fig. 3). Small changes in the ratio F690/F730 during the fluorescence-induction kinetics as well as larger changes due to a lower chlorophyll content and/or a decline of photosynthesis are recorded and documented with the LICAF system. Changes in the ratio F690/F730 during the induction kinetics are usually not found when the photosynthetic electron transport is blocked by the herbicide diuron. The LICAF system also permits resolution of the fast fluorescence-induction rise to fmax in 5 µs measuring periods every 200 µs. Further details and advantages of the LICAF system see Kocsanyi et al., 1988. With certain precautions the LICAF system can also be applied in field experiments.

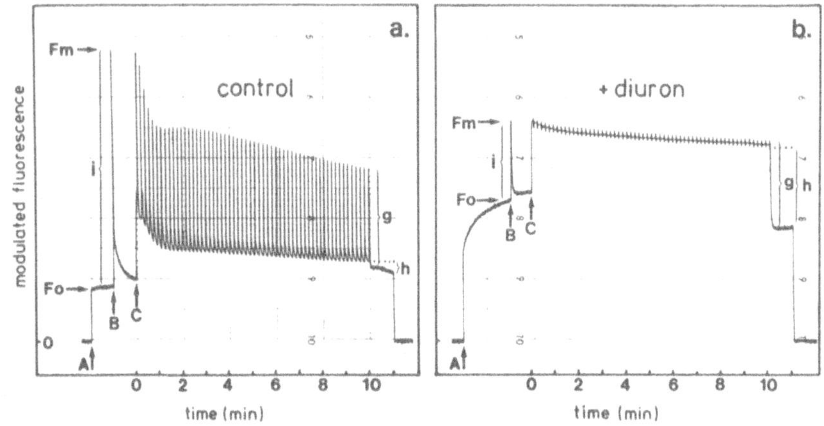

Figure 4. Original tracing of a modulated chlorophyll fluorescence induction kinetic with repetitive saturating light pulses in green predarkened (20 min) bean leaves (Phaseolus vulgaris). A. measuring light pulses (to detect the Fo level), B. saturation pulse 1s to detect Fmax, C. actinic light (+saturating light pulses every 10s). g= (Fv)s = maximal variable fluorescence, h= Fv = variable fluorescence. Distance (g-h) = saturation pulse-induced fluorescence spikes which indicate the amount of Q_A reoxidation capacity per leaf area unit (after Schreiber et al. 1986). **a)** green photosynthetically active leaf, **b)** leaf treated with the herbicide diuron (10^{-3}M) 40 min before the measurement.

A very powerful instrument is the recently developed **PAM-fluorometer** (pulse amplitude modulation) which provides additional and complementary information to the other fluorescence methods such as fluorescence induction kinetics, Rfd-values, Ap-index, fluorescence ratio F690/F735 and the LICAF system. The PAM fluorometer developed by Schreiber (for description see Schreiber et al. 1986) permits exact determination of the ground fluorescence fo, the maximum fluorescence fmax (using a saturating light pulse) as well as the photochemical (qQ) and non-photochemical quenching (qE) of the in vivo chlorophyll fluorescence during the fluorescence-induction kinetics and the steady state fluorescence (Fig. 4a). The non-photochemical quenching qE increases during stress conditions e.g. water or herbicide stress (Fig. 5 and Table 2), whereas the photochemical Q-quenching is little affected. A very fast changing fluorescence parameter is the **height of the saturating fluorescence spikes** (distance g-h in Fig. 4), which represents the capacity for reoxidation of the primary quencher Q_A of photosystem II. We found that the higher the capacity for photosynthetic quantum conversion (net CO_2-assimilation) per leaf area unit, the higher the saturation pulse-induced fluorescence spikes and the higher the Q_A-reoxidation capacity of the leaf (Lichtenthaler and Rinderle, 1988). The height of the saturation pulse-induced fluorescence spikes (g-h) decreases very rapidly to extremely low values upon stress treatment, as is shown in Table 2 (water-stress condition). The fluorescence spikes (g-h) also dissappear when the electron transport and the Q_A-reoxidation is blocked by the herbicide diuron (Fig. 4b).

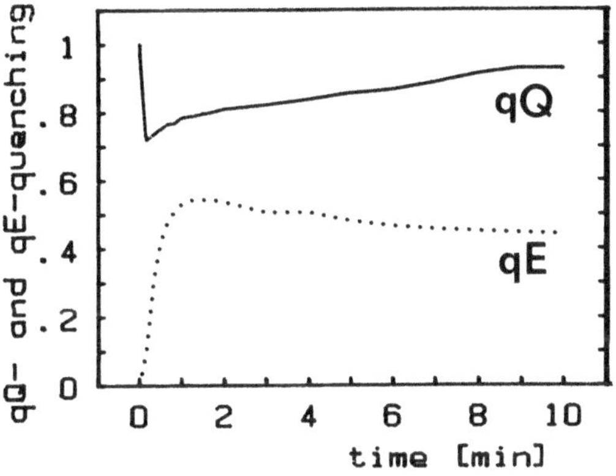

Figure 5. Time course of the quenching coefficients qQ (photochemical quenching) and qE (non-photochemical quenching) during the light-induced chlorophyll fluorescence induction kinetics in normal green bean leaves, as calculated from the tracings in Fig. 4a (applying the equations of Schreiber et al., 1986 as given by Lichtenthaler and Rinderle,1988a). In the diuron-treated bean leaf qE = 0 and qQ < 0.1.

The PAM-fluorometer is certainly a very powerful instrument to further describe and localize stress to the photosynthetic apparatus, once this had been detected by fluorescence screening via Rfd-values. That the PAM-fluorometer can also be applied for the determination of Rfd-values by using its saturation light is shown here in the contribution Haitz and Lichtenthaler, 1988.

In contrast to the room temperature fluorescence spectra, the **low temperature chlorophyll fluorescence emission spectra** are not only determined by the photosystem II fluorescence, but also by chlorophyll fluorescence of the photosystem I. One can differentiate between the fluorescence contribution of photosystems I and II and the light-harvesting antenna. Details are found in the review article of Murata and Satoh, 1986 and the contribution of Siffel and Szesták, 1988 in this volume. The problem is that the low-temperature fluorescence signals of frozen leaves are also influenced by partial reabsorption of the emitted fluorescence by the chlorophyll. This requires certain precautions. Either one works with isolated and diluted chloroplast preparations, or one dilutes the ground, deep-frozen leaves with ice (Weis, 1984). The method of registration of low temperature chlorophyll fluorescence spectra is more complex and takes more time expense than the fluorescence induction and saturation pulse methods mentioned before. Yet it provides additional information on the size of the photosynthetic units and of the pigment antennae and their change due to stress. Low temperature fluorescence spectroscopy can be an interesting complementary instrumentation once a stress condition or damage has been detected by the fluorescence induction kinetics, but is not a routine method of stress screening.

CONCLUSION: With the registration of several chlorophyll fluorescence parameters such as laser-induced fluorescence induction kinetics at two wavelengths, determination of the fluorescence-decrease ratios (Rfd 690 and Rfd 730), the stress-adaptation index Ap and the relative variable fluorescence fmax/fo as well as registration of the chlorophyll fluorescence spectra and the ratio F690/F735, the application of the new LICAF system and the PAM fluorometer (Q- and E-quenching, fluorescence spikes g-h) it is possible to detect, follow and define short-term and long-term stress conditions in plants and damage to the photosynthetic apparatus as well as regeneration once the stressor is removed. Examples of various applications of chlorophyll fluorescence in stress physiology are found in this book and in the review of Lichtenthaler and Rinderle, 1988.

ACKNOWLEDGEMENTS
Part of the work described here was sponsored by a grant from the PEF (Projekt Europäisches Forschungszentrum) Karlsruhe which is gratefully acknowledged. I wish to thank Mr. M. Haitz, Mrs. U. Prenzel, U. Rinderle and I. Schürer for assistance.

REFERENCES

HAITZ M and LICHTENTHALER HK, 1988. The measurement of Rfd-values as plant vitality index with the portable field fluorometer and the PAM-fluorometer. In: Applications of Chlorophyll Fluorescence, Lichtenthaler HK ed., pp.233-238, Kluwer Academic Publishers, Dordrecht (this volume).

KOSCANYI L, HAITZ M and LICHTENTHALER HK, 1988. Measurement of laser-induced chlorophyll fluorescence kinetics using a fast acoustooptic device. In: Applications of Chlorophyll Fluorescence, Lichtenthaler HK ed., pp. 91 - 93, Kluwer Academic Publishers Dordrecht (this volume).

KRAUSE GH and WEIS E, 1984. Chlorophyll fluorescence as a tool in plant physiology. II. Interpretation of fluorescence signals, Photosynth. Res. 5: 139-157.

LARCHER W, 1987. Stress bei Pflanzen. Naturwissenschaften 74: 158-167.

LEVITT J, 1980. Responses of plants to environmental stresses. Vol. 1, Academic Press, New York.

LICHTENTHALER HK, 1984. Influence of environmental factors on composition and function of the photosynthetic apparatus. In: Advances in Photosynthesis Research, Sybesma C ed., Vol. IV, pp. 241-244, Martinus Nijhoff/Dr.W. Junk Publisher, The Hague.

LICHTENTHALER HK, 1986. Laser-induced chlorophyll fluorescence of living plants. In: Proceed. Remote Sensing Sensing Symp. IGARSS Zürich Vol. III, pp. 1587-1590, ESA Publications Division, Noordwijk.

LICHTENTHALER HK, 1987. Chlorophyll fluorescence signatures of leaves during the autumnal chlorophyll breakdown. J. Plant Physiol. 131: 101-110.

LICHTENTHALER HK and BUSCHMANN C, 1984a. Das Waldsterben aus botanischer Sicht, G. Braun Verlag Karlsruhe.

LICHTENTHALER HK and BUSCHMANN C, 1984b. Photooxidative changes in pigment composition and photosynthetic activity of air-polluted spruce needles (Picea abies L.). In Advances in Photosynthesis Research Vol. IV, Sybesma C ed., pp. 245-250, Martinus Nijhoff Publisher, The Hague.

LICHTENTHALER HK and RINDERLE U, 1988. The role of chlorophyll fluorescence in the detection of stress conditions in plants. CRC Critical Reviews in Analytical Chemistry 19, Suppl. I: S 29-S 85.

LICHTENTHALER HK, BUSCHMANN C, DÖLL M, FIETZ H-J, BACH T, KOZEL U, MEIER D and RAHMSDORF U, 1981. Photosynthetic activity, chloroplast ultrastructure, and leaf characteristics of high-light and low-light plants and of sun and shade leaves. Photosynthesis Res. 2: 115-141.

LICHTENTHALER HK, KUHN G, PRENZEL U, BUSCHMANN C and MEIER D, 1982. Adaptation of chloroplast-ultrastructure and of chlorophyll-protein levels to high-light and low-light growth conditions. Z. Naturforsch. 37c: 464-475.

LICHTENTHALER HK, MEIER D and BUSCHMANN C, 1984. Development of chloroplasts at high and low quanta fluence rates. Israel J. Botany 33: 185-194.

LICHTENTHALER HK, SCHMUCK G, DÖLL M and BUSCHMANN C, 1985. Untersuchungen über die Funktionsfähigkeit des Photosyntheseapparates bei Nadeln gesunder und geschädigter Koniferen. In: PEF Bericht KfK-PEF2, pp. 81-105, Kernforschungszentrum, Karlsruhe.

LICHTENTHALER HK, BUSCHMANN C, RINDERLE U and SCHMUCK G, 1986. Application of chlorophyll fluorescence in ecophysiology. Radiation Environmental Biophysics 25: 297-308.

LICHTENTHALER HK, RINDERLE U, KRITIKOS G and ROCK B, 1987. Classification of damaged spruce stands in the Northern Black Forest by airborne reflectance and terrestrial chlorophyll fluorescence measurements. In: 2nd DFVLR Status Seminar: Untersuchung und Kartierung von Waldschäden mit Methoden der Fernerkundung pp. 228-252, DFVLR Oberpfaffenhofen/München.

MURATA N and SATOH K, 1986. Absorption and fluorescence emission by intact cells, chloroplasts, and chlorophyll-protein complexes. In: Light Emission by Plants and Bacteria. Govindjee, Amesz N and Fork DC eds., pp. 137-159, Academic Press Inc., Orlando.

RINDERLE U and LICHTENTHALER HK, 1988. The chlorophyll fluorescence ratio F690/F735 as a possible stress indicator. In: Applications of Chlorophyll Fluorescence, Lichtenthaler HK ed., pp. 176 - 183, Kluwer Academic Press, Dordrecht (this volume).

ROCK BN, HOSHIZAKI T, LICHTENTHALER HK and G SCHMUCK, 1986. Comparision of in situ spectral measurements of forest decline symptoms in Vermont (USA) and the Schwarzwald (FRG). In: Proc. Internat. Geoscience Remote Sensing Symposium, IGARSS Zürich, Vol. III, pp. 1667-1672, ESA Publications Division, Noordwijk.

SCHREIBER U, SCHLIWA U and BILGER W, 1986. Continuous recording of photochemical and non-photochemical chlorophyll fluorescence quenching with a new type of modulation fluorometer. Photosynth. Res. 10:51-62.

SCHMUCK G and LICHTENTHALER HK, 1986. Application of laser-induced chlorophyll a fluorescence in the forest decline research. In: Proc. Internat. Geoscience and Remote Sensing Symposium, IGARSS Zürich, Vol. III, pp. 1587-1590, ESA Publications Division, Noordwijk.

ŠIFFEL LP and ŠESTÁK Z, 1988. Low temperature fluorescence spectra of chloroplasts: methodical aspects and possible applications. In: Applications of Chlorophyll Fluorescence, Lichtenthaler HK ed., pp. 50 - 56 , Kluwer Academic Press, Dordrecht (this volume).

STRASSER RJ, 1986. Laser-induced fluorescence of plants and its application to environmental research. In: Proc. Internat. Geoscience Remote Sensing Symposium, IGARSS Zürich, Vol. III, pp. 1581-1585, ESA Publications Division, Noordwijk.

STRASSER R, SCHWARZ B and BUCHER J, 1987. Simultane Messung der Chlorophyll-Fluoreszenz Kinetik bei verschiedenen Wellenlängen als rasches Verfahren zur Frühdiagnose von Immissionsbelastungen an Waldbäumen. Ozoneinwirkung auf Buchen und Pappeln. Europ. J. Forest Pathology 17: 149-157.

WEIS E, 1984. Short term acclimation of spinach to high temperatures. Effect on chlorophyll fluorescence at 293 and 77 Kelvin in intact leaves. Plant Physiol. 74:402-407.

WHITMORE EM and FREER-SMITH PH, 1982. Growth effects of SO_2 and/or NO_2 on woody plants and grasses during spring and summer. Nature 300: 55-57.

CHLOROPHYLL FLUORESCENCE SIGNATURES AS VITALITY INDICATOR IN FOREST DECLINE RESEARCH

Hartmut K. LICHTENTHALER and Ursula RINDERLE

Botanisches Institut II (Plant Physiology and Plant Biochemistry), University of Karlsruhe, Kaiserstr. 12, D-7500 Karlsruhe, F.R.G.

KEY WORDS: Chlorophyll fluorescence, forest decline, Kautsky effect, PAM-fluorometer, photosynthesis, Rfd-values, vitality index.

ABSTRACT

The chlorophyll-fluorescence signatures of the needles of healthy (damage class 0/1, Althof, 450 m above sea level) and damaged spruce trees (damage class 3/4, Mauzenberg 700 m altitude) in the northern Black Forest were determined by using a self-constructed portable field fluorometer as well as the PAM-fluorometer. These new instruments allow measurement of the red-laser-induced chlorophyll fluorescence induction kinetics (Kautsky effect) at 690 and 730 nm, the determination of the Rfd-values (Rfd 690 and Rfd 730) as vitality index, and the stress-adaptation index A_p as well as the photochemical Q-quenching and the non-photochemical E-quenching of the spruce needles. The differences in the fluorescence signatures and photosynthetic activity between the needles of healthy and damaged trees are presented. In the needles of the damaged trees the Rfd-values, the capacity for photosynthesis and the reoxidation of Q_A are strongly reduced, whereas the non-photochemical E-quenching is increased.

The chlorophyll fluorescence is a very powerful non-destructive method for fast outdoor screening of tree physiology and for ground-truth measurements in remote sensing of forest decline as well as detection of damage to the photosynthetic apparatus.

INTRODUCTION

The forest decline observed since the beginning of the 80's in Central Europe (Black Forest, Vosges, Alpes, Bavarian Forest and also the mountains in Czechoslovakia) as well as in United States of America (e.g. Vermont) is caused by a combined action of natural and anthropogenic stress factors (Lichtenthaler and Buschmann 1984a, b; Lichtenthaler, 1984). The conifer trees were at first much more affected than the broadleaf trees. The damage was visualized for example by a considerable needle loss (up to 80 or 90 %), by smaller needles, lower chlorophyll content (lower accumulation rate of chlorophyll a+b) and by a yellowish discolouring of the needles. The latter could occur spotwise or start from the needle's end and develop to the middle part of the needle or primarily concern the upper leaf side (photo-oxidative bleaching) of the outer sun-exposed needle years or preferentially act on the older needle, inner years (Lichtenthaler et al., 1985). The various damage symptoms in the growth of trees, needle content of branches and the colour of the needle years are shown in the colour plates No. 4 to 9 at the end of this book. The in vivo chlorophyll fluorescence is a very suitable tool to document and screen the decline of forest trees and their

H. K. Lichtenthaler (ed.) Applications of Chlorophyll Fluorescence, 143–149.

photosynthesis in the differently discoloured needles of damaged conifers.

In photosynthetically active green-plant tissue (leaves, needles) the largest part of the light energy absorbed by the pigments (chlorophylls, carotenoids) is used for photosynthesis (photosynthetic quantum conversion). A minor part is re-emitted as chlorophyll fluorescence, the spectrum of which exhibits maxima near 690 and 735 nm (Lichtenthaler, 1986; Lichtenthaler et al., 1986; Lichtenthaler and Rinderle, 1988). In damaged forest trees and stressed plants as well as in leaves treated with photosystem II herbicides, such as diuron or bentazon, the photosynthetic activity declines (Lichtenthaler and Rinderle, 1988; Nagel et al., 1987; Schmuck and Lichtenthaler, 1986) and the chlorophyll fluorescence emission increases. The variable fluorescence-decrease ratio (Rfd-value), which indicates the potential photosynthetic activity, has been established with great success as a vitality index and stress indicator in plants (Lichtenthaler and Rinderle, 1988). With a new self-constructed portable field fluorometer one can simultaneously measure the chlorophyll-fluorescence induction kinetics in the 690 and 730 nm regions and determine the values for Rfd 690 and Rfd 730 (Lichtenthaler and Rinderle, 1988) as well as the stress-adaptation index A_p (Strasser et al, 1987). With another new device, the PAM-fluorometer (Schreiber et al., 1986), it is possible to determine the photochemical Q-quenching and the non-photochemical E-quenching. Here we report on the differences in several fluorescence parameters between needles from healthy and damaged spruce trees of the northern Black Forest (see also the colour plates No 4-9 at the end of this book).

MATERIALS AND METHODS

The fluorescence signatures of different needle years, of mainly healthy (Althof, damage class 0/1) and of damaged spruce trees (Mauzenberg, damage class 3/4, colour plates No 7 and 9) were measured using three different fluorescence methods: **A.** The red laser-induced chlorophyll fluorescence kinetics (determination of Rfd- values as vitality index of needles) measured near 690 and near 730 nm in a portable field fluorometer (Lichtenthaler and Rinderle, 1988). **B.** The chlorophyll fluorescence emission spectra at room temperature induced by blue light (470±30 nm) recorded with a Shimadzu MPS 5000 spectrometer under steady-state conditions of the chlorophyll fluorescence 5 min after onset of illumination. **C.** The differentiation between photochemical Q-quenching and non-photochemical E-quenching using the new PAM-fluorometer of A. Walz, Effeltrich (Schreiber et al., 1986; Lichtenthaler and Rinderle, 1988). In this new fluorometer, the excitation light to measure chlorophyll fluorescence is separately applied to the actinic light, which drives the photosynthetic reactions. Ground fluorescence fo is excited repetitively by 1 µs pulses of low intensity. The photosynthetic prenyl pigments (chlorophylls and carotenoids) were extracted with 100% acetone and the pigments quantitatively determined using the newly established extinction coefficients of Lichtenthaler, 1987 . The CO_2-assimilation rates were determined using the CO_2/H_2O-porometer system of Walz (s. Nagel et al., 1987).

RESULTS AND DISCUSSION

Green and yellowish-green needles. The red-laser-induced chlorophyll-fluorescence induction kinetics (Kautsky effect: fast and slow component) of spruce needles of healthy and damaged trees, measured in the 690 nm region, are shown in Fig. 1 for controls and highly damaged yellowish-green needles. The relative variable fluorescence, i.e. the increase of the fluorescence from the ground fluorescence fo to the maximum fluorescence fmax and slow fluorescence decrease (min range) from fmax to fs, is lower in the needles of damaged trees. In addition, the fo and fs (steady state fluorescence) is increased in this particular needle sample. Consequently the Rfd-values, as an indicator of the potential photosynthetic activity of a leaf (Lichtenthaler, 1986; Lichtenthaler et al., 1986; Lichtenthaler and Rinderle, 1988), and the ratio fmax/fo are considerably lower in yellowish-green needles of the damaged spruces (Fig. 1).

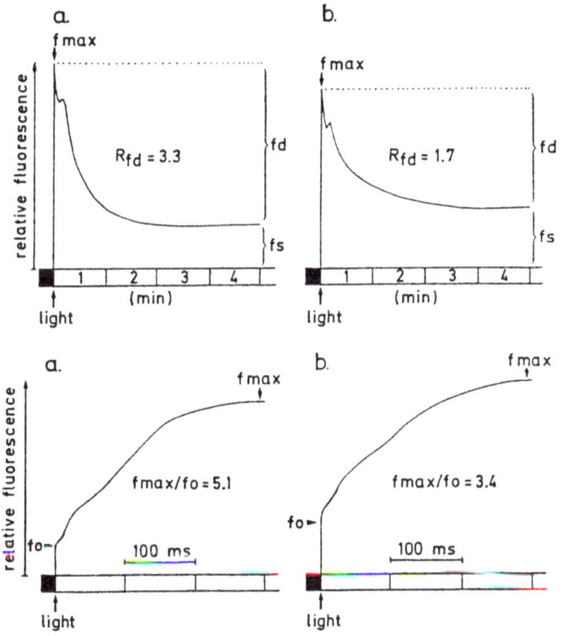

FIGURE 1. Red-laser-induced chlorophyll fluorescence induction kinetics of green needles (healthy spruce) and yellowish-green needles (damaged spruce) measured in July 1987 (needle year 1985). The fast fluorescence rise via fo (ground fluorescence) to fmax (maximum fluorescence) proceeds in ca. 300 ms, the slow fluorescence decrease fd from fmax to fs (steady state fluorescence) is accomplished after 4 to 5 min. The Rfd-ratio is defined as Rfd=fd/fs. **a**) Althof (damage class 0/1) and **b**) Mauzenberg (damage class 3/4) in the northern Black Forest.

Green and light-green needles. In the further investigation we used green (healthy trees) and olive-green to light-green needles of damaged trees. All needle-years of the damaged trees exhibit a much lower chlorophyll content, whereas the Rfd-values are only slightly reduced (Tab. 1). This indicates that the remaining chlorophyll of the needles is physiologically active in photosynthetic quantum conversion, which is also demonstrated by rather normal values for the stress adaptation index A_p. Due to the lower chlorophyll content, the net CO_2-assimilation per needle area is, however, much lower in the needles of damaged than healthy trees. On a chlorophyll basis the net CO_2-assimilation of the needles of the damaged spruces is slightly higher in all needle years (Tab. 1). Though the chlorophyll of the needles of the damaged spruces is physiological fully active, the lower values for the pigment ratio chlorophylls to carotenoids (a+b/x+c) indicates that there is stress on the photosynthetic apparatus of the damaged spruces.

TABLE 1. Chlorophyll content (μg a+b·cm^{-2}), pigment ratios (chlorophylls/carotenoids = a+b/x+c) and values for the variable chlorophyll fluorescence ratio (Rfd-values at 690 and 730 nm), for the stress adaptation index A_P and net CO_2-assimilation (P_N) in green needles of healthy (Althof) and green to light-green needles from damaged spruces (Mauzenberg). Mean of 3 determinations from 2 trees at each site. The measurements were made in August 1987 with light-exposed N1 and N2 needles.

Picea abies (needle-year)	a+b	a+b / x+c	Rfd-values 690nm	Rfd-values 730nm	A_P-index	P_N* (per m^2)	P_N** (per a+b)	transpiration***	gH_2O ***
class 0/1									
1984	112	5.3	4.7	3.2	0.262	5.4	0.7	0.6	50
1985	134	5.3	5.2	3.4	0.299	6.7	0.8	1.4	54
1986	110	5.3	5.4	3.6	0.274	7.0	1.0	1.1	45
1987	71	5.1	5.4	3.6	0.284	7.2	1.6	1.0	48
class 3/4									
1984	50	4.7	4.4	3.0	0.263	2.6	0.8	0.82	35
1985	37	4.3	4.0	2.9	0.224	2.4	1.0	0.45	24
1986	47	4.6	4.4	3.2	0.232	4.0	1.3	0.73	30
1987	46	4.7	5.9	3.9	0.295	5.5	1.9	0.95	38

* P_N in $\mu mol\ CO_2 \cdot m^{-2} \cdot s^{-1}$; ** P_N in mg $CO_2 \cdot$ mg a+b$^{-1} \cdot h^{-1}$

*** transpiration and stomata conductivity gH_2O in mmol $H_2O \cdot m^{-2} \cdot s^{-1}$

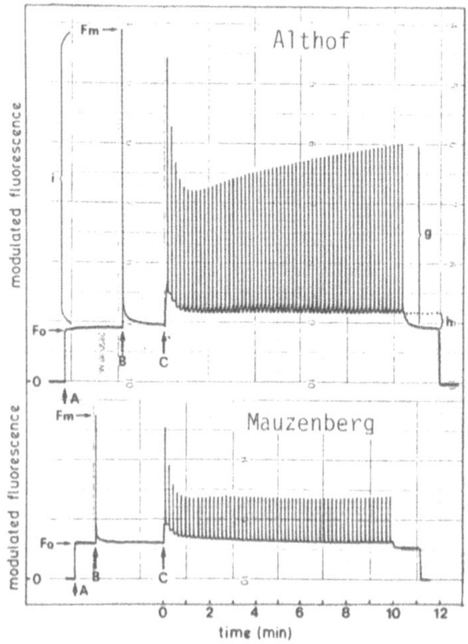

FIGURE 2. Chlorophyll fluorescence induction kinetics measured with a PAM-fluorometer. Original tracing with repetitive application of saturation light pulses of green needles of the needle-year 1985 from a healthy (Althof, damage class 0/1) and olive-green needles from a damaged spruce (Mauzenberg, damage class 3/4). The leaves were dark-adapted (20 min).
A: measuring light pulses (to detect the Fo level), **B**: saturation pulse (1s to detect maximum fluorescence Fm), **C**: actinic light (+ saturating light pulses every 10s). The needles were measured in August 1987 . Fo= ground fluorescence; g= (Fv)s = maximal variable fluorescence at any given time upon application of a saturation pulse; h= Fv = variable fluorescence at any given time during the induction period; i= (Fv)m = fluorescence rise from Fo to Fm induced by the saturation pulse, according to Schreiber et al., 1986.

With the PAM-fluorometer one can differentiate between the photochemical (qQ) and non-photochemical quenching (qE) of the in vivo chlorophyll fluorescence and determine fo, fmax (Fm) as well as the rate of Q_A-reoxidation (Schreiber, 1986). A typical example is shown in Fig. 2. Saturating light pulses, given every 10s, will fully reduce Q_A in the photosynthetic electron transport chain. In photosynthetically active leaf tissue the reduced Q_A will be reoxidized and can be reduced by the next light pulse. **The higher the saturation pulse-induced fluorescence spikes, the larger the amount of reoxidized Q_A per needle area and the higher the photosynthetic capacity per needle-area unit.** The comparison between the green needles (needle-year 1985) of a healthy and light-green needles of a damaged spruce tree (damage class 3/4) shows that the capacity for Q_A-reoxidation is much lower in the needles of the damaged spruce, as is the chlorophyll content (Fig. 2 and Tab. 1). This is further evidenced by slightly lower Rfd-values as well as lower rates of CO_2-fixation per needle-area unit (Table 1). From the tracings shown in Fig. 2 one can calculate the Q- and E-quenching as is shown in Fig. 3.

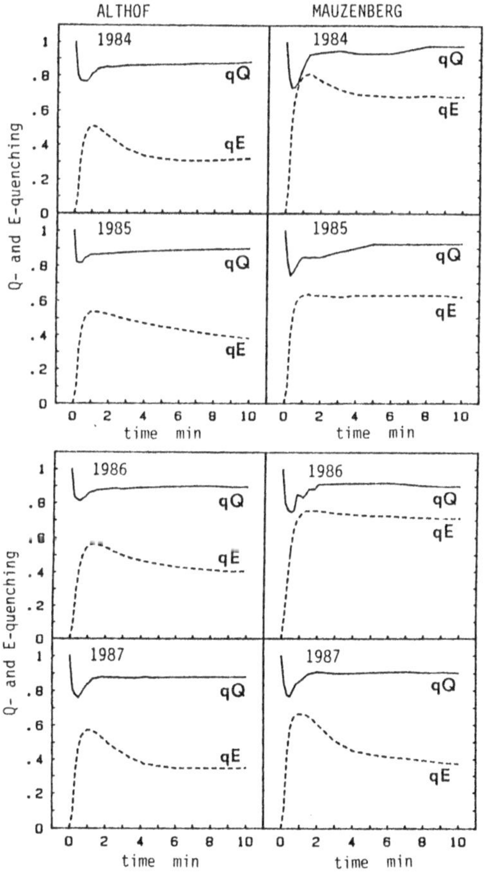

FIGURE 3. Time course of the quenching coefficients qQ (photochemical quenching) and qE (nonphotochemical quenching) during the chlorophyll fluorescence-induction kinetics. The values were calculated from the original tracings obtained in a PAM-fluorometer in August 1987. Althof: green needles of 1984 to 1987 of a healthy spruce and Mauzenberg: olive-green needles of damaged spruce (damage class 3/4).

In needles of damaged trees the non-photochemical E-quenching is significantly higher than in fully green needles of healthy spruces. This mainly applies to the needle years 1984-1986. The difference in the youngest needle year in summer 1987 was quite small (Fig. 3) but became larger and significant in November (Table 2). In contrast the Q-quenching, as measured in August 1987, is practically the same in needles of healthy and damaged spruces (Fig. 3 and Tab. 2). This is evidence that the chlorophyll present in the needles of healthy and of damaged trees is organized in functional photosynthetic units, even in the needles with lower chlorophyll content of the stressed trees at the Mauzenberg site. The higher qE-values as well as the reduced values for fmax and the ratio fmax/fo (Tab. 2) indicate, however, that the photosynthetic apparatus of the damaged tree is not under fully normal operational conditions and affected by stress.

TABLE 2. Values for the fluorescence-quenching coefficients qQ and qE after 10 min of a saturation-pulse kinetic as well as for the ground fluorescence fo, the maximum fluorescence fmax and the ratio fmax/fo of needles of healthy (Althof, damage class 0/1) and damaged spruces (Mauzenberg, damage class 3/4) determined from measurements with the PAM-fluorometer. Mean of 6 determinations in each case from August to October 1987. fo and fmax are given in relative units.

Picea abies (needle-year)	qQ	qE	fo	fmax	fmax/fo
class 0/1					
1985	0.90	0.48*	28	164*	5.9*
1986	0.91	0.54*	29	173*	6.0*
1987	0.92	0.49	36	199*	5.9
class 3/4					
1985	0.94	0.60*	21	106*	4.9*
1986	0.91	0.68*	24	93*	3.9*
1987	0.92	0.47	27	143	5.4

* These differences between needles of healthy and damaged trees are significant ($P \leq 0.02$).

CONCLUSION: The portable field fluorometer (Rfd-values) and the PAM-fluorometer offer very good possibilities for ground-truth measurements to exactly define - in addition to measurements of P_N, chlorophyll content and pigment ratios - the physiological state of trees. These chlorophyll fluorescence methods which permit detection of stress to the photosynthetic apparatus should be applied in all future ground-truth and ecophysiological investigations, since they allow a faster outdoor screening of the vegetation than other more time-consuming ecophysiological methods.

ACKNOWLEDGEMENTS
This work was sponsored in part by a grant from the PEF Karlsruhe, which is gratefully acknowledged. We wish to thank Ms. U. Prenzel, B. Weigelt, I. Schürer and Mr. M. Haitz for assistance.

REFERENCES

LICHTENTHALER HK, 1984. Luftschadstoffe als Auslöser des Baumsterbens. Naturwiss. Rundschau 37: 271-277.

LICHTENTHALER HK, 1986. Laser-induced chlorophyll fluorescence of living plants. In: Proc. Internat. Geoscience and Remote Sensing Symposium IGARSS Zürich 1986, Vol. III, pp. 1571-1579, ESA Publications Division, Noordwijk.

LICHTENTHALER HK, 1987. Chlorophylls and carotenoids, the pigments of photosynthetic biomembranes. Methods in Enzymology: 148, 350-382.

LICHTENTHALER HK and BUSCHMANN C, 1984a. Beziehungen zwischen Photosynthese und Baumsterben. Allgem. Forstzeitschrift 39: 12-16.

LICHTENTHALER HK, and BUSCHMANN C, 1984b. Das Waldsterben aus botanischer Sicht. G. Braun Verlag, Karlsruhe.

LICHTENTHALER HK and RINDERLE U, 1988. The role of chlorophyll fluorescence in the detection of stress conditions in plants. CRC Critical Reviews in Analytical Chemistry 19, Suppl. I: S 29-S 85.

LICHTENTHALER HK, SCHMUCK G, DÖLL M and BUSCHMANN C, 1985. Untersuchungen über die Funktionsfähigkeit des Photosyntheseapparates bei Nadeln gesunder und geschädigter Koniferen. PEF-Bericht KfK-PEF 2, PEF Projektleitung ed., pp. 81-105, Kernforschungszentrum Karlsruhe.

LICHTENTHALER HK, BUSCHMANN C, RINDERLE U and SCHMUCK G, 1986. Application of chlorophyll fluorescence in ecophysiology. Radiation and Environmental Biophysics 25: 297-308.

NAGEL EM, BUSCHMANN C and LICHTENTHALER HK, 1987. Photoacoustic spectra of needles as an indicator for the activity of the photosynthetic apparatus of healthy and damaged conifers. Physiol. Plantarum 70: 427-437.

SCHMUCK G and LICHTENTHALER HK, 1986. Application of laser-induced chlorophyll a fluorescence in the forest decline research. In: Proc. Internat. Geoscience and Remote Sensing Symposium IGARSS Zürich 1986, Vol. III, pp. 1587-1590, ESA Publications Division, Noordwijk.

SCHREIBER U, 1986. Detection of rapid induction kinetics with a new type of high-frequency modulated chlorophyll fluorometer. Photosynth. Res.: 9, 261-272.

SCHREIBER U, SCHLIWA U and BILGER W, 1986. Continuous recording of photochemical and non-photochemical chlorophyll fluorescence quenching with a new type of modulation fluorometer. Photosynth. Res. 10: 51-62.

STRASSER R, SCHWARZ B and BUCHER J, 1987. Simultane Messung der Chlorophyll Fluoreszenz-Kinetik bei verschiedenen Wellenlängen als rasches Verfahren zur Frühdiagnose von Immissionsbelastungen an Waldbäumen. Ozoneinwirkung auf Buchen und Pappeln. European Journal of Forest Pathology: 17, 149-157.

APPLICATION OF THE PAM FLUOROMETER IN STRESS DETECTION

SCHREIBER U., BILGER W., KLUGHAMMER C. and NEUBAUER C.
Lehrstuhl Botanik I, Universität Würzburg,
Mittlerer Dallenbergweg 64, D-8700 Würzburg, FRG

ABSTRACT

Different parameters of chlorophyll fluorescence contain a wealth of information on the state of the photosynthetic apparatus, which is affected in more or less direct ways by environmental stress. The recently developed PAM Fluorometer provides the means to harvest this information. Essential aspects for practical applications in ecophysiology are portability, tolerance of unfiltered sun light and diagnostic techniques for localization of stress induced changes. Additional insights are possible by means of a new emitter-detector unit converting the fluorometer into a system for measuring P700 absorbance changes. Some examples of application are given, relating to *Fusarium* infection, photoinhibition and drought stress.

INTRODUCTION

Chlorophyll fluorescence is a complex indicator of photosynthesis (for recent reviews, see Briantais et al 1986, Renger and Schreiber 1986, Schreiber and Bilger 1987). On one hand quantitative evaluation of the complex information in not always easy, on the other hand it may be safely stated that hardly any change in photosynthetic reactions will escape from being reflected in one or the other fluorescence parameter. This feature makes fluorescence an ideal indicator for early detection and detailed analysis of stress effects on plants. To make full use of the large potential of this versatile indicator, a fluorometer with considerable flexibility is needed. The recently introduced PAM Fluorometer meets the requirements for a thorough analysis of fluorescence parameters. Details on technical aspects have been described elsewhere (Schreiber 1986, Schreiber et al 1986, Schreiber and Bilger 1987, Schreiber et al 1988). The following properties of the PAM Fluorometer are of particular interest with respect to the detection and analysis of stress effects:

1) The systems is insensitive to ambient day light (even full sun light), thus allowing fluorescence measurements under field conditions.
2) Even under conditions of steady state illumination, information on photochemical efficiency and non-photochemical energy dissipation can be obtained by the so-called saturation pulse method.
3) Suitable accessory devices (clocks, timers, pulse modulation frequency control, trigger circuitry, saturation pulse and flash lamps, data acquisition systems) are available in compact design for fluorescence analysis in the field.
4) At a modulation frequency of 100 kHz even the fastest induction and relaxation kinetics can be recorded.
5) With a recently developed emitter-detector unit for P700 measurements, the same instrument can also assess properties of PS I, complementing the fluorescence information which primarily relates to properties of PS II.

H. K. Lichtenthaler (ed.) Applications of Chlorophyll Fluorescence, 151–155.

RESULTS

Fig. 1 shows an example of application of the saturation pulse method to characterize changes in photosynthetic performance caused by the infection of a tomato plant with the phytopathogenic fungus *Fusarium oxysporum*. Saturating light pulses are applied repetitively, to determine photochemical and non-photochemical quenching components (see Schreiber et al 1986, Schreiber and Bilger 1987). It is apparent that *Fusarium* infection considerably slows down the induction kinetics of fluorescence and of photochemical and non-photochemical quenching. A detailed discussion of the presented results would be out of scope of this presentation. Briefly, it may be remarked that the observed changes point towards a limitation of electron transport between PS II and NADP, possibly caused by some toxic substance set free by the fungus.

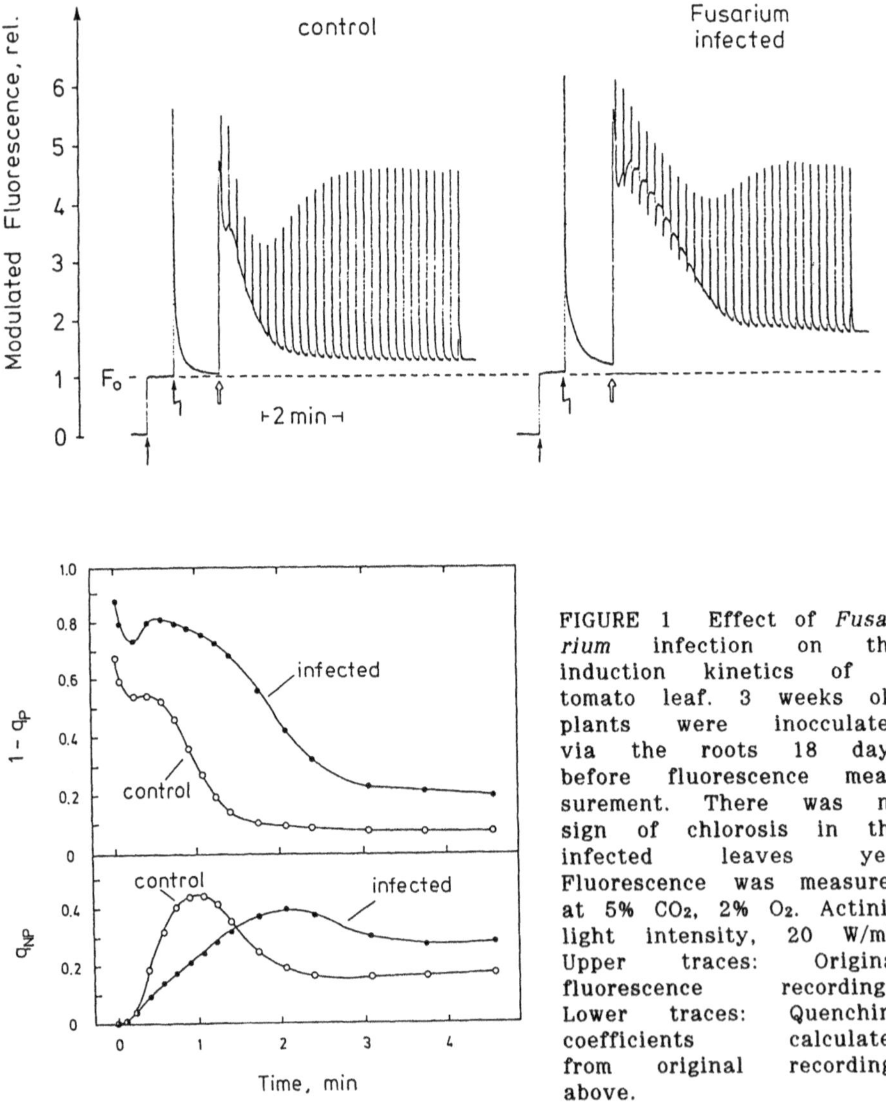

FIGURE 1 Effect of *Fusarium* infection on the induction kinetics of a tomato leaf. 3 weeks old plants were inocculated via the roots 18 days before fluorescence measurement. There was no sign of chlorosis in the infected leaves yet. Fluorescence was measured at 5% CO_2, 2% O_2. Actinic light intensity, 20 W/m^2. Upper traces: Original fluorescence recordings. Lower traces: Quenching coefficients calculated from original recordings above.

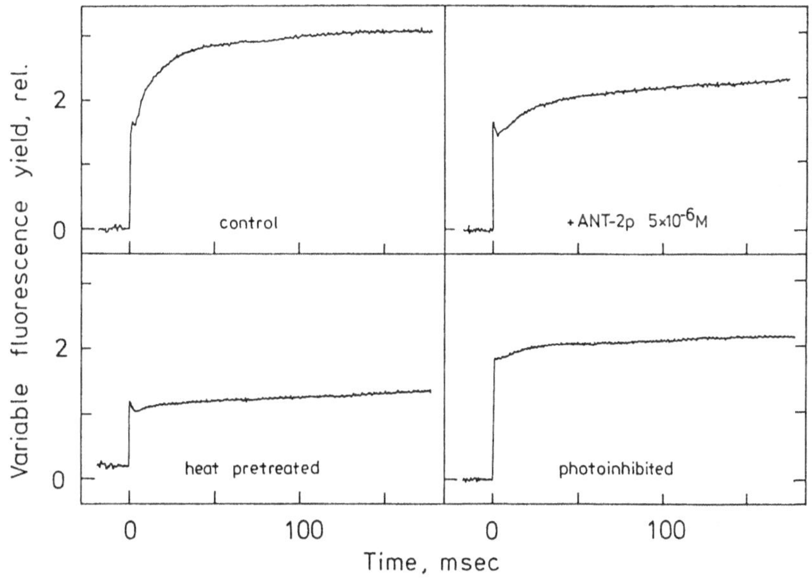

FIGURE 2 Polyphasic rise of chlorophyll fluorescence upon illumination with 2000 W/m^2 white light. Spinach leaf discs; treatments: infiltration with $5x10^{-6}M$ ANT-2p; 5 min exposure to 48°C; 30 min exposure to 2000 W/m^2, followed by 60 min dark recovery.

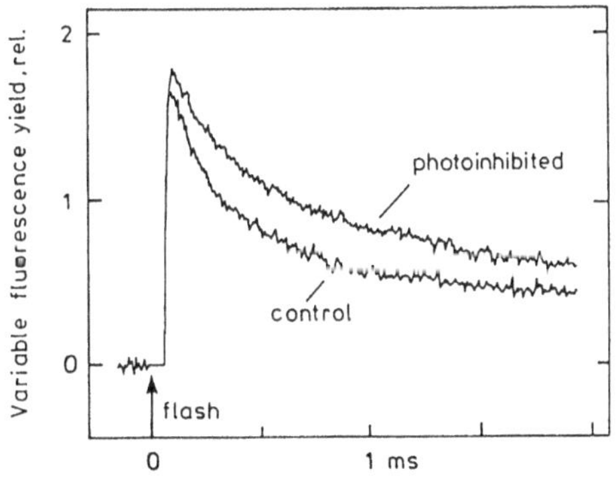

FIGURE 3 Fluorescence relaxation kinetics following single turnover saturating flash in control and photoinhibited spinach leaf sample. Control and photoinhibited samples identical to those of Fig. 2. Application of XST 103 flash lamp (Walz).

Fig. 2 shows the effects of the ADRY reagent ANT-2p and of stress treatments by heat and photoinhibition on the polyphasic fluorescence rise kinetics induced by saturating light. All three treatments lead to preferential suppression of the so-called I_1-I_2 phase, indicating loss of PS II donor side activity (Schreiber and Neubauer 1987). While such loss is not unexpected for ANT-2p (Renger 1972) and for heat treatment (Yamashita and Butler 1968), in the case of photoinhibition it contrasts the hypothesis of the Q_B-protein being the primary site of photoinhibitory damage (Kyle et al 1984). Actually, as shown in Fig. 3, there is only a relatively small effect on the $Q_A \longrightarrow Q_B$ transfer monitored via the relaxation kinetics following a flash, after the same photoinhibitory treatment which strongly affected the donor side.

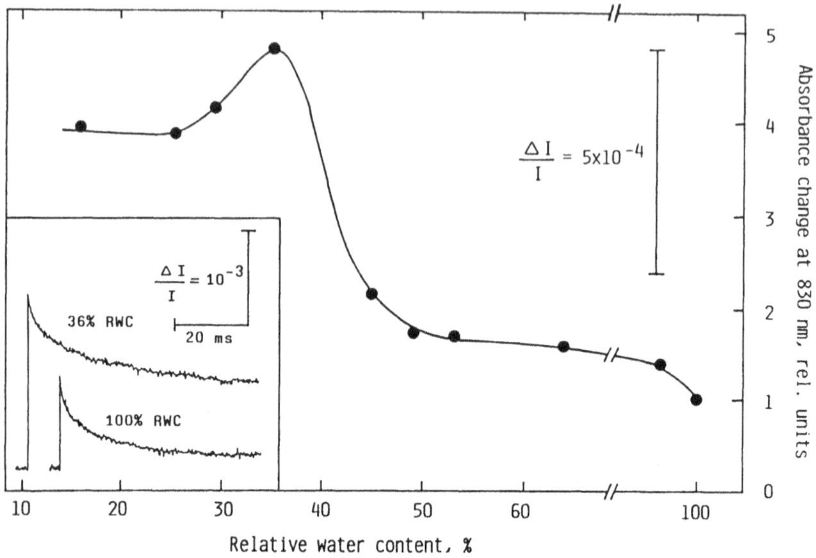

FIGURE 4 Relationship between a slow phase of $P700^+$ re-reduction following a single turnover flash and the relative water content of *Arbutus unedo* leaf discs. The absorbance increase (due to $P700^+$) observed 50 ms following a flash is plotted.

Finally the application of the PAM Fluorometer, modified to monitor P700 absorbance changes (Schreiber et al 1988), shall be demonstrated for detection of drought stress in *Arbutus unedo* (Fig. 4). P700 was measured via the absorbance of the cation radical $P700^+$. Leaf disc samples with different relative water contents were illuminated by single turnover saturating flashes, producing transient oxidation of P700, the rate of re-reduction being affected by desiccation (see inset of Fig. 4). It is shown that in *Arbutus* at a critical relative water content of appox. 42%, $P700^+$ re-reduction (within 50 ms following a flash) becomes severely slowed down. In a control sample, plastocyanin re-reduces most of $P700^+$ with half-times of 20 µs and 140 µs (Haehnel 1977) (not resolved in Fig. 4). Dessiccation obviously interferes strongly with the interaction of plastocyanin and P700. In this context, the observation of Hincha et al (1987) of a loss of plastocyanin in thylakoids isolated from leaves, which had lost more than 50% of their water, appears important. Via P700 measurements the modified PAM Fluorometer may provide a rapid, non-destructive screening test for drought tolerance.

ACKNOWLEDGEMENT

This work was supported by the Deutsche Forschungsgemeinschaft (Forschergruppe Ökophysiologie and SFB 176). We wish to thank Ulrich Heber for stimulating discussions.

REFERENCES

Briantais J-M, Vernotte C, Krause GH and Weis E (1986) Chlorophyll fluorescence of higher plants: Chloroplasts and leaves. In: Govindjee, Amesz J, Fork DC. Light emission by plants and photosynthetic bacteria. pp 539-577, New York, Academic Press.

Haehnel W (1977) Electron transport between plastoquinone and chlorophyll a_I in chloroplasts. II. Reaction kinetics and the function of plastocyanin. Biochim Biophys Acta 459: 418-441

Hincha DK, Höfner R, Schwab KB, Heber U and Schmitt JM (1987) Membrane rupture is the common cause of damage to chloroplast membranes in leaves injured by freezing or excessive wilting. Plant Physiol 83, 251-253

Kyle DJ, Ohad I and Arntzen CJ (1984) Membrane protein damage and repair: Selective loss of quinone-protein function in chloroplast membranes. Proc Natl Acad Sci USA 81: 4070-4074

Renger G (1973) Studies of the mechanism of destabilization of the positive charges trapped in the photosynthetic watersplitting enzyme Y by a deactivation accelerating agent. Biochim Biophys Acta 314: 390-402

Renger G and Schreiber U (1986) Practical applications of fluorometric methods to algae and higher plant research. In: Govindjee, Amesz J, Fork DC. Light emission by plants and bacteria. pp 587-619, New York, Academic Press.

Schreiber U (1986) Detection of rapid induction kinetics with a new type of high-frequency modulated chlorophyll fluorometer. Photosynth Res 9: 261-272

Schreiber U, Bilger W and Schliwa U (1986) Continuous recording of photochemical and non-photochemical chlorophyll fluorescence quenching with a new type of modulation fluorometer. Photosynth Res 10: 51-62

Schreiber U and Bilger W (1987) Rapid assessment of stress effects on plant leaves by chlorophyll fluorescence measurements. In: Tenhunen et al. Plant response to stress. pp 27-53, Berlin, Springer Verlag

Schreiber U and Neubauer C (1987) The polyphasic rise of chlorophyll fluorescence upon onset of strong continuous illumination: II. Partial control by the photosystem II donor side and possible ways of interpretation. Z Naturforsch 42c: 1255-1264

Schreiber U, Klughammer C and Neubauer C (1988) Measuring P700 absorbance changes around 830 nm with a new type of pulse modulation system. Z Naturforsch, in press

Yamashita T and Butler WL (1968) Inhibition of chloroplasts by UV-irradiation and heat-treatment. Plant Physiol 43: 2037-2040

CHANGES IN CHLOROPHYLL FLUORESCENCE RELATED TO PHOTOINHIBITION OF PHOTOSYNTHESIS AND COLD ACCLIMATION OF GREEN PLANTS

Susanne Somersalo[1] and G. Heinrich Krause[2], Institute for Biochemistry of Plants, University of Düsseldorf, Universitätsstraße 1, D-4000 Düsseldorf 1, Germany (FRG)

Key words: cold acclimation, fluorescence of chlorophyll, photoinhibition, photosynthesis, quantum yield, Spinacia oleracea L.

ABSTRACT

Fluorescence induction at 20°C and 77K was applied in this study to characterize the photoinhibition of spinach leaves (Spinacia oleracea L.) induced in moderate light at chilling temperature. When plants acclimated to +18°C and 260-300 μmol photons m^{-2} s^{-1} were exposed to 550 μmol m^{-2} s^{-1} at +4°C, they developed a strong photoinhibition within a few hours. This was manifested by a decline in variable fluorescence (F_V) of both photosystems (PS) and in quantum yield of O_2 evolution. The changes in fluorescence characteristics were more pronounced in the signal recorded at 20°C that at 77K. Moreover, the ratio of variable to maximum fluorescence (F_V/F_M) measured at 20°C was more closely related to quantum yield than F_V/F_M at 77K. The F_V/F_M ratio recorded at 20°C is thus regarded as the more sensitive indicator of photoinhibition. The effects related to photoinhibition of unhardened spinach leaves, including a pronounced increase of initial fluorescence (F_O), were fully reversible at +18°C in dim light. The reversible photoinhibition is viewed as a protective mechanism serving for thermal dissipation of excess light energy. Cold acclimation of the plants increased their resistance to photoinhibition. The mechanism of photoinhibition seemed to be altered in cold acclimated leaves, as indicated by constant F_O. Cold acclimation apparently enforces other protective mechanisms, and photoinhibition then occurs only in higher light.

INTRODUCTION

Photoinhibition of photosynthesis is defined as a decrease in efficiency (optimal quantum yield) and capacity of photosynthetic O_2 evolution or CO_2 assimilation, caused by excessive light energy (for reviews see Kyle et al., eds., 1987). Various environmental stress factors (e.g., chilling) promote photoinhibition. In ecophysiological research, photoinhibition therefore has gained increased interest.

Photoinhibition is initiated in PS II and consistently is related to a quenching of variable chlorophyll fluorescence. This 'photoinhibitory' fluorescence quenching is usually expressed as a decrease in the F_V/F_M ratio of PS II, which according to Butler and coworkers (Butler 1977) is a measure of

1) Permanent address: Department of Biology, University of Turku, SF-20500 Turku Finland.
2) To whom reprint request should be sent.

Abbreviations used: F_O, initial fluorescence; F_M, maximum fluorescence; F_V, variable fluorescence (F_M-F_O); PFD, photon flux density; PS, photosystem.

H. K. Lichtenthaler (ed.) Applications of Chlorophyll Fluorescence, 157–164.

the efficiency of PS II in the primary photochemistry. F_V/F_M ratios can be determined either at room temperature or at 77K from fluorescence induction signals. Further changes in fluorescence parameters, e.g. the yield of F_o, may serve to characterize the mechanism of photoinhibition.

In the present study, a comparison of fluorescence data obtained at 20°C and 77K is given. The fluorescence induction is analysed in relation to photoinhibition occurring in spinach leaves at low temperature. In this chilling-resistant plant, temperatures around +4°C induce a reversible photoinhibition at light fluxes similar to those of the natural habitat. Cold acclimation of the plants strongly decreased their susceptibility to photoinhibition. The reversible photoinhibition is discussed as a protective mechanism for thermal energy dissipation preventing damage of the photosynthetic apparatus until other protective systems become effective during cold acclimation.

MATERIAL AND METHODS

Material: 5 weeks old spinach plants (*Spinacia oleracea* L.) were acclimated for 10 days to +18°C (unhardened plants). Cold acclimation of the plants was carried out similar to Klosson and Krause (1981). PFD in both cases was 260-300 μmol m^{-2} s^{-1} and the light period 8 h.

Pretreatments: Detached spinach leaves were exposed for different times to white light, PFD 550 to 1400 μmol m^{-2} s^{-1} in normal air at +4°C. The temperature of the lower leaf side was +5.5 to +6°C. For recovery, the leaves were kept at +4°C or +18°C, PFD 2.5 to 5.0 μmol m^{-2} s^{-1}.

77K fluorescence: Fluorescence induction was measured at 77K from leaf discs, similar to Ögren and Öquist (1984) using blue exciting light. The leaf discs were kept for 5 minutes in complete darkness at room temperature before freezing.

20°C fluorescence: Fluorescence induction at 20°C was measured in the far red region from leaf discs with a Hansatech fluorometer (Hansatech, Kings Lynn, Norfolk, UK). Red exciting light was provided from a set of photodiodes (Hansatech). Before recording the fluorescence, the leaf discs were kept 10 minutes in complete darkness in CO_2-free air at 20°C.

Optimal quantum yield of O_2 evolution was determined according to Björkman and Demmig (1987).

RESULTS

Photoinhibition of unhardened leaves at chilling temperature

Figure 1a illustrates the decrease in the F_V/F_M ratio at 77K caused by photoinhibition treatment at +4°C of leaves acclimated to +18°C. The decline in F_V/F_M of PS II was accompanied by a smaller but significant decrease in F_V/F_M of PS I. As seen from Figure 1b,c, the change in F_V/F_M of PS II was based on a decrease in F_V and an increase in F_o. In PS I fluorescence, F_V was lowered, while F_o was not significantly altered. The data are consistent with the view that photoinhibition is induced by a transformation of PS II reaction centers to fluorescence quenchers (*cf.* Cleland et al. 1986). Such altered centers would efficiently trap excitation energy and convert this to heat.

Table 1 and Figures 2 and 3 provide a comparison of the effects of photoinhibition treatment on fluorescence emitted by PS II at 77K and 20°C and on

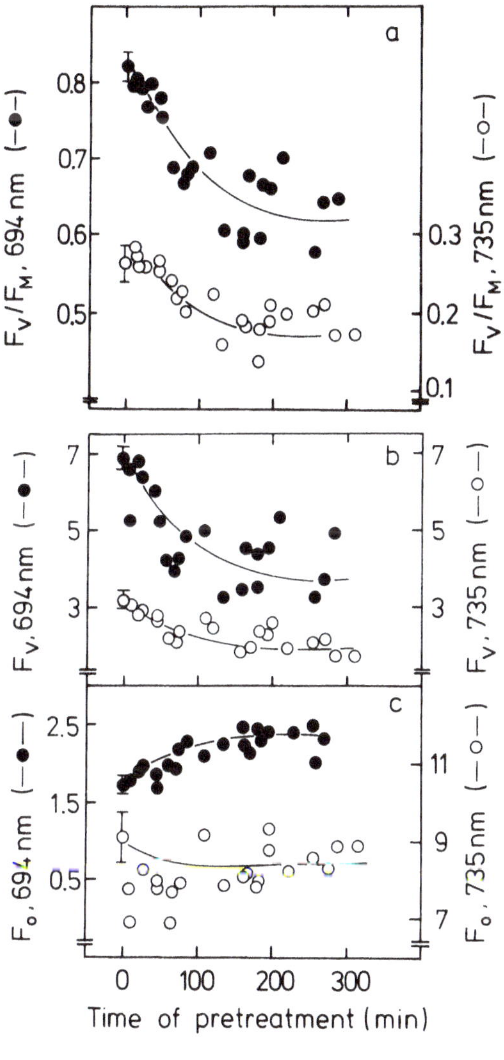

Fig. 1. Effects of photoinhibition treatment of unhardened spinach leaves (550 µmol photons m^{-2} s^{-1} at +4°C) on 77K fluorescence characteristics. Fluorescence (relative units) of PS II (●) was recorded at 694 nm and of PS I (○) at 735 nm. a) F_V/F_M ratio, b) F_V, c) F_o. Standard deviations of the controls are given, n = 8 to 9.

optimal quantum yield of O_2 evolution. Notably, larger changes in fluorescence parameters were always observed at 20°C as compared to 77K (Table 1a). In the usual technique, fluorescence is excited and measured from the upper leaf side. At 20°C a high difference was seen between F_V/F_M ratios measured from upper and lower sides of photoinhibited leaves, respectively (Table 1). There was considerably less difference between upper and lower side in corresponding signals obtained at 77K. In the mean F_V/F_M values of upper and lower sides, a relatively small but still significant difference between 20°C and 77K was seen. The data show that the more severely inhibited upper leaf layers make a higher contribution to the signal at 20°C that at 77K.

Figure 2 demonstrates that photoinhibition, expressed as a decline in F_V/F_M or as a decrease in quantum yield, appears as an approximate first order reaction. The calculated rate constant of the decline in F_V/F_M at 20°C (but not at 77K) was close to that of quantum yield. Figure 3 shows a linear correlation between decrease in quantum yield and F_V/F_M ratio, both at 77K and 20°C. However, the regression line of the 77K data exhibits a steeper slope, and the extrapolation deviates from the origin. As the upper leaf layers particularly influence the quantum yield, as well as the 20°C fluorescence signal, we view the F_V/F_M ratio at 20°C as the more suitable measure of the decline of the quantum yield in photoinhibition.

Table 1. Effects of irradiation (3h,550 µmol photons m^{-2} s^{-1} at +4°C) on fluorescence induction of unhardened spinach leaves excited and measured at 20°C and 77K from the upper and lower leaf sides, respectively. Standard deviations are given; n = 3 to 6.

Conditions		20°C	77K
a) Upper leaf side			
F_V	control	3.54 ± 0.11	5.74 ± 0.40
	irradiated	0.78 ± 0.10	3.68 ± 0.22
F_o	control	0.67 ± 0.07	1.29 ± 0.14
	irradiated	0.95 ± 0.08	1.69 ± 0.08
F_V/F_M	control	0.84 ± 0.01	0.83 ± 0.02
	irradiated	0.45 ± 0.04	0.69 ± 0.02
b) Lower leaf side			
F_V/F_M	control	0.82 ± 0.01	0.79 ± 0.01
	irradiated	0.80 ± 0.02	0.75 ± 0.02
Mean of upper and lower sides			
F_V/F_M	control	0.83	0.81
	irradiated	0.62	0.72

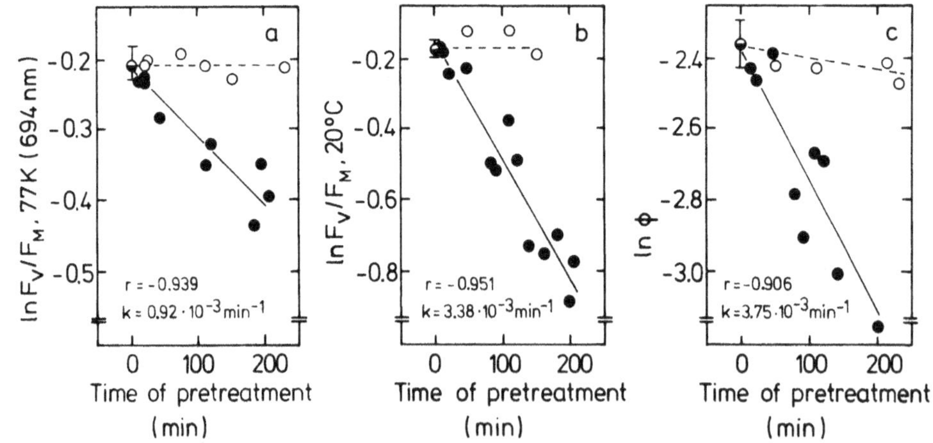

Fig. 2. Semilogarithmic plots of F_V/F_M ratios and quantum yield of unhardened spinach leaves versus time of pretreatment; (●), light treatment (550 µmol m^{-2} s^{-1}) at +4°C and (○), leaves kept in darkness at +4°C. a) F_V/F_M of PS II (77K), b) F_V/F_M (20°C), c) quantum yield (Φ). Standard deviations of the controls are indicated; n = 5 to 9. Correlation coefficients and rate constants obtained from the regression lines are given.

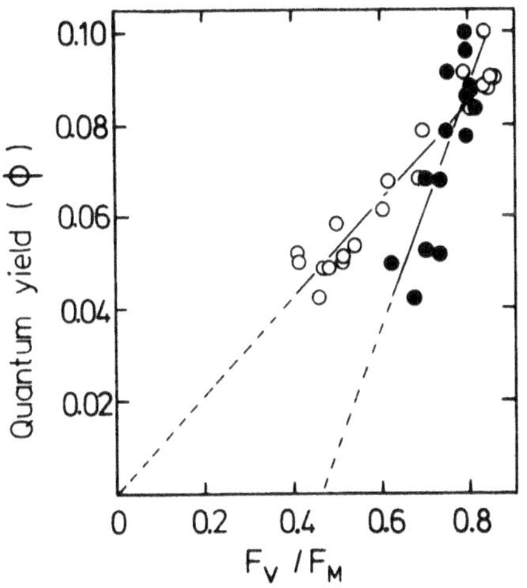

Fig. 3. Quantum yield of O_2 evolution plotted versus F_V/F_M ratio, recorded at 77K (●) and at 20°C (○) with unhardened spinach leaves.

Reversibility of photoinhibition in unhardened leaves

When photoinhibited leaves were kept in low light, recovery from photoinhibition was observed. This is depicted in Figure 4 for fluorescence parameters recorded at 20°C. The reversion of changes related to photoinhibition was complete after about 3 h at +18°C. In contrast, only incomplete recovery was seen at +4°C, even after 5 h. Figure 4a shows at +4°C a fast partial reversion of the F_V/F_M decrease. This resulted from a full reversion of changes in F_o (Fig.4c) and partial recovery of F_V (Fig.4d), so that F_M did not significantly increase. It appears that these recovery effects reveal two mechanism of photoinhibition; one is characterized by increase in F_o, decrease in F_V and fast recovery at +4°C, the other by decrease in F_V and recovery at higher temperatures only.

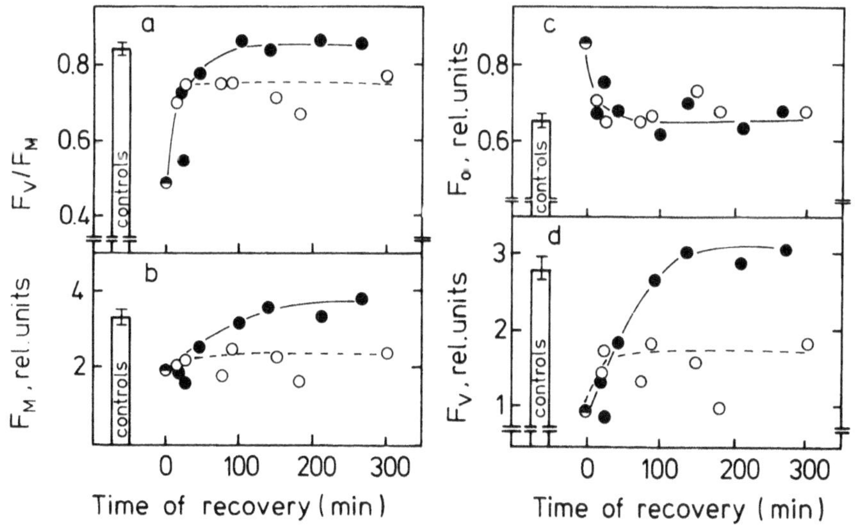

Fig. 4. Fluorescence characteristics at 20°C of unhardened spinach leaves during recovery treatment at +4°C (○) and at +18°C (●), PFD 2.5 to 5.0 μmol m^{-2} s^{-1}. The recovery treatment was started after exposing the leaves for 3 h to 550 μmol m^{-2} s^{-1}. a) F_V/F_M, b) F_M, c) F_o, d) F_V. Standard deviations of the controls are indicated; n = 9.

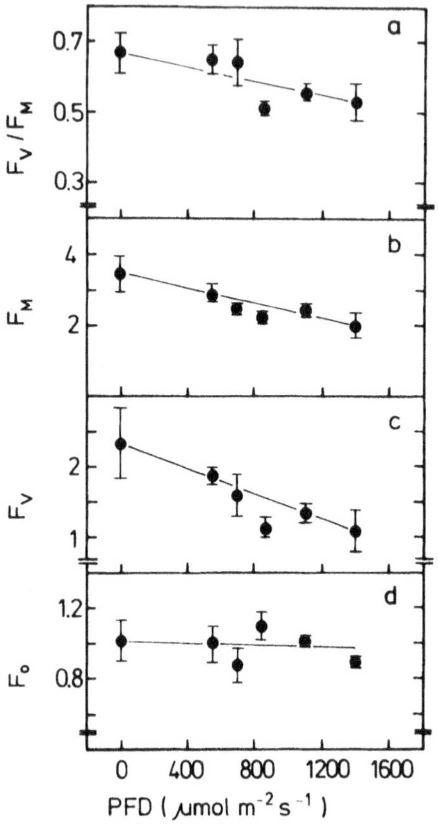

Fig. 5. Fluorescence parameters at 20°C (relative units) of cold acclimated (hardened) spinach leaves as a function of PFD during photoinhibition pretreatment (130 min at +4°). a) F_V/F_M, b) F_M, c) F_V, d) F_o. Standard deviations are given; n = 3 to 8.

Photoinhibition in cold-acclimated leaves

After long-term acclimation of spinach plants, resulting in increased frost tolerance (Klosson and Krause 1981), the susceptibility of the leaves to photoinhibition at chilling temperatures was considerably decreased. Figure 5 shows the effect of photoinhibition treatments on fluorescence (20°C) of such hardened leaves. The slightly lowered F_V/F_M of the controls supposedly indicates a limited light stress during the acclimation procedure. No significant further photoinhibition could be observed at a PFD of 550 µmol m^{-2} s^{-1}, which strongly inhibited unhardened leaves. Light above 800 µmol m^{-2} s^{-1} caused photoinhibition; the decrease in F_V/F_M and in F_M (Fig. 5a,b) resulted from lowered F_V. The pronounced increase in F_o, seen in unhardened leaves, was absent. We presently investigate,whether the decreased susceptibility to photoinhibition and absence of increase in F_o in hardened leaves is based on higher rates of 'repair' (cf. Greer et al. 1986). Preliminary experiments (not shown) indicate that hardened leaves are, indeed, capable of faster recovery after photoinhibition treatment.

DISCUSSION

Chlorophyll fluorescence measured at 20°C and 77K is applied here to characterize a reversible photoinhibition, which occurs in a chilling-resistant plant at temperatures close to +4°C. The effect of chilling may be based on decreased energy utilization in carbon metabolism (see Krause and Cornic 1987) and on lowered rates of continuous 'repair' processes (Greer et al. 1986). The decline in variable fluorescence at 77K can be observed in emission bands of both PS II and PS I (Fig. 1), indicating that the major effect of photoinhibition is an increase in the rate constant of thermal deactivation of photosynthetic pigments, in agreement with other investigations (Powles and Björkman 1982, Ögren and Öquist 1984, Bárenyi and Krause 1985, Demmig and Björkman 1987).

The decline in F_V/F_M, both at 20°C and at 77K, represents a first order reaction (Fig. 2) and is linearly related to the decrease in quantum yield (Fig. 3). This confirms recent reports on similar correlations (Demmig and Björkman 1987, Leverenz and Öquist 1987). Our data show a rate constant of photoinhibition of quantum yield close to that of the F_V/F_M decline at 20°C, but not at 77K. The plot of quantum yield versus F_V/F_M extrapolates to the origin only for 20°C fluorescence (Fig. 3). Obviously, at room temperature, fluorescence emission from the upper, more strongly inhibited layers of chloroplasts is pronounced (Table 1). In contrast, F_V/F_M at 77K represents effects closer to the average of all chloroplasts. Thus we regard the F_V/F_M ratio at 20°C as the more sensitive indicator. Measurements of the induction signal at 20°C may serve as a routine method to detect photoinhibition.

Unhardened leaves exhibited an increase in F_0 on photoinhibition treatments at chilling temperature. Similar increases in F_0 related to photoinhibition have been interpreted as an indication of permanent damage (Demmig and Björkman 1987). In contrast, our data show a fast reversion of this effect in dim light both at +4°C and +8°C (Fig. 4). The incomplete recovery of F_V at 4°C indicates involvement of different mechanisms, but the full reversion seen at +18°C excludes permanent destructions.

Cold acclimation of the plants decreased their susceptibility to photoinhibition. Moreover, it appeared to alter the mechanism of photoinhibition, as seen by constancy of F_0 (Fig. 5). The reduced sensitivity may partly be due to increased rates of recovery, taking place already during pretreatment (Greer et al. 1986). In addition, long-term acclimation may lead to increased activities of protective reaction systems, e.g. of scavengers for active oxygen species (S. Schöner and G.H. Krause, unpublished).

In conclusion, we interprete the reversible photoinhibition as a controlled mechanism that protects the photosynthetic apparatus by means of thermal energy dissipation. It is effective within short time periods after changes in light conditions, however, includes a (reversible) decrease in rates of photosynthesis. Reversible photoinhibition has to be viewed in context with other protective mechanisms. An even faster response to excess energization of thylakoid membranes is the energy-dependent (ΔpH-related) fluorescence quenching. This has been postulated to represent a regulated process that balances thermal de-excitation with photosynthetic energy conversion according to energy utilization in carbon metabolism (Krause and Behrend 1986, Krause and Laasch 1987, Weis and Berry 1987). When in high PFD this effect becomes light-saturated, reversible photoinhibition may be induced, opening a further pathway for energy dissipation. During long-term acclimation, probably additional protective systems become effective, so that photoinhibition occurs only under more extreme conditions.

ACKNOWLEDGEMENTS

S. Somersalo acknowledges a grant of the Academy of Finland. The study was supported by the Deutsche Forschungsgemeinschaft.

REFERENCES

Barényi B and Krause GH 1985. Inhibition of photosynthetic reactions by light. A study with isolated spinach chloroplasts. Planta 163: 218-226.

Björkman O and Demmig B 1987. Photon yield of O_2 evolution and chlorophyll fluorescence characteristics at 77K among vascular plants of diverse origins. Planta 170: 489-504.

Butler WL 1977. Chlorophyll fluorescence: a probe for electron transport and energy transfer. In: Trebst A and Avron M. Encyclopedia of plant physiology, new series. Springer, Berlin.

Cleland RE, Melis A and Neale PJ 1986. Mechanism of photoinhibition: Photochemical reaction center inactivation in system II of chloroplasts. Photosynth. Res. 9: 79-88.

Demmig B and Björkman O 1986. Comparison of the effect of excessive light on chlorophyll fluorescence (77K) and photon yield of O_2 evolution of leaves of higher plants. Planta 171: 171-184.

Greer DH, Berry JA and Björkman O 1986. Photoinhibition of photosynthesis in intact bean leaves: a role of light and temperature, and requirement for chloroplast-protein synthesis during recovery. Planta 168: 253-260.

Klosson RJ and Krause GH 1981. Freezing injury in cold acclimated and unhardened spinach leaves. I. Photosynthetic reactions of thylakoids isolated from frost-damaged leaves. Planta: 339-346.

Krause GH and Behrend U 1986. ΔpH-dependent chlorophyll fluorescence quenching indicating a mechanism of protection against photoinhibition of chloroplasts. FEBS Lett. 200: 298-302.

Krause GH and Cornic C 1987. CO_2 and O_2 interactions in photoinhibition. In: Kyle DJ, Osmond CB and Arntzen CJ. Photoinhibition. Topics in photosynthesis, vol. 9, Elsevier, Amsterdam

Krause GH and Laasch H 1987. Energy-dependent chlorophyll fluorescence quenching in chloroplasts correlated with quantum yield of photosynthesis. Z. Naturforsch. 42c: 581-584.

Kyle DJ, Osmond CB and Arntzen CJ (eds.) 1987. Photoinhibition. Topics in photosynthesis, vol. 9. Elsevier, Amsterdam.

Leverenz JW and Öquist G 1987. Quantum yields of photosynthesis at temperatures between -2°C and 35°C in a cold-tolerant C_3 plant (_Pinus sylvestris_) during the course of one year. Plant Cell Environ. 10: 287-295.

Ögren E and Öquist G 1984. Photoinhibition of photosynthesis in _Lemna gibba_ as induced by the interaction between light and temperature. III. Chlorophyll fluorescence at 77K. Physiol. Plant. 62: 193-200.

Powles SB and Björkman 1982. Photoinhibition of photosynthesis: effect on chlorophyll fluorescence at 77K in intact leaves and in chloroplast membranes of _Nerium oleander_. Planta 165: 97-107.

Weis E and Berry JA 1987. Quantum efficiency of photosystem II in relation to 'energy'-dependent quenching of chlorophyll fluorescence. Biochim. Biophys. Acta 894: 198-207.

SCREENING FOR PHOTOINHIBITION OF PHOTOSYNTHESIS IN THE FIELD USING A PORTABLE FLUORIMETER

Erling Ögren and Gunnar Öquist, Dept of Plant Physiology, Univ. of Umeå, S-901 87 Umeå, Sweden.

Abstract

A portable, microprocessor operated instrument for measuring chlorophyll fluorescence induction kinetics is described. It is particularly useful for measuring the cardinal points of the rapid phase of the induction curve. Using this instrument we show that photosynthesis of field-grown willow leaves may be partially photoinhibited (max. 20 %) under clear and changeable summer days in Northern Sweden. The inhibition, residing in photosystem II as deduced from the fluorescence analyses, was light dependent, showed recovery kinetics on the hours scale and was not associated with unfavourable water and temperature conditions. Hence, photoinhibition may occur under field conditions that are not associated with any other stress than sunlight.

Introduction

Laboratory studies have proved that chlorophyll fluorescence is a powerful analytical tool in studies on stress effects in photosynthesis (Ögren and Öquist 1984a; Baker and Horton 1988; Strand and Öquist 1988). Because the fluorescence technique is sensitive, rapid and non-destructive, it has a great potential in diagnosing plants growing in the field for stress effects before visible damages occur. For this purpose we constructed a portable fluorimeter designed to allow rapid measurements of chlorophyll fluorescence kinetics both in the field and the laboratory.

This instrument was applied in a study on photoinhibition of photosynthesis, i.e. inactivation of photosynthesis caused by excessive light absorption. The majority of studies on photoinhibition has been conducted under laboratory conditions. Only a few studies have considered the ecological significance of photoinhibition. Those available have demonstrated that plants growing in the field become more susceptible to photoinhibition under conditions unfavorable for growth and photosynthesis. Bright sunlight can cause photoinhibition under stresses of drought (Björkman and Powles 1984), chilling temperatures (Farage and Long 1987), freezing temperatures (Öquist and Ögren 1985; Strand and Lundmark 1987), high temperatures (Ludlow and Björkman 1984; Adams et al. 1987), limiting nitrogen supply (Osmond 1983) and long term shade conditions (Powles and Critchley 1980). These studies leave one question unanswered: Does photoinhibition occur in the field under conditions that are not associated with any other stress

H. K. Lichtenthaler (ed.) Applications of Chlorophyll Fluorescence, 165–172.

than full sunlight? This question was addressed in the present study using a willow stand growing under the moderate conditions of light and temperature that prevail during the summers in Northern Sweden. A full account of the results has been presented in (Ögren 1988).

Material and methods

The plant material studied was a stand of a single clone of *Salix* sp. growing in a field in Umeå, Northern Sweden. If not otherwise stated the leaves investigated were oriented towards the sector SE-SW; one leaf per shoot was selected among the third- to fifth-uppermost fully expanded ones.

Chlorophyll fluorescence at the prevailing temperatures was measured with a fibre-optic based, microprocessor operated, portable instrument. A detailed technical description of the instrument is published elsewhere (Öquist and Wass 1988). A block diagram of the instrument is given in Fig. 1. The actinic light (330 to 660 nm with a peak at 500 nm), provided by a halogen lamp and a Schott glass filter Bg 39, is guided to the leaf via one branch of a bifurcated optical fibre. The PPFD of the actinic light can be increased in four steps (max. 600 µmol m^{-2} s^{-1}). The emitted chlorophyll fluorescence is carried back to the photodetector, a photodiode, via the other branch. A 690-nm interference filter prevents the actinic light to reach the photodetector. A fluorescence induction measurement is initiated by opening an electromagnetic shutter positioned in front of the actinic lamp. The instrument is operated by 12 V obtained from rechargable batteries or from 110/220 V AC mains. The program is written in TURBOPASCAL in a read only memory, EPROM. A built in CMOS microprocessor unit (CPU), with two attached read and write memories, stores the fluorescence induction curve and calculates the fluorescence parameters F_0, F_M ($F_P = F_M$ when the actinic light is strong enough to close all photosystem II reaction centers at the P-peak), $F_V = F_M - F_0$ and $t_{1/2}$ (the half rise time from F_0 to F_M at the P-peak). All functions including the microprocessor is commanded from a keyboard. A recorded induction curve can be plotted onto a chart recorder or an oscilloscope. Since the instrument operates with continuous excitation and fluorescence detection, the leaf sample has to be well dark adapted prior to measurement. Simple, light weight clip-on cuvettes are used for this purpose. The cuvettes are equipped with a gate allowing the optical fibre to be inserted without daylight reaching the sample. The fluorimeter is manufactured and sold under the name of Plant Stress Meter by BioMonitor S.C.I. AB, Box 7034, S-900 07 Umeå, Sweden. In the present study the protocol of measurements were as follows: the leaf was excited by a 5-s period of a PPFD of 100 µmol m^{-2} s^{-1} after 45-55 min of dark-adaptation.

Chlorophyll fluorescence at 77K was measured using an apparatus described elsewhere (Ögren and Öquist 1984a). The sample, dark adapted for 45 min prior to freezing, was excited at 433 nm with a PPFD of 1-1.5 µmol m^{-2} s^{-1}. Fluorescence was detected at 689 nm.

The quantum yield of O_2 evolution was measured at CO_2 saturation and at 25° C using a leaf disc electrode (LD2/CB1, Hansatech

Ltd.) as described in (Ögren 1988). Limiting PPFDs were provided with the Hansatech lamp and a set of neutral density filters. The leaf absorptance was measured with a light integrating sphere.

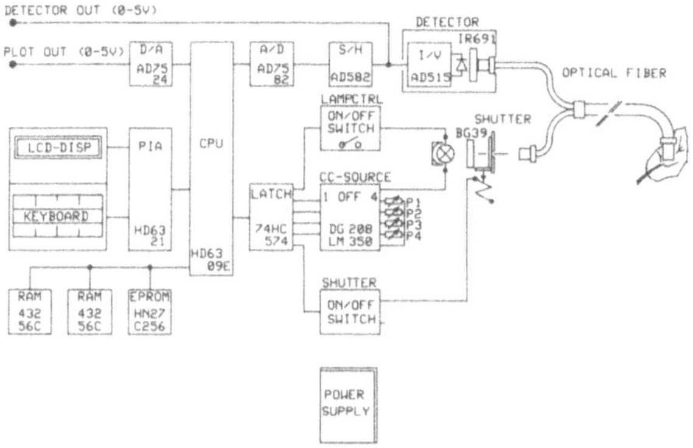

FIGURE 1. A block diagram of the fluoresence instrument. Explanations are given in **Materials and methods**. From Öquist and Wass 1988.

Results

Diurnal changes in fluorescence characteristics of attached leaves of a willow stand were followed on several days in July 1987. The cardinal points of the rapid phase of the fluorescence induction curve were recorded in the morning and in the afternoon. The ratio of F_V/F_P, where F_P is the transient peak fluorescence and $F_V=F_P-F_0$, at 9:00 h (open) and at 14:30 h (closed symbols) is plotted versus the average PPFD in between these hours (Fig. 2). For leaves on peripheral shoots (triangles) the F_V/F_P ratio dropped roughly 15% on clear days with an average PPFD around 1500 μmol m^{-2} s^{-1}. On cloudy days (PPFD<500 μmol m^{-2} s^{-1}) the F_V/F_P ratio stayed constant, whereas an intermediate decrease was observed on changeable days. Hence, the effect seems to be light dependent. The other environmental factors could not explain the diurnal variation in F_V/F_P. The leaf temperature in the afternoon was within the optimal range of 20-27° C (data not shown). Neither did water stress occur since the leaf water potential did not drop below -0.85 MPa. When shade acclimated leaves were exposed to full sunlight a larger decrease in F_V/F_P ratio was observed (Fig. 2, squares). Thus, there seems to be a differential sensitivity of shade and sun acclimated leaves.

The depression in F_V/F_P ratio was fully reversible. As shown in Fig. 3 the leaves recovered from most of the effect on the same day.

To examine the mechanisms underlying the depression in F_V/F_P ratio, a comparison was made with fluorescence induction at 77K. At this low temperature, the ratio of F_V/F_M measured at around 690 nm, where F_M is the steady-state maximum level and $F_V=F_M-F_0$, is known to be a measure of the efficiency of the primary photochemical reaction of photosystem II (Butler 1978). Fig. 4

shows that there was a nearly 1:1-relationship between the room temperature ratio of F_V/F_P and the 77K ratio of F_V/F_M when willow leaves had experienced various light conditions at their natural positions. It can therefore be inferred that a damage within the primary photochemistry of photosystem II underlies the diurnal change in F_V/F_P ratio.

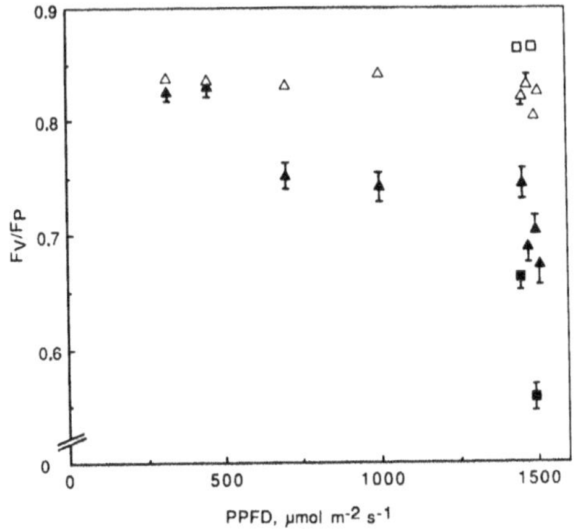

FIGURE 2. The F_V/F_P ratio of attached willow leaves at 9:00 (*open*) and at 14:30 h (*closed symbols*) versus the average, external PPFD of the intervening time period. Leaves were naturally exposed on peripheral shoots (*triangles*), or developed in the shade but exposed to direct sunlight by trimming of the stand (*squares*).Measurements were done in July 1987. Mean values and SE>0.007 are given for 25 replicate leaves (triangle) and 10 replicate measurements on a single leaf (square). From Ögren 1988.

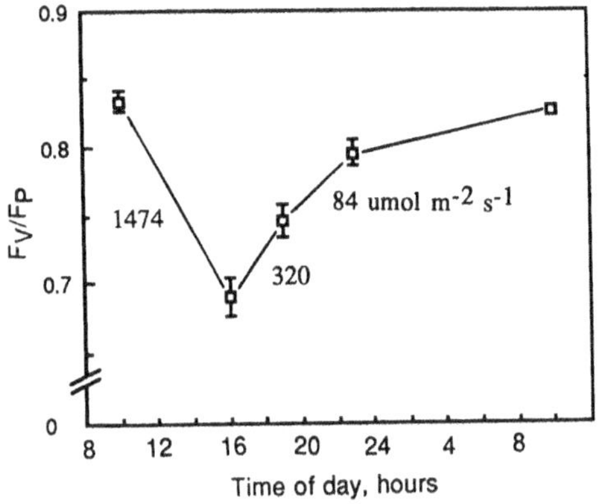

FIGURE 3. The time course of recovery of the F_V/F_P ratio in naturally exposed leaves on peripheral shoots. The mean values of the external PPFD of the time periods in between measurements are indicated. Mean values and SE>0.006 of 25 replicate leaves are given. From Ögren 1988.

To determine to what extent the damage on photosystem II-photochemistry is expressed at the leaf level, the maximal quantum yield of O_2 evolution and the ratio of F_V/F_P were compared (Fig. 5). This was done for leaves that had experienced natural (closed) or artificial (open symbols) high-light conditions. Irrespective of the leaves being shade (triangles) or sun acclimated (squares), or the type of treatment, the damage on photosystem II-photochemistry as probed by the F_V/F_P ratio correlated fairly well with the damage on the quantum yield of photosynthesis. Also, the relationship is roughly 1:1 suggesting that the extent of inhibition of the F_V/F_P ratio can be taken as an estimate of the extent of inhibition of the quantum yield of photosynthesis.

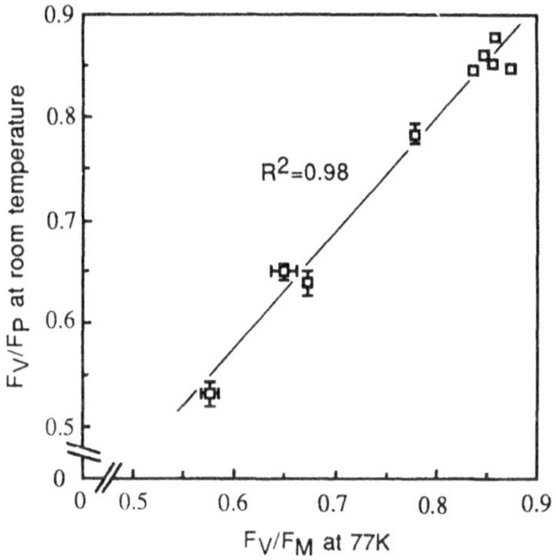

FIGURE 4. Relationship between the F_V/F_P ratio at room temperature and the F_V/F_M ratio at 77K in willow leaves that at their natural positions had experienced various PPFDs. The mean values and SEs > 0.006 of 3 (77K) or 13 (room temperature) replicate measurements of one leaf are given. From Ögren 1988.

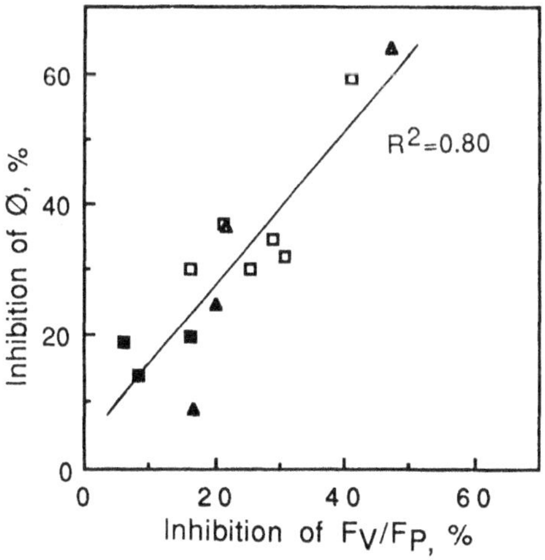

FIGURE 5. Relationship between inhibition of the quantum yield of O_2 evolution and inhibition of the F_V/F_P ratio in sun (*squares*) and shade (*triangles*) acclimated willow leaves after exposure to natural (*closed*) and artificial (*open symbols*) light conditions. From Ögren 1988.

In the studies mentioned so far the leaves examined were those most liable to receive bright light (positioned at the shoot tops, facing SE-SW). In another study all fully expanded and active leaves of a peripheral shoot were examined. On three successive days of bright light the F_V/F_P ratio was measured at 9:00 h (open) and at 15:30 h (closed symbols) and plotted versus the leaf position numbered downwards (Fig. 6). Though the majority of the leaves showed a depression in the F_V/F_P ratio there was a great leaf to leaf variability. This can presumably be ascribed to the variability in the intercepted light among the leaves (data not shown).

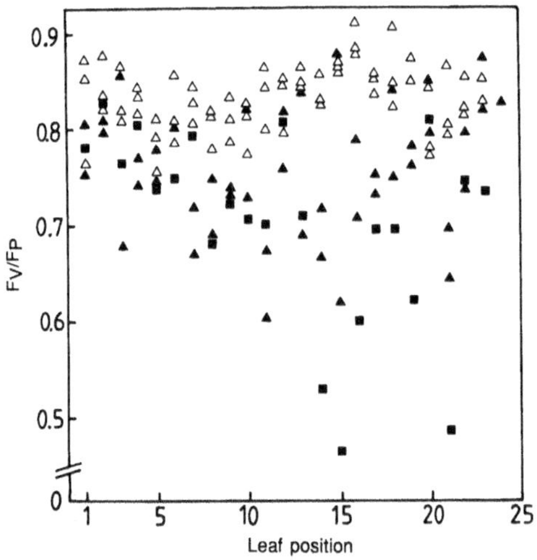

FIGURE 6. The F_V/F_P ratio at 9:00 (*open*) and at 15:30 h (*closed symbols*) on three successive days as a function of the position of the leaf (numbered from the uppermost fully expanded) within a single peripheral shoot. The average value of external PPFD of the intervening time period was about 1000 and 1300 µmol m^{-2} s^{-1} for, respectively, the first two days (triangles) and the 3rd day (squares). From Ögren 1988.

Discussion

Willow leaves with exposed positions in a field-grown stand often showed a decrease in the F_V/F_P ratio in the afternoon. This effect is suggested to reflect photoinhibition of photosynthesis on the basis of the following arguments:

(i) The depression in the room temperature ratio of F_V/F_P was paralleled by an equal depression in the 77K ratio of F_V/F_M (Fig. 4), and a similar depression in the maximal quantum yield of photosynthesis.(Fig. 5). These effects are consistent with the general picture of photoinhibition, namely a damage residing within the primary photochemistry of photosystem II that is fully expressed by the quantum yield of photosynthesis (Ögren and Öquist 1984a,b; Ögren et al.1984; Demmig and Björkman 1987).

(ii) The decline in the F_V/F_P ratio was most pronounced on clear days, less so on changeable days and absent on cloudy days (Fig. 2).

(iii) The recovery of the F_V/F_P ratio following a decline took several hours to go to completion. This rate agrees well with other reports on photoinhibition (Ögren et al. 1984). It rules out various regulatory mechanisms such as state transitions. These mechanisms require much less time for relaxation, probably the 45-min period of darkness that preceded the measurement is enough.

(iv) Leaves developed in the shade underwent a larger decline in the F_V/F_P ratio on exposure to full sunlight as compared to leaves developed on peripheral shoots (Fig. 2). This finding is consistent with the general feature of photoinhibition, that shade leaves are more sensitive than sun leaves (Powles and Chritchley 1980).

The evidences given above leave little doubt that photoinhibition frequently affected the willow leaves. However, the importance of this photoinhibition for the carbon gain and the growth of the plants can only very tentatively be discussed. From the climate data and the observation that photoinhibition was observed on clear and changeable days, one can estimate that on roughly one-third of the days of the growing season (1987), up to 20% photoinhibition occurred in leaves of peripheral willow shoots. Since photoinhibition is more detrimental for carbon uptake at limiting than at saturating light levels one might guess that loss of potential carbon gain occurred during late hours of dim light when the photosynthetic quantum yield was not fully reactivated. However, one can not exclude the possibility that these data underestimate the importance of photoinhibition because some recovery might have occurred during the 45-min period of dark adaptation that preceded the measurement. If the instantaneous photoinhibition was larger than what is presented here, the carbon uptake even at saturating light levels might have been affected.

Preliminary results using other species suggest that photoinhibition under non-stress conditions is not restricted to willow. For instance in plants of *Trientalis europaea* growing in an open spot, a diurnal change in the F_V/F_P ratio was observed on clear but not on cloudy days (data not shown).

Acknowledgement: The instrument was developed with a grant from the National Swedish Board for Technical Development and the Swedish Natural Science Research Council supported the experimental work.

References

Adams III, W. W., Smith, S. D., Osmond, C. B. (1987) Photoinhibition of the CAM succulent *Opuntia basilaris* growing in Death Valley: evidence from 77K fluorescence and quantum yield. Oecologia 71, 221-228

Baker, N. R., Horton, P. (1988) Chlorophyll fluorescence quenching during photoinhibition. In: Photoinhibition. Topics in Photosynthesis, D. J. Kyle, C. B. Osmond and C. J. Arntzen, eds., Vol. 9. pp.145-168, Elsevier Sci. Publ. Amsterdam

Björkman, O., Powles, S.P. (1984) Inhibition of photosynthetic reactions under water stress: interaction with light level. Planta 161, 490-504
Butler, W.L. (1978) Energy distribution in the photochemical apparatus of photosynthesis. Ann. Rev. Plant Physiol. 29, 345-378
Demmig, B., Björkman, O. (1987) Comparison of the effect of excessive light on chlorophyll fluorescence (77K) and photon yield of O_2 evolution in leaves of higher plants. Planta 171, 171-184
Farage, P.K., Long, S.P. (1987) Damage to maize photosynthesis in the field during periods when chilling is combined with high photon fluxes. In: Progress in photosynthesis research. Proceedings of the VIIth international congress on photosynthesis, vol. 4, pp. 139-142, Biggins, J., ed., Martinus Nijhoff Publishers, Dordrecht
Ludlow, M.M., Björkman, O. (1984) Paraheliotropic leaf movement in Siratro as a protective mechanism against drought-induced damage to primary photosynthetic reactions: damage by excessive light and heat. Planta 161, 505-518
Ögren, E. (1988) Photoinhibition of photosynthesis in willow leaves under field conditions. Planta (in press)
Ögren, E., Öquist, G. (1984) Photoinhibition of photosynthesis in *Lemna gibba* as induced by the interaction between light and temperature. II. Photosynthetic electron transport. Physiol. Plant. 62, 187-192
Ögren, E., Öquist, G. (1984) Photoinhibition of photosynthesis in *Lemna gibba* as induced by the interaction between light and temperature. III. Chlorophyll fluorescence at 77 K. Physiol. Plant. 62, 193-200
Ögren, E., Öquist, G., Hällgren, J.-E. (1984) Photoinhibition of photosynthesis in *Lemna gibba* as induced by the interaction between light and temperature. I. Photosynthesis in vivo. Physiol. Plant. 62, 181-186
Öquist, G., Ögren, E. (1985) Effects of winter stress on photosynthetic electron transport and energy distribution between the two photosystems of pine as assayed by chlorophyll fluorescence kinetics. Photosynth. Res. 7, 19-30
Öquist G, Wass R (1988) A portable, microprocessor operated instrument for measuring chlorophyll fluorescence kinetics in stress physiology. Physiol. Plant. (in press)
Osmond, C.B. (1983) Interactions between irradiance, nitrogen nutrition, and water stress in the sun-shade responses of *Solanum dulcamara*. Oecologia 57, 316-321
Powles, S.B., Critchley, C. (1980) Effect of light intensity during growth on photoinhibition of intact attached bean leaflets. Plant Physiol. 65, 1181-1187
Strand, M., Lundmark, T. (1987) Effects of low night temperature and light on chlorophyll fluorescence of field-grown seedlings of Scots pine (*Pinus sylvestris* L.) Tree Physiol. 3, 211-224
Strand, M., Öquist, G. (1988) Effects of frost hardening, dehardening and freezing stress on *in vivo* chlorophyll fluorescence of Scots pine seedlings (*Pinus sylvestris* L.). Plant Cell Environ. (in press)

WINTER STRESS AND CHLOROPHYLL FLUORESCENCE IN NORWAY SPRUCE (Picea abies, L., Karst.)

H.R. BOLHÀR-NORDENKAMPF and E. LECHNER
Dept. for Horticultural Plant Physiology, Institute for Plant Physiology. University of Vienna, Althanstr. 14, A-1091 Vienna.

KEY WORDS: fluorescence induction, winter stress, photoinhibition, Picea abies.

ABSTRACT

Winter stress induces loss of photosynthetic capacity in at least two ways: 1.) At freezing temperatures below -4° C the photosynthetic membranes were disintegrated. 2.) Chilling temperatures enhance the sensitivity for photoinhibition. Fast kinetics of chlorophyll fluorescence induction (FV/Fm = Fm-FO/Fm; Fm = P) provide adequate means to assess the photosynthetic capacity of a plant.

Samples from the experimental garden (Vienna) and from three sites in the alpine region (Zillertal) were measured with a microscopic fluorimeter and additionally with a portable fluorimeter. Freezing temperatures induce a marked loss of FV/Fm independently of incident light-intensities. FV/Fm also declines if chilling temperatures are combined with high light-intensities. Chilling temperatures in interaction with low light had no photoinhibitory effect. Therefore, shaded and snow covered branches are protected against light stress.

Chlorophyll fluorescence is a sensitive tool to detect stress phenomena as well as a measuring device which is manageable even under unfavourable climatic conditions.

INTRODUCTION

In the alpine region evergreen conifers are exposed to unfavourable environmental conditions for a rather long period in the year, particularly during winter (Öquist, 1983). Under normal conditions, the plants are subjected to the following stresses: light, temperature, drought, limitations in mineral nutrition, pests and diseases. Plants affected by one of these stresses react more sensitive to the occurrence of an additional stress, e.g. temperature and light (Bongi and Long, 1987). It seems likely that plants exposed to natural stresses are predisposed to be injured by anthropogenic stresses, i.e. air pollution. During the ontogenesis the extent of effects triggered off by stresses is subject to further changes. Therefore, a knowledge of tree physiology under natural stress conditions is the most important prerequisite if metabolic reactions of trees are to indicate damage and particularly the cause of the damage.

Light and temperature exert a direct influence upon the primary processes of photosynthesis. Drought and biotic stresses can affect the photosynthetic capacity indirectly. Thus, the determination of the photo-

H. K. Lichtenthaler (ed.) Applications of Chlorophyll Fluorescence, 173–180.

synthetic capacity provides an adequate means to assess the plant's response to these stress factors.

The photosynthetic activity can either be determined in the traditional way, i. e. by means of gas exchange measurements, or by chlorophyll fluorescence measurements (Ireland et al., 1988). During the fast induction phase of photosynthesis, variable fluorescence (Fm-FO; Fm = P), which apart from heat is one of the mechanisms of deactivation of the excitation energy is proportional to the photochemical capacity of Photosystem II (Butler and Kitajima, 1975). The slow kinetics of chlorophyll fluorescence is increasingly influenced by the activity of the Calvin Cycle, with heat becoming the more prominent mechanism of energy dissipation (Buschmann, 1986). This goes to show that measurements of chlorophyll fluorescence provide information on disturbances of photosynthesis (Renger and Schreiber, 1988).

For a field-study it is absolutely necessary to use an easily manageable measuring equipment. During winter, many measuring devices cannot be used because temperatures are too low. Furthermore problems may occur because of unfavourable conditions, such as snow, rain, high athmospheric humidity or wind. Compared to measurements of CO_2-gas exchange, chlorophyll fluorescence measurements is, because of its lower technological expenditure a suitable means of measurement, which additionally permits to obtain a larger amount of data for a better statistical reliability.

The aim of the current investigations is to determine the photosynthetic capacity of spruce needles with respect to their light .and temperature pre-history over the winter period.

MATERIAL AND METHODS

The first investigation was started in order to examine the basic winter stress phenomena. This study was made in the Experimental Garden of the Institute of Plant Physiology in Vienna. The findings obtained served as a basis for a follow-up study in the Zillertal, in the Tyrolian Alps.

Samples of branches were collected weekly from spruce trees in the Experimental Garden from January to June 1987. The air pollution expressed in maximal day mean values was: 0.15 mg SO_2/m^{-3}, 20 ppb NO_x and 60 ppb O_3.

After a two-hour pretreatment with low light of about 10 $\mu mol.m^{-2}.s^{-1}$ at 15° C, the fluorescence induction kinetics were measured with a microscopic fluorimeter (Bolhàr-Nordenkampf and Lechner, 1988). The fast kinetics of the induction curve was used to characterize the PS-II activity using the value FV/Fm (Fm-FO/Fm). For detection of a disturbance in the Calvin Cycle slow kinetics were recorded and a characteristic value F_{dr} calculated: Fm-FT/Fm (cp. Lichtenthaler et al., 1986). All confidence limits and the t-tests were calculated with a probability level of 10 %.

A "level of disturbance" for the detection of an impairment of the photosynthetic apparatus was determined by treatment with the photosynthetic inhibitor DCMU (3-(3,4-dichlorphenyl)-1,1-dimethylurea).

For the "Zillertal"-study starting in January 1988 branches were collected every month from three sites situated at different heights. On the respective days the fast kinetics of the fluorescence induction was determined after two hours of thermic adaptation at 20° C and a thirty-minute predarkening phase. At least 30 measurements were done using the portable fluorescence device (Plant Stress Meter, PSM, Biomonitor Umea, Sweden).

Additionally, samples of branches were cold-stored and transported to the laboratory in Vienna. After a standardized pretreatment in a climatic chamber, the fluorescence induction of single needles was measured with a microscopic fluorimeter.

Site 1 is located just above the bottom of the valley (720 m above sea level), the major air pollutants there being SO_2 and NO_x; the limits established for these pollutants are exceeded only on very few days of the year. However, when foehn weather occurs, fairly large amounts of ozone enter the valley for several hours, exceeding maximal day mean values of 0.060 $mg.m^{-3}$.

Site 2 is located at 1,000 metres, just above the main inversion layer. The ozone concentrations are constantly high reaching maximal day mean values of 0.120 $mg.m^{-3}$. The mean temperature values of one month are some 2° C lower and the irradiation intensity 5 % higher than at site 1.

Site 3 is situated at a height of 1,480 m in a "Reinluftgebiet", i. e. an area of relatively low primary air-pollution. The maximal day mean values for SO_2 are 0.050 $mg.m^{-3}$, those for NO_x are 0.025 $mg.m^{-3}$. The pollutant concentrations are to a high degree influenced by the direction of the wind. Temporarily, the ozone stress sinks in correspondence with the NO-concentration, but still shows maximal day mean values of 0.160 $mg.m^{-3}$. The monthly temperature average is about 4-5° C lower, the irradiation intensity some 20 % higher than at site 1.

RESULTS and DISCUSSION

Although the process of frost hardening induced by low temperatures and short photoperiod (Sakai and Larcher, 1987) starts already in September, the fluorescence values of spruce needles from the Experimental Garden were virtually normal until December. In the course of January, both FV/Fm and F_{dr} showed a pronounced decline. This decline was caused by a substantial decrease of the maximal value Fm. To a minor extent, FO and the steady state level T were declining during winter (Fig. 1).

In February, a relaxation of the fluorescence response with a tendency to reach normal values was observable. At the beginning of March a decline of FV/Fm and F_{dr} was once more detectable. These changes in the fluorescence values during winter occur mainly as a consequence of the photoinhibition caused by the interaction of light and temperature (Lundmark and Strand, 1987). Chilling morning temperatures cause a pronounced photoinhibition only in combination with moderate high light (+ 3.4°C; 9.0 $mol.m^{-2}$, from dawn to noon) (see March in fig. 1). At relatively low light intensities (+ 3.3° C; 5.0 $mol.m^{-2}$, from dawn to noon) in combination with chilling temperatures, the impairment of the photosynthetic apparatus is much lower and the values FV/Fm and F_{dr} are therefore virtually normal (see December and February in fig. 1). If the temperatures fall clearly below zero, freezing temperatures may cause an alteration in the membrane system of the thylakoids (Senser and Beck, 1979). Irrespective of the light intensity, an impairment of the entire electron transport chain (Martin et al., 1978) may cause the decline of FV/Fm and F_{dr}. With temperatures rising clearly above +5° C, even high light intensities no longer cause a permanent photoinhibition. Temperatures and light intensities of the days immediately preceding the measurements were observed to be predominantly responsible for the short term fluctuations of the fluorescence values.

During the period of flushing, however, a disturbance of F_{dr} was observable indicating a disturbance situated in the Calvin Cycle.

Fig. 1

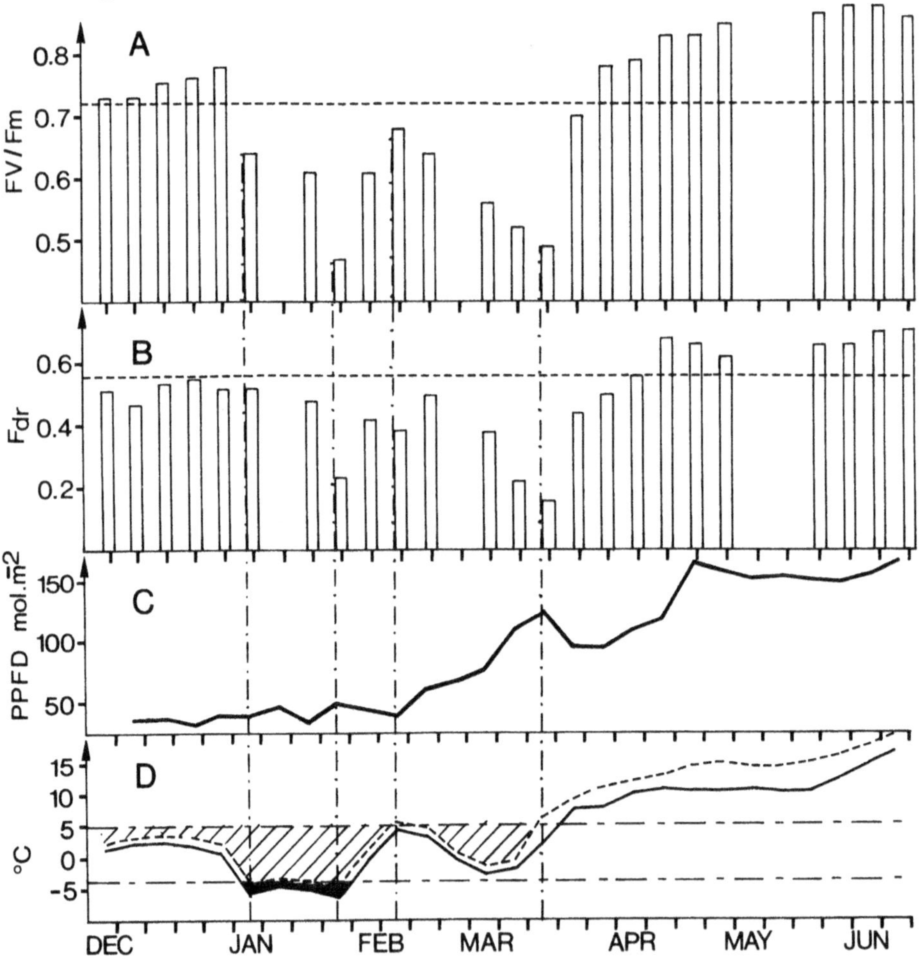

Fig. 1: Chlorophyll fluorescence of spruce needles during winter. Abscissa: months of the year. A) Mean values of FV/Fm, B) Mean values of F_{dr}. --------- = "level of disturbance". C) Photosynthetic photon flux density (400-700 nm): sum of irradiation in mol.m^{-2} and week. D) Week mean values of the temperatures calculated from day mean values in °C. --------- = day temperature, ———— = night temperature.

Fig. 2

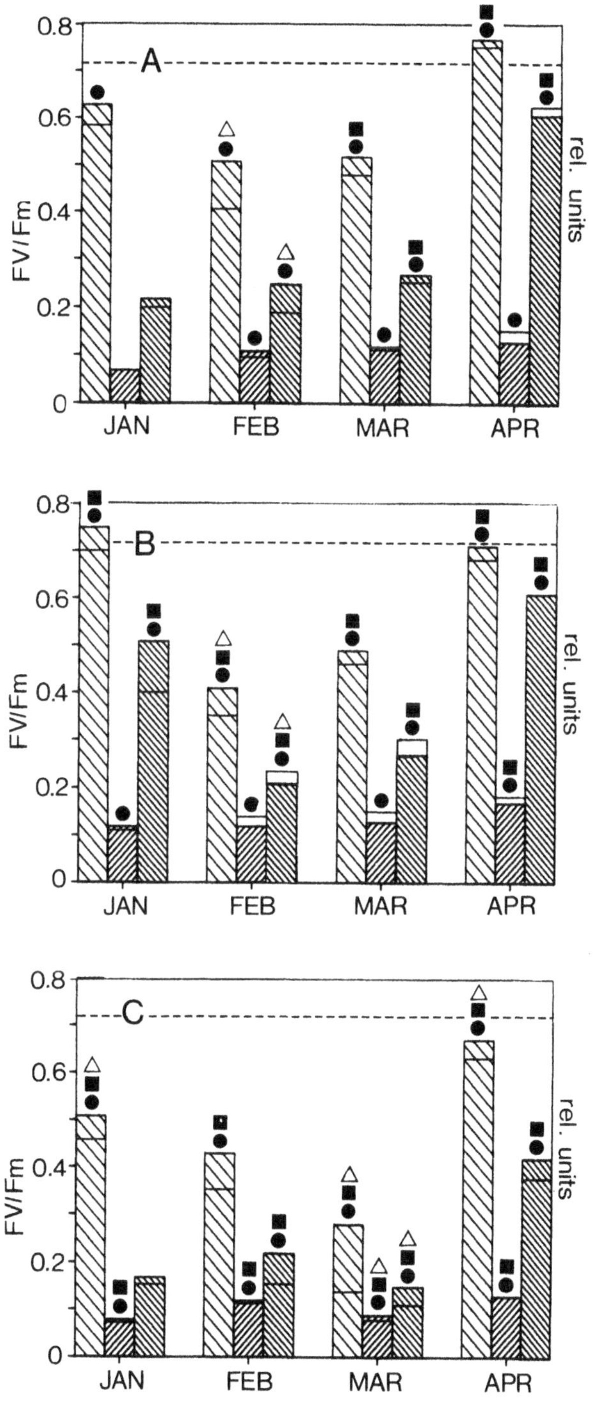

Fig. 2:

Characteristics of fast fluorescence kinetics during winter.

Mean values:

▧ = FV/Fm
▨ = FO
▧ = Fm

Line at 0.72 indicates "level of disturbance" for FV/Fm.

Bars indicate values from the needle set of 1986. Line across bars indicates values from needle set of 1987.

Significant differences ($p < 0.1$):

between successive months:

● = needle set 1987,

■ = needle set 1986.

between needle set 1987 and 1986:

△

A) site 1,
B) site 2,
C) site 3.

From the results already available, it is obvious that the extent of winter stress even in the alpine region fluctuates due to the climatic pre-history. As a result of a relatively mild winter in January no pronounced decrease of FV/Fm was observable. However, during February at all sites a significant decline of variable fluorescence occurred, which corresponded with a marked decrease of temperature.

The sharpest decline of FV/Fm clearly occurs at site 3. Fig. 2A shows a remarkable decline of FV/Fm of the samples from January to March. Even in April the values were evidently below "the level of disturbance". The decline of FV/Fm is mainly caused by a decrease of the maximal level Fm. Obviously, the needle set of 1987 exhibits significantly lower values of FV/Fm. It seems likely, that young needles are more sensitive to winter stress, possibly because frost hardening is less developed than in the needle set of 1986.

The same phenomenon is observable at site 2, however, in that case the more expressed decrease of FV/Fm of the needle set 1987 is mainly caused by an increase of FO giving evidence to the occurrence of photoinhibition. Contrary to site 3, in March FV/Fm exhibits a distinct increase compared to February. A comparable course of FV/Fm during winter is detectable at site 1, yet the decline of FV/Fm is mainly caused by decrease of P. Even there, variable fluorescence in March is higher than in February.

Fig. 3:

FV/Fm from shaded, sun exposed and snow covered branches: February, site 3. For explanations of symbols see Fig. 2. Significant difference ($p < 0.1$):

1986 1987

△ ▲ = shaded/sun exposed

■ = sun exposed/snow covered

○ ● = shaded/snow covered

* = sets 86/87.

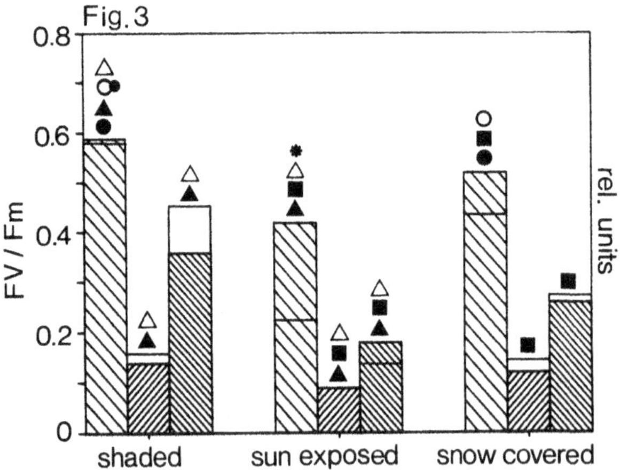

Fig. 3 shows a comparison between shaded, sun exposed and snow covered branches from site 3 in February. FV/Fm from sun exposed needle set 1987 exhibits an unambiguously low variable fluorescence provoked by a marked decrease of Fm. Contrary to this snow covered branches display fairly higher FV/Fm. Shaded branches show the highest FV/Fm values due to high Fm levels. FV/Fm of the needle set from 1986 have the same characteristics, however, the differences between shaded and sun exposed branches are less prominent. The decrease of FV/Fm is a result of light stress enhanced by chilling (Krause and Somersalo, 1988). This is a sign of photoinhibition. The consequence of this is that photoinhibition in needles protected from sunlight by a snow cover or by permanent shading is lower.

These results show clearly that the measurement of fluorescence induction is a sensitive tool to detect diverse stress phenomena. In addition, the fluorescence measuring equipment can be used even under harmful climatic conditions.

ACKNOWLEDGEMENTS

We appreciate the permission by the Forstliche Bundesanstalt to use the site in Zillertal. Thanks are due to D.I. Glattes for providing us with climatic data and to B. Dresel for excellent help in translating the manuscript. The project was supported by the Ministries for Science and Agriculture.

REFERENCES

Bolhàr-Nordenkampf, HR and Lechner, E, 1988: Temperature and light dependent modifications of chlorophyll fluorescence kinetics in spruce needles during winter. Photosyn Res, in press.

Bongi, G and Long, SP, 1987: Light-dependent damage of photosynthesis in olive leaves during chilling and high light treatment. Plant, Cell Environ 10: 241-249.

Buschmann, C, 1986: Fluoreszenz- und Wärmeabstrahlung bei Pflanzen. Naturwissenschaften 73: 691-699.

Butler, WL and Kitajima, M, 1975: A tripartite model for chloroplast fluorescence. In: Avron, M. Proceedings of the 3rd international congress on photosynthesis. Elsevier, Amsterdam.

Ireland, CR, Telfer, A, Covello, PS, Baker, NR and Barber, J., 1988: Studies on the limitations to photosynthesis in leaves of the atrazine resistant mutant of Senecio vulgaris L. Planta 173: 459-467.

Krause, GH and Somersalo, S, 1988: Fluorescence as a tool in photosynthesis research: application in studies of photoinhibition, cold acclimation and freezing stress. Proc R Soc Lond, in press.

Lichtenthaler, HK, Buschmann, C, Rinderle, U and Schmuck, G, 1986: Application of chlorophyll fluorescence in ecophysiology. Radiat Environ Biophys 25: 297-308.

Martin, B, Martensson, O and Öquist, G, 1978: Effects of frost hardening and dehardening on photosynthetic electron transport and fluorescence properties in isolated chloroplasts of Pinus silvestris. Physiol Plant 43: 297-305.

Öquist, 1983: Effects of low temperature on photosynthesis. Plant Cell Environ 6: 281-300.

Renger, G and Schreiber, U, 1986: Practical applications of fluorimetric methods to algae and higher plant research. In: Govindjee, Amesz, A and Fork, DC. Light emission by plants and bacteria. Academic Press, Orlando.

Sakai, A and Larcher, W, 1987: Mechanisms of frost survival. In: Mechanisms of frost survival. Responses and adaptation to freezing stress. Springer Verlag, Berlin.

Senser, M and Beck, E, 1979: Kälteresistenz der Fichte. II Einfluß von Photoperiode und Temperatur auf die Struktur und photochemischen Reaktionen von Chloroplasten. Ber Dt Bot Ges, Bd. 92: 243-259.

Strand, M. and Lundmark, T, 1987: Effects of low night temperature and light on chlorophyll fluorescence of field-grown seedlings of Scots pine. Tree Physiol 3: 211-224.

FLUORESCENCE ROUTINE TESTS TO DESCRIBE THE BEHAVIOUR OF A PLANT IN ITS ENVIRONMENT

R.J.Strasser, B.Schwarz, P.Eggenberg

Bioenergetics Laboratory, University of Geneva
Station de Botanique Expérimentale, CH 1254 Lullier, Geneva

INTRODUCTION:

Easy methods have to be developed for the description of the behaviour of a plant in its environment. Fluorescence techniques are very useful because the measurements can be done directly on the sample (leaf, needle, algae etc.) without any further extraction. The sample can be analysed in two different ways.
1) In vivo method: Different variations of fluorescence induction curves are measured directly on the living sample. For this procedure it is important that each individual test run is short, so that in a reasonable time many samples can be measured.
2) Sampling method: Small pieces of samples are fixed (e.g. one needle or a disc of 1cm diameter) between two transparent tapes, wraped in aluminium foil and collected in a dewar containing liquid nitrogen. Many samples can therefore be collected and stored for a long time. The samples can be brought to the laboratory where detailed fluorescence analysis is possible. The information which can be obtained by these techniques are summarized elsewhere (Strasser, Graf 1988).
In this paper we would like to propose some fluorescence tests based on the measurements of twin channel fluorescence induction kinetics (Lombard, Strasser 1984; Strasser, Schwarz, Bucher 1987).
Empirical indexes are defined for sample activity and sample architecture criteria. Many experiences have shown so far that the response of plants to stress is reflected in these empirical indexes.

MATERIALS AND METHODS:

Any equipment which is able to measure an usual fluorescence induction kinetics can be used to calculate the VITALITY index. Two fluorescence signals have to be measured simultaneously at two different wavelengths (e.g. $710 < F' < 750$ and $680 < F'' < 700$ nm) to obtain the ADAPTATION index which is very sensitive to early detection of stress effects (Strasser, Schwarz, Bucher 1987). Usually fluorescence peak and terminal fluorescence values are necessary for the calculation of fluorescence ratios. Fluorimeters with a modulated light source (AC-method) bring many very convenient advantages to the method with a DC light source only. HANSATECH LIMITED, King's Lynn, Norfolk, Great Britain builts a low cost twin-channel modulated fluorimeter which can be highly recommended for routine measurements, even in the field. The data presented here are done with such an instrument. Sample activity and architecture criteria lead to the following definitions.

H. K. Lichtenthaler (ed.) Applications of Chlorophyll Fluorescence, 181–187.

Sample activity criteria: (Only fluorescence signals are needed)

P : Productivity index: It is derived of delayed fluorescence signals (not shown here) with a set up as published earlier (Strasser 1974).

V : Vitality index : It is the ratio of a high fluorescence signal $F_{(max)}$ or $F_{(peak)}$ and a reference fluorescence signal. This reference fluorescence signal can be the terminal fluorescence F_T or the initial fluorescence F_0. The subscript ' or '' refers to the two different fluorescence wavelengths. If F_T and $F_{(peak)}$ are used, the vitality index corresponds to the R_{fd}- values found in the literature.

$$V_i = \frac{F_i}{F_{(ref)}}$$

A : Adaptation index : It correlates the vitality indexes of two different wavelengths.

$$A_i = 1 - \frac{F'_i \ / \ F'_{ref}}{F''_i \ / \ F''_{ref}}$$

Fig.1 shows two experimental routine test runs, one with (top) one without (bottom) detection of the maximum fluorescence intensity. The vitality and adaptation indexes of these curves are shown in Fig.2.

Sample architecture criteria: (fluorescence and reflection measurements are needed).
The reflection signal can be obtained with the fluorescence equipment just by using an additional light source (e.g. LED) which emits at the wavelength where the fluorescence is measured.
The measured (apparent) fluorescence signal of a green sample undergoes reabsorption events. This is demonstrated in Fig.3 where pea leaves were frozen to 77 K and "diluted" with different amounts of icepowder. Emission and reflection spectra of these samples are shown. A simplified equation can be derived which correlates the real fluorescence ratio $(F_i/F_{ref})_{real}$ with the apparent observed fluorescence ratio $(F_i/F_{ref})_{apparent}$ and the reflection ratio R_i/R_{ref}. The theoretically derived equation shown in Fig.4 appears to be practically confirmed with experimental data. This leads to the following definitions shown in Fig.4 top.

S : Structure index
D : Density index of pigments (mainly due to the chlorophylls)
M : Morphology index

The structure index contains all the reabsorption events: it is a combination of the pigment content and the morphology of the sample.

The density index of pigments classifies the sample on a scale between zero and unity.

The morphology index indicates the scatter properties and therefore the morphology of the sample. M can be calculated with additional reflection data (not shown) or with the data of the leaf and the diluted leaf powder.

Fig.1 : two test runs of TWIN CHANNEL fluorescence induction kinetics.
Top : with initial and terminal saturating light pulse
Bottom : without detection of F_M

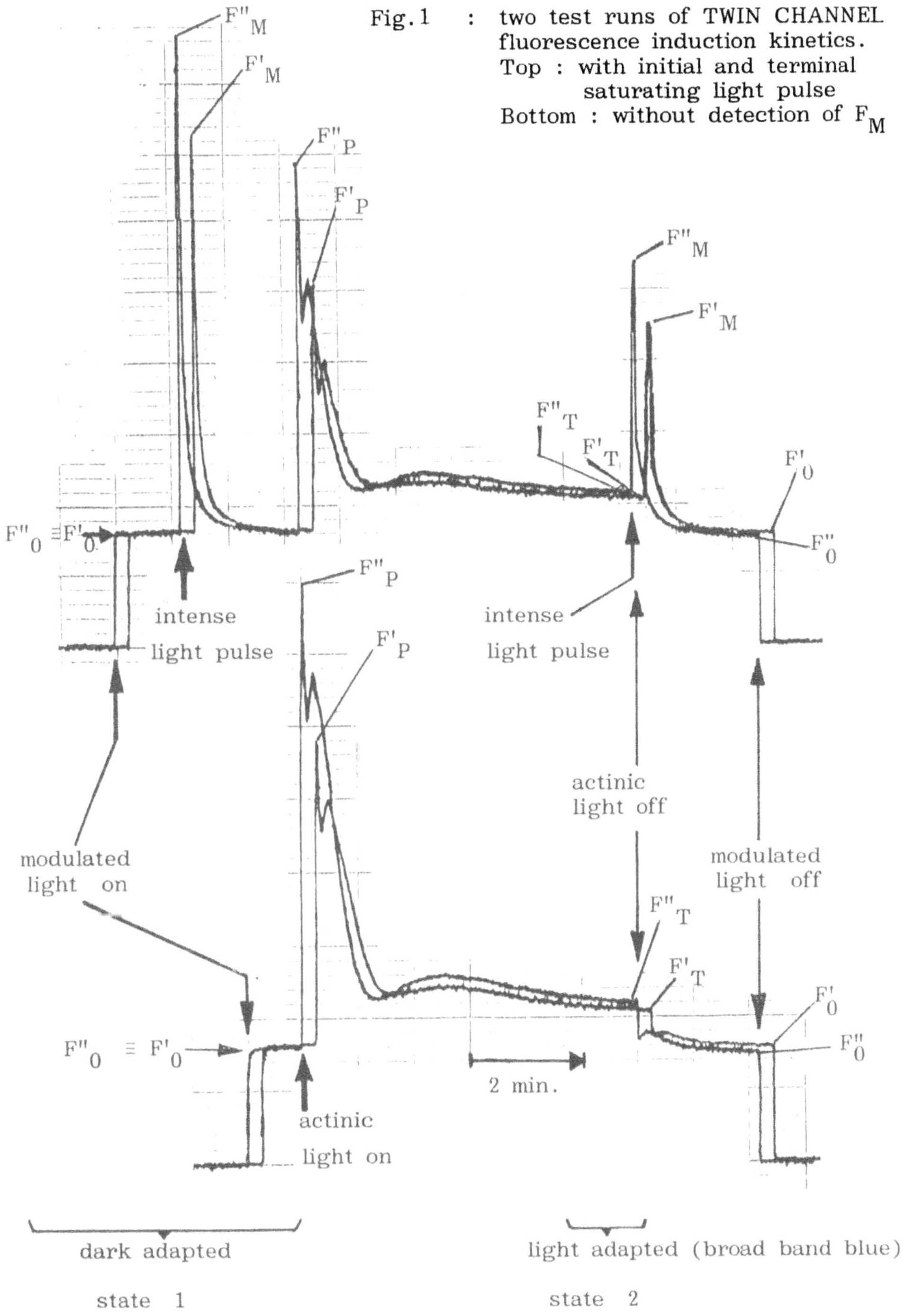

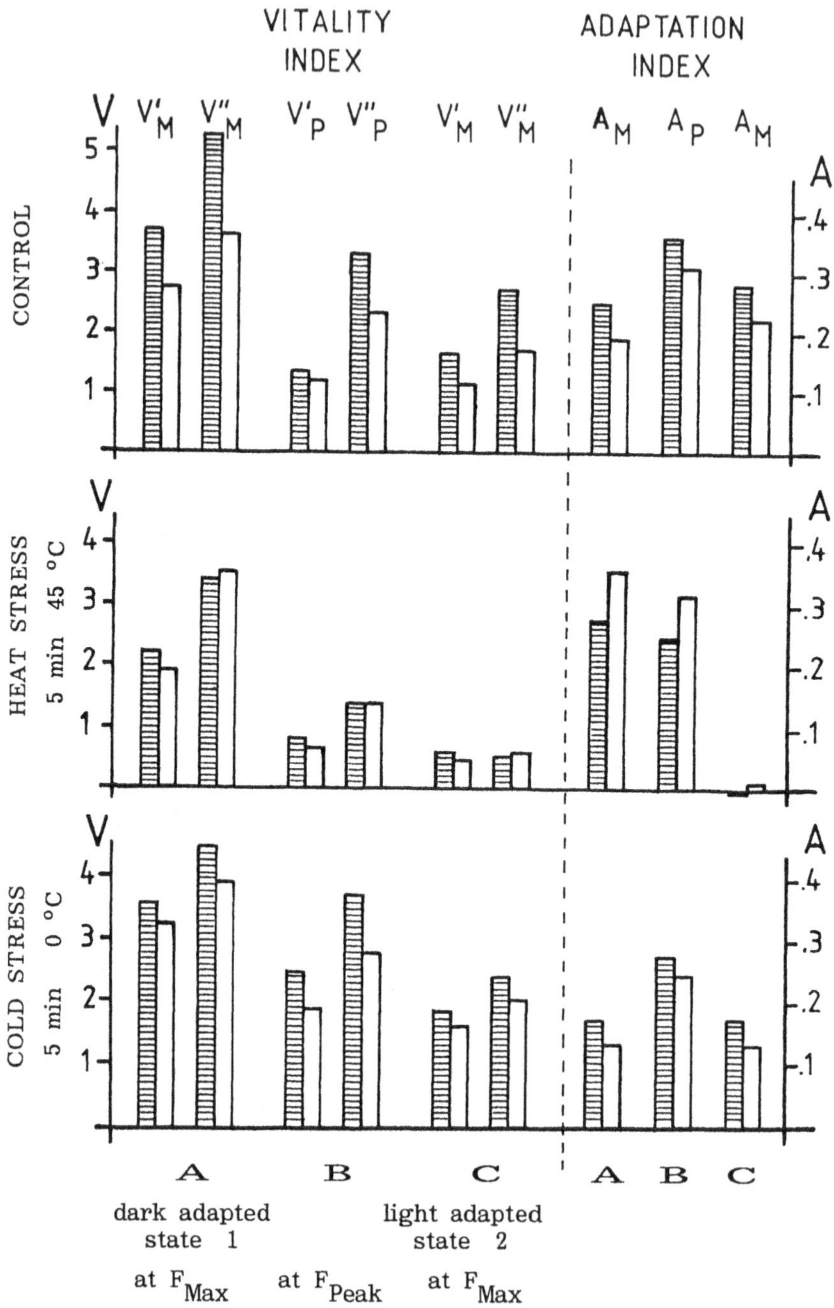

Fig. 2 : Comparison of VITALITY and ADAPTATON INDEX using (▤) F-terminal (F_T) or (□) F-initial (F_0) as reference. The subscript ' refers to the fluorescence signal at 730 nm, the subscript " to the fluorescence signal at 685 nm.

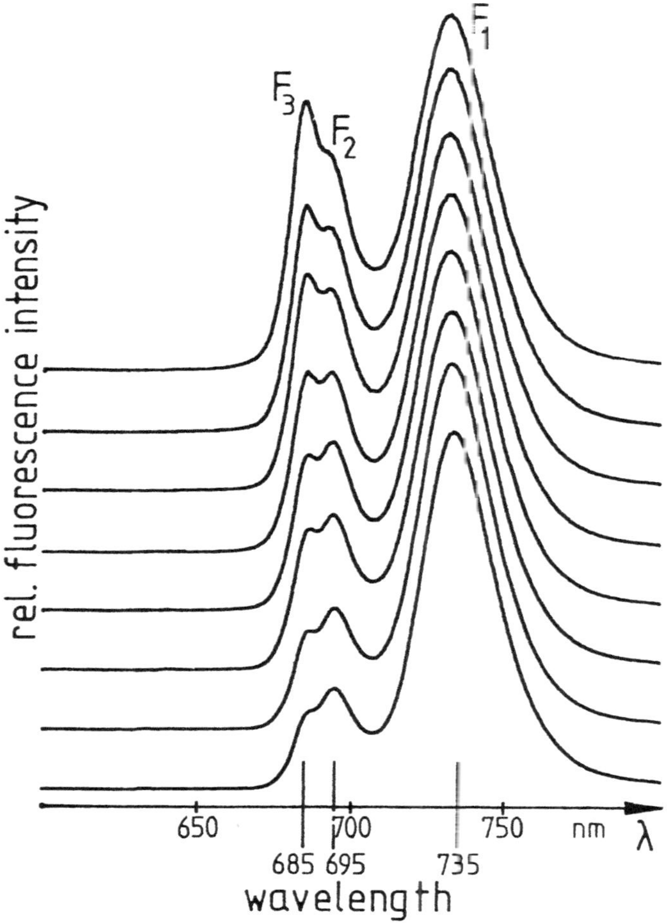

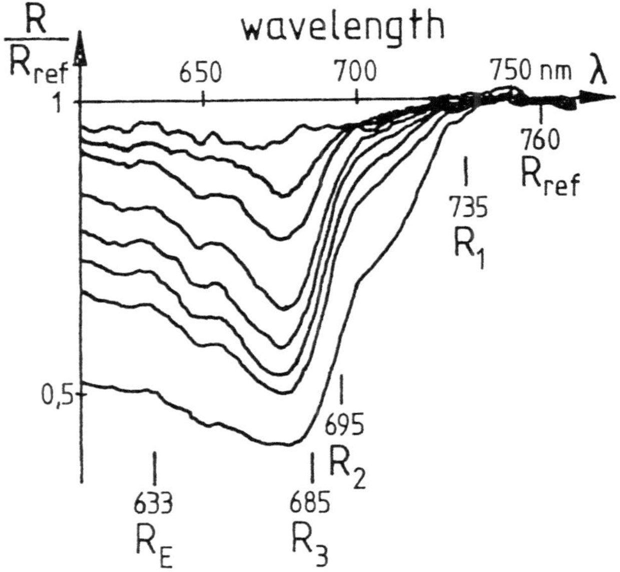

Fig.3 : Fluorescence (left) and reflection (top) spectra of pea leaves (lowest curve) and leave powder "diluted" with different amounts of ice. All manipulations are done at 77 K.
Excitation light: He-Ne-Laser 633 nm

Theoretical Equation:

$$\frac{(F_i/F_{ref})_{apparent}}{(F_i/F_{ref})_{real}} = \left(\frac{R_i}{R_{ref}}\right)^M$$

Definitions:

S = Structure index

D = Pigment density index

M = Morphology index

Empirical Equation:

$$\log\left(\frac{F_2}{F_1}\right)_{apparent} = \log\left(\frac{F_2}{F_1}\right)_{real} - M\cdot\log\left(\frac{R_{ref}}{R_2}\right)$$

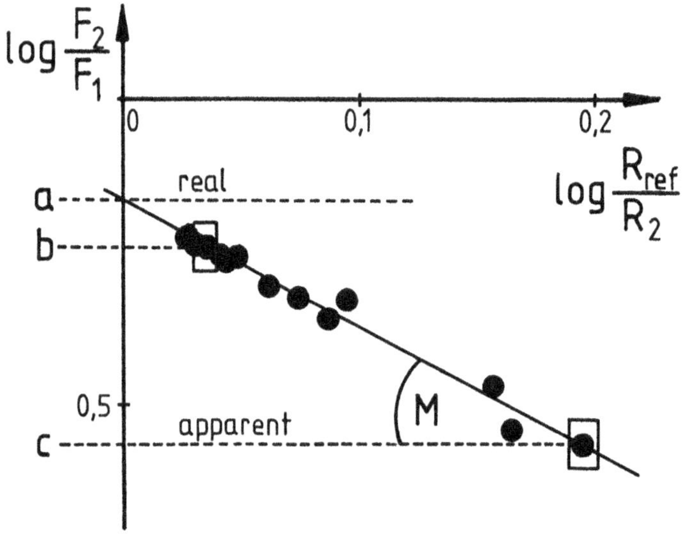

Fig. 4 : Correlation of fluorescence and reflection at wavelength i and reference wavelength ref.
Pea leaves were frozen to 77 K, diluted with ice and the fluorescence and reflection spectra were measured at 77 K.

$F_i = F_2 = F_{695}$ $\quad$ $R_i = R_2 = R_{695}$

$F_{ref} = F_1 = F_{735}$ $\quad$ $R_{ref} = R_{760}$

M is the ratio of the light paths of fluorescence and incident light at the same wavelength ($d_{F_i}/d_{I_i} = M$).

a) corresponds to the real fluorescence ratio
b) corresponds to the fluorescence ratio of a highly diluted sample
c) corresponds to the apparent fluorescence ratio of a leaf

The rectangles at b and c indicate the standard deviation of ten different samples. Rectangle at c for ten different pea leaves. Rectangle at b for diluted leaf powder of the same ten leaves.

DISCUSSION:

Already with very simple fluorescence signals it is possible to define indexes which characterize a sample.
Reflection and fluorescence kinetics can be measured with the same instrument. The calculation of the mentioned indexes are possible with the very simple equipment consisting of a light source, two filters and two photocells. However a low cost TWIN-CHANNEL modulated fluorimeter brings a lot of very handy advantages.
If many samples have to be measured the experimental time for organisational reasons becomes very important. However the detection of the terminal fluorescence signal needs normally more than 5 minutes.
Fig.2 shows that the trend of all signals is the same when the initial fluorescence is taken as a reference signal instead of the terminal fluorescence. Therefore less than one minute of experimental time is needed per sample to get all the data which allow the calculation of the different indexes. Fig.2 also shows that the adaptation index brings other informations than the vitality index. It is less affected by daily and seasonal variations and it reacts (as experimented so far) earlier upon stress than the vitality index or the R_{fd}-values.

REFERENCES:

All references can be found in the review:

- Lichtenthaler, H.K. , Rinderle, U.
  THE ROLE OF CHLOROPHYLL FLUORESCENCE IN THE DETECTION OF STRESS CONDITIONS IN PLANTS
  Critical Reviews in Analytical Chemistry 1988, Vol.19, p 29-85
  CRC Press, Inc., Boca Raton, Florida

Summary of fluorescence tests:

- Lombard, F. , Strasser, R.J.
  EVIDENCE FOR SPILL OVER CHANGES DURING STATE 1 TO STATE 2 TRANSITION IN GREEN LEAVES
  In: Progress in Photosynthesis Research, Vol.III, 271-274, (1984)
  C.Sybesma (Ed.), Martinus Nijhoff, The Hague, Netherlands
- Strasser, R.J. , Schwarz, B. , Bucher, J.B.
  SIMULTANE MESSUNG DER CHLOROPHYLL FLUORESZENZ-KINETIK BEI VERSCHIEDENEN WELLENLÄNGEN ALS RASCHES VERFAHREN ZUR FRÜHDIAGNOSE VON IMMISSIONSBELASTUNGEN AN WALDBÄUMEN: OZONEINWIRKUNGEN AUF BUCHEN UND PAPPELN
  In: Eur.J.Forest Pathol. 17/3, 149-157, (1987)
- Strasser, R.J. , Graf, J.A.
  FLUORESCENCE MEASUREMENTS – A MEANS OF EARLY DIAGNOSIS OF AIR POLLUTION DAMAGES TO HIGHER PLANTS
  In: Int. Symp. Air Pollut. Plant Metab., Blauk, L.,Barell, N., Schulte-Hostede, S. and Wellburn, A. (Eds.), De Gruyter, Berlin, 1988 in press
- Weis, E.
  CHLOROPHYLL FLUORESCENCE AT 77 K IN INTACT LEAVES: CHARACTERIZATION OF A TECHNIQUE TO ELIMINATE ARTIFACTS RELATED TO SELF-ABSORPTION
  In: Photosynthesis Research 6, 73-86, (1985),
  Martinus Nijhoff / Dr.W.Junk Publishers, Dordrecht, Netherlands

THE CHLOROPHYLL FLUORESCENCE RATIO F690/F735 AS A POSSIBLE STRESS INDICATOR

Ursula RINDERLE and Hartmut K. LICHTENTHALER

Botanisches Institut II (Plant Physiology and Plant Biochemistry),
University of Karlsruhe, Kaiserstr. 12, D-7500 Karlsruhe 1, F.R.G.

KEY WORDS: Chlorophyll fluorescence emission spectra, fluorescence ratio F690/F735, laser-induced fluorescence, Rfd-values, remote sensing of vegetation, stress detection in plants, vitality index.

ABSTRACT

The chlorophyll fluorescence emission spectra, as excited by blue and red light including laser light, were determined depending on the chlorophyll content and the photosynthetic activity of leaves and needles. The fluorescence-emission spectra exhibit two maxima or a shoulder in the 690 nm and the 735 nm region. The corresponding fluorescence ratio F690/F735 for green leaves is higher (values of 0.8 to 1.1) when excited with blue light (range 400 to 500 nm) than excited with light in the yellow to red wavelength region (525 to 633 nm), which only yields values for F690/F735 of 0.5 to 0.7. The values for the ratio F690/F735 are drastically increased with decreasing chlorophyll content of leaves and to a lower degree also by a decline of photosynthetic activity (e.g. herbicide treatment, needles of damaged forest trees) no matter whether the fluorescence is excited by red or blue light. Since stress induces a lower chlorophyll content as well as lower rates of photosynthesis, the ratio F690/F735 can be taken as indicator of stress to plants. The application of the ratio F690/F735 in detecting stress to terrestrial vegetation via remote sensing of the chlorophyll fluorescence is discussed.

INTRODUCTION

The light-induced chlorophyll fluorescence of intact green leaves and needles provides general information on the functional integrity of the photosynthetic apparatus and on the degree of photosynthetic quantum conversion of the absorbed light. Under normal photosynthetic conditions, when no stress is applied to the plant, the absorbed light is primarily used for photosynthesis, whereas only a small proportion is transformed into heat or chlorophyll fluorescence. Under environmental or anthropogenic stress conditions the intensity of the chlorophyll fluorescence is increased (Lichtenthaler and Pfister, 1978; Krause and Weis, 1984), which also changes the shape of chlorophyll-fluorescence spectra taken at room temperature (Lichtenthaler, 1986 and 1987a; Lichtenthaler et al., 1986; Lichtenthaler and Rinderle, 1988). This can be seen in increased values of the fluorescence ratio F690/F735. Stress and damage to the plant and the photosynthetic apparatus will not only reduce the rate of photosynthesis but in most cases also result in a lower chlorophyll content of leaves being brought about by either a degradation or a lower accumulation of chlorophylls. Here we describe how the values of the fluorescence ratio F690/F735 are modified depending on the chlorophyll content and photosynthetic activity, and demon-

H. K. Lichtenthaler (ed.) Applications of Chlorophyll Fluorescence, 189–196.

strate that the ratio F690/F735 is a suitable indicator of stress to terrestrial plants. This has great bearings for the remote sensing and stress detection of vegetation via chlorophyll fluorescence from active laser-equipped airborne systems.

MATERIALS AND METHODS

The chlorophyll-fluorescence emission spectra were recorded at the steady state of fluorescence (5 min after onset of illumination) applying a Shimadzu MPS 5000 spectrometer. Excitation and detection of the chlorophyll fluorescence was performed on the same leaf side (reflection measuring mode), or the fluorescence was excited from the lower leaf side and sensed from the upper leaf side (transmission measuring mode). The fluorescence was comparatively excited by light of the blue region (470±30 nm) and of the red region (620±20 nm). In some cases a He/Ne laser (Spectra Physics, 5mW, 632.8 nm) and a He/Cd laser (Laser 2000 GmbH, model 456XL, 10 mW, 442 nm) were applied. The light intensity of the two lasers, though different in energy, was about the same in the density of light quanta (ca. 500 $\mu E \cdot m^{-2} \cdot s^{-1}$).
The ratio of the two fluorescence maxima F690/F735 was determined at the actual maxima measured near 690 and 735 nm, irrespective of whether the maximum was shifted a few nm towards longer or shorter wavelengths. Only in cases of a shoulder was the fluorescence intensity taken from the exact wavelength position at 690 or 735 nm.

The variable fluorescence ratio (Rfd-values) as vitality index was determined from the chlorophyll-fluorescence induction kinetics of predarkened leaves (20 min) near 690 and 730 nm using a portable, self-constructed field fluorometer (Lichtenthaler and Rinderle, 1988) with a He/Ne laser as excitation light. From the ratios Rfd 690 and Rfd 730 the stress-adaptation index A_p was calculated after Strasser et al., 1987. The chlorophyll content was determined in acetone (100 %) using the new coefficients of Lichtenthaler, 1987b.

RESULTS AND DISCUSSION

Excitation and sensing of chlorophyll fluorescence at the same leaf side
The shape of the blue-light-induced chlorophyll-fluorescence emission spectra of intact green leaves is characterized by two maxima in the 690 and 735 nm region, when the fluorescence is excited and sensed at the same leaf side (reflection measuring mode). The two fluorescence maxima of fully green leaves are in most plants of about equal height (values for the fluorescence ratio F690/F735 of ca. 0.85 to 1.1) when sensed from the upper leaf side (Fig. 1A, solid line). When the blue-light-induced chlorophyll fluorescence is excited and sensed at the lower side of the leaf, the fluorescence maximum near 690 nm is considerably higher, as are the values of the ratio F690/F735 (ca. 1.2 to 1.7), as is shown in Fig. 1A (dotted line) and in Table 1. Many agricultural and forest plants are C_3-plants, which possess bifacial leaves, with densely packed cells in the upper leaf half (palisade parenchyma) and much fewer cells, which are separated by large aerial intercellular spaces, in the lower leaf half (spongy parenchyma). Consequently the upper leaf half contains more chlorophyll than the lower leaf half, and the probability of reabsorption of the emitted fluorescence is much higher. This is the explanation why the fluorescence peak in the 690 nm region and the ratio F690/F735 is lower when excited and sensed at the upper than at the lower leaf side. Similar differences between upper and lower leaf sides are found for the leaf of the C_4-plants maize and sugar cane, which exhibit

a concentric arrangement of their mesophyll cells around the vascular bundles (Tab. 1 and Fig. 1B). In the case of equifacial leaves (Avena sativa), the differences in the chlorophyll-fluorescence emission spectra between the upper and lower leaf side are very small (Tab. 1). Small differences in the ratio F690/F735 are also found between sun and shade leaves (Table 1), which are known to have a different chlorophyll content and photosynthetic activity per leaf area unit (Lichtenthaler et al. 1981). The result of this investigation show that the ratio F690/F735 is very much determined by the chlorophyll content of the leaves and leaf halves.

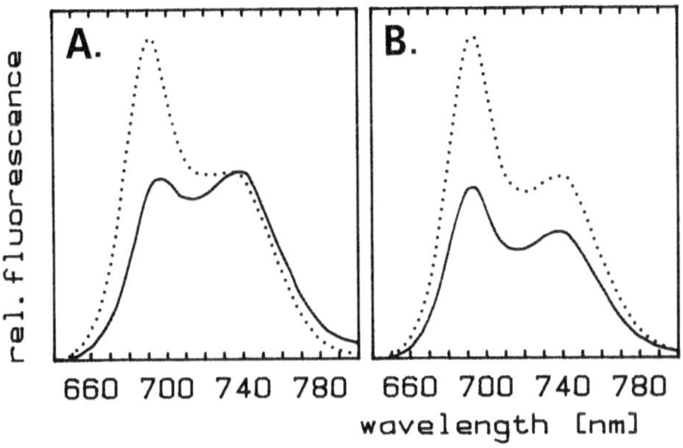

FIGURE 1. Chlorophyll fluorescence emission spectra of the upper (———) and lower leaf side (·····) **A.** of a C_3- plant (Nicotiana tabacum) and **B.** of a C_4-plant (Zea mays). The fluorescence was excited and sensed from the same leaf side (reflection measuring mode). Excitation light 470±30 nm.

TABLE 1. Differences in the fluorescence ratio F690/F735 between upper and lower leaf side of different plants. Mean values of 12 leaves in each case (with standard deviation). Excitation at 470±30 nm.

	F690/F735 upper side*		F690/F735 lower side*	
C_3-plants (bifacial leaves)				
Carpinus betulus	0.87	± 0.04	1.30	± 0.09
Fagus sylvatica				
sun leaf	0.92	± 0.11	1.39	± 0,11
shade leaf	1.06	± 0.07	1.48	± 0.06
Nicotiana tabacum	0.96	± 0.11	1.71	± 0.08
Phaseolus vulgaris	0.92	± 0.07	1.70	± 0.04
Quercus robur				
sun leaf	0.90	± 0.06	1.24	± 0.08
shade leaf	0.87	± 0.05	1.41	± 0.05
Raphanus sativus	0,98	± 0.06	1.59	± 0.11
Avena sativa (equifacial leaves)	1.20	± 0.12	1.32	± 0.10
C_4-plants				
Saccharum officinarum	0.92	± 0.05	1.32	± 0.07
Zea mays	1.38	± 0.08	1.75	± 0.11

* the differences between upper and lower leaf side are highly significant in all plants (P = 0.001) with the exception of the equifacial Avena leaf (P = 0.02).

Differences between blue and red-light excitation. The chlorophyll-fluorescence spectra of leaves look quite different when the excitation is performed with light in the range of 525 to 633 nm instead of blue light. In these cases the shorter-wavelength chlorophyll-fluorescence peak near 690 nm appears to be suppressed as compared to the longer-wavelength fluorescence peak near 735 nm. This is demonstrated by comparing the fluorescence spectra of the same leaf in Fig. 2A and B (solid lines), where excitation and sensing of the fluorescence was performed at the upper leaf side (reflection measuring mode). The value for the ratio F690/F735 is much higher for blue-light (0.85) than for red-light excitation (0.55).

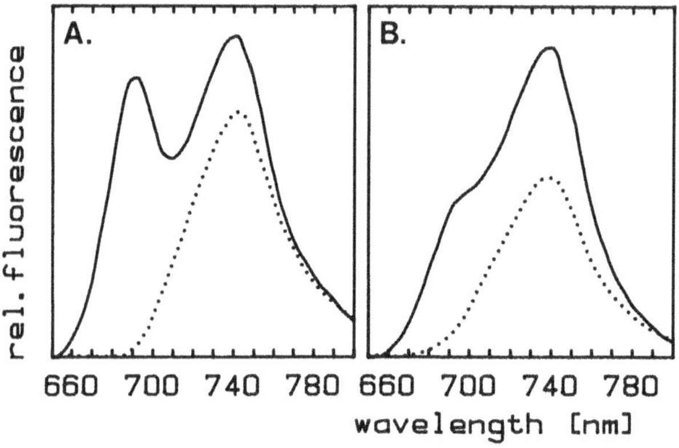

FIGURE 2. Chlorophyll fluorescence emission spectra of a dark-green Ficus leaf (Ficus benjamini), **A.** excited by a blue laser (442 nm) and **B.** by a red laser (632.8 nm). Solid lines: Fluorescence was excited and sensed at the upper leaf side (reflection measuring mode). Dotted lines: excitation was applied from the lower, but sensed from the upper leaf side (transmission measuring mode).

The reason for the differences in the chlorophyll-fluorescence spectra and the ratio F690/F735 between blue-light and yellow-to-red-light excitation (excitation and sensing via the upper leaf side) is the fact that blue light (range 400-500 nm) is absorbed by the photosynthetic accessory pigments of the pigment antenna (chlorophyll a, chlorophyll b, carotenoids) and the greater part does not penetrate into deeper leaf layers. Hence the blue-light-induced chlorophyll fluorescence predominantly derives from the upper leaf half. In contrast, excitation light of 525 to 633 nm, which does not correspond to the main absorption bands of chlorophylls and which can no longer be absorbed by carotenoids, penetrates into much deeper leaf layers than blue light. As a consequence the red-light-induced chlorophyll fluorescence emanates from deeper leaf layers and its short-wavelength form near 690 nm therefore is reabsorbed to a much larger degree by the chlorophylls than the blue-light-excited fluorescence (Lichtenthaler, 1986; Lichtenthaler et al. 1986; Lichtenthaler and Rinderle, 1988).These results indicate that the chlorophyll-fluorescence maxima are of about equal height when the fluorescence:

1) is excited and sensed from the upper leaf side and
2) when the excitation wavelength is in the range of 390 to 500 nm

When the excitation light is applied from below (transmission measuring mode) there is little difference between the shape of the fluorescence spectra excited with blue and red light and only one fluorescence peak in the 735 nm region can be seen (Fig. 2A and B, dotted lines). The shorter wavelength chlorophyll fluorescence maximum in the 690 nm region is suppressed by reabsorption of the emitted fluorescence by the chlorophylls. The absorption bands of the in vivo chlorophylls strongly overlap with this emitted short-wavelength chlorophyll fluorescence, whereas the fluorescence in the second maximum near 735 nm is little affected. The explanation for this behaviour is the fact that the excitation of the fluorescence at the lower leaf side (transmission measuring mode) the emitted fluorescence comes from deeper leaf layers and is reabsorbed to a larger degree by the chlorophylls than when it is excited at the upper leaf side (reflection measuring mode), where the major part of the chlorophyll fluorescence is emitted from the upper leaf half, where reabsorption is not as high.

Role of chlorophyll content and photosynthetic activity. That the shape of the chlorophyll fluorescence spectra and the height of the ratio is determined by the chlorophyll content of the leaf is shown in Fig. 3. The spruce needles of the damaged tree, which possess a ca. 30 % lower chlorophyll content, exhibit a higher short-wavelength fluorescence maximum than the fully green needles from a healthy spruce. These differences in the shape and the ratio F690/F735 are evident when exciting them with either blue light (Fig. 3A) or with red light (Fig. 3B). Though the absolute values for the ratio F690/F735 are different for blue and red-light excitation, the increase of the ratio with decreasing chlorophyll content is seen in both cases. The dependence of the ratio on the chlorophyll content was further investigated during the autumnal loss of chlorophyll in Parthenocissus leaves, where the ratio F690/F735 increased up to values of ca. 8 (Fig. 4).

We have also measured blue-light and red-light-induced chlorophyll-fluorescence spectra and the ratio F690/F735 for leaves of different chlorophyll content and different rates of photosynthesis (Table 2). Intactness or damage of the photosynthetic function was determined by measuring the variable fluorescence in form of the fluorescence decrease ratio (Rfd-values) in the 690 and 735 nm range (Rfd 690 and Rfd 730) as vitality index (Lichtenthaler 1986, 1987a, Lichtenthaler et al. 1986) as well as the stress adaptation index A_p (after Strasser et al., 1987).

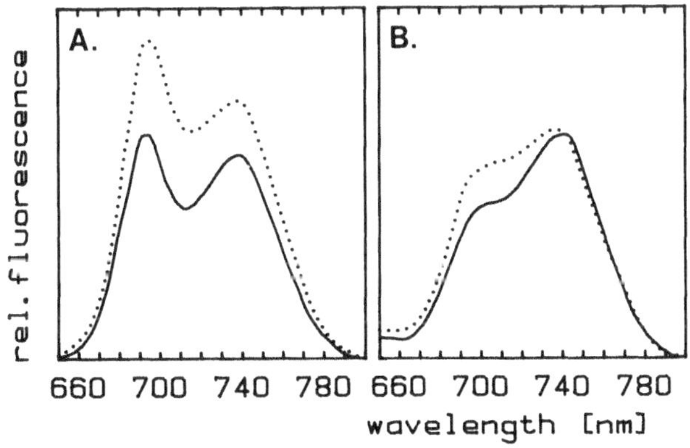

FIGURE 3. Chlorophyll fluorescence emission spectra of spruce needles (Picea abies Karst.L.) from a healthy tree (damage class 0/1;——) and light-green needles (with small yellow spots) from a damaged tree (damage class 3/4;·····) of the northern Black Forest. **A.** excitation by blue light (470±30 nm) and **B.** by red light (620±20 nm).

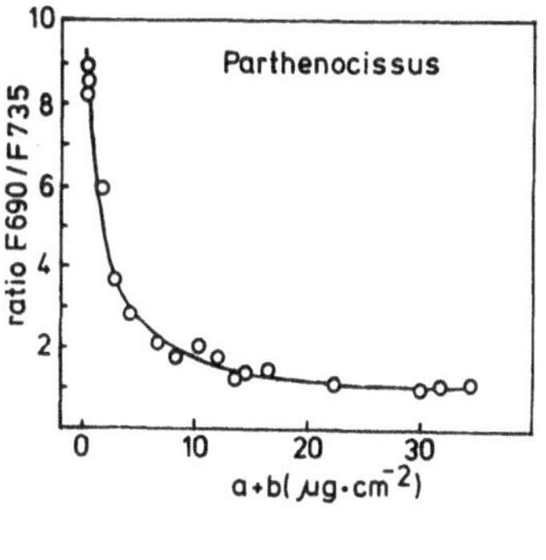

FIGURE 4. Dependence of the chlorophyll fluorescence ratio F690/F735 in Parthenocissus leaves on the chlorophyll content (a+b) per leaf area unit. Excitation light 470±30 nm (taken from Lichtenthaler, 1987a)

TABLE 2. Values of the **chlorophyll-fluorescence ratio F690/F735** (excited at 470 and 620 nm), **Rfd-values** (fluorescence vitality index measured in the 690 and 730 nm region) and **stress-adaptation index A_p** of leaves of different chlorophyll content and physiological state from several plants. Mean values of at least 3 to 5 determinations per leaf-type. Rfd-values higher than 2.5 indicate a good to very good photosynthetic activity of the leaf.

Plant	a+b ($\mu g \cdot cm^{-2}$)	ratio F690/F735 (470 nm)	(620 nm)	Rfd-values 690 nm	730 nm	A_p-index
Ilex aquifolium						
dark-green leaf:	70	0.98	0.5	4.0	2.4	0.33
light-green senescent leaf:	21	1.91	1.68	1.2	1.0	0.09
frost-injured senescent leaf:	6	3.5-6	2-2.8	0.7	0.5	0.12
Picea abies						
damage class 0/1 green NI-needles						
needle year 1987:	50	1.21	0.90	6.0	3.8	0.31
needle year 1986:	75	1.16	0.75	6.1	3.9	0.31
needle year 1985:	90	1.09	0.72	5.1	3.5	0.26
needle year 1984:	86	1.12	0.70	5.0	3.6	0.23
damage class 3/4 olive-green NI-needles*						
needle year 1987:	52	1.10	0.86	3.9	2.8	0.22
needle year 1986:	45	1.21	0.94	2.1	1.8	0.09
needle year 1985:	28	1.60	1.28	1.2	1.1	0.05
needle year 1984:	32	1.26	0.83	2.5	2.3	0.06
Phaseolus vulgaris						
green control leaf:	63	0.90	0.61	2.8	2.1	0.18
diuron-treated:	63	1.15	0,78	0	0	0
Zea mays						
light green leaf:	42	1.32	0.85	2.5	1.9	0.17
diuron-treated:	42	1.65	1.10	0	0	0
Fagus sylvatica **						
sun leaf:	56	1.01	0.63	3.7	2.6	0.23
shade leaf:	44	1.14	0.79	1.8	1.4	0.14
Nicotiana tabacum**						
green leaf (su/su)	46	1.10	0.76	3.3	2.2	0.26
aurea leaf (Su/su)	16	1.90	1.27	4.0	2.6	0.28

* the olive-green needles with some yellow spots and tips are from spruces of damage class 3/4 (ca. 85% needle loss)
** red excitation light provided by a He/Ne-laser (5 mW, 632.8 nm).

The ratio F690/F735 increases with decreasing chlorophyll content e.g. in senescent and frost-injured Ilex leaves. The ratio is significantly higher in the needles of damaged spruce trees as compared to the needles of healthy trees (except for the youngest needle year 1987). This is paralleled by a decreased chlorophyll content and lower Rfd-values. In diuron-treated (DCMU) bean leaves (Phaseolus) and maize, where the photosynthetic function is lost (Rfd-values=0), we also observed an increase of the ratio F690/F735, though the chlorophyll content remained the same before and after treatment with a 10^{-3} molar solution of diuron (Table 2). This indicates that not only a lower chlorophyll content but also a decline or loss of the photosynthetic function will increase the values of the ratio F690/F735. The block of the photosynthetic electron transport by diuron causes increased chlorophyll-fluorescence emission (Fig. 5), but the increase in the 690 nm region is higher than that in the 735 nm region. This increase of the ratio F690/F735 to higher values can be seen after fluorescence excitation by blue light or red light (Fig. 5A and B).

The lower photosynthetic rates per leaf area unit of shade as compared to sun leaves of the beech (Lichtenthaler et al. 1981), which is also seen in lower Rfd-values (Table 2) is also documented by slightly, yet significantly higher values for the ratio F690/F735. The somewhat lower chlorophyll content of shade leaves cannot alone account for this difference.

In the case of aurea leaves, which possess a much lower chlorophyll content than normal green tobacco leaves, the ratio F690/F735 is increased and there solely due to the lower chlorophyll content, since the Rfd-values indicate in both aurea and green tobacco an efficient photosynthetic function (Table 2).

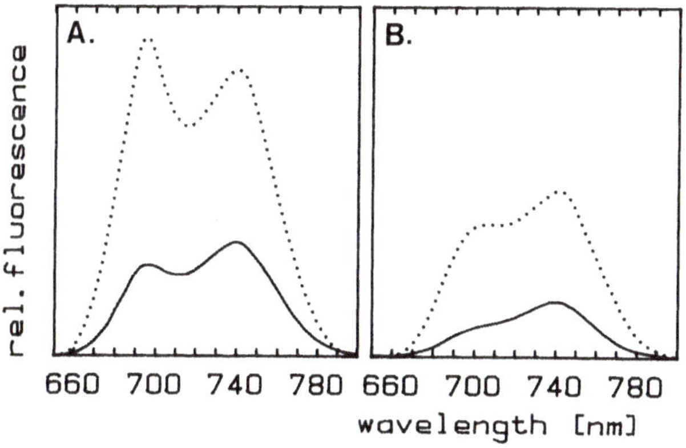

FIGURE 5. Chlorophyll fluorescence spectra of green bean leaves (Phaseolus vulgaris;———) and bean leaves treated with the herbicide diuron ($2x10^{-3}$M; ·····). Excitation in **A.** by blue light and **B.** by red light. Mean of 5 leaves per condition.

CONCLUSION: The shape of the chlorophyll-fluorescence emission spectra is determined mainly by two parameters, the chlorophyll content and the photosynthetic activity of the leaf. The height of the fluorescence ratio F690/F735 increases not only with decreasing chlorophyll content, but also when the process of photosynthesis declines e.g. in damaged forest trees or is blocked by herbicides or other stress factors. The increase of the ratio proceeds no matter whether the chlorophyll fluorescence is excited by blue or red light. The results demonstrate that the ratio F690/F735 is a good stress indicator for plants and may be applied in remote sensing of the physiological state of terrestrial vegetation (crop plants, forest trees) via laser-induced chlorophyll fluorescence from airborne systems. Blue-light and red-light lasers appear to be equally suitable for application in the remote sensing of the ratio F690/F735. In the case of blue-light lasers one could dispense with special precautions against overlapping of the excitation light with the chlorophyll fluorescence, which are required using red-light lasers. This would favour a blue-light laser for remote sensing. That the ratio F690/F735 is in fact a suitable stress indicator in remote sensing was shown in a provisional experiment with an active airborne system equipped with a pulsed excimer blue laser (450 nm) as excitation light, where differences in the ratio F690/F735 could be sensed for trees of different physiological state (Zimmermann and Günther, 1986).

ACKNOWLEDGEMENTS
The work described was sponsored by a grant from the European Joint Research Centre (JRC), Ispra which is gratefully acknowledged. We wish to thank Ms. B. Weigelt and Ms. K. Führich as well as Mr. M. Haitz and S. Lichtenthaler for technical assistance and the Laser 2000 GmbH, München for short-term lending of a blue laser.

REFERENCES

KRAUSE GH and WEIS E, 1984. Chlorophyll fluorescence as a tool in plant physiology. Photosynth. Res. 5: 139-157.

LICHTENTHALER HK, 1986. Laser-induced chlorophyll fluorescence of living plants. In: Proc. Internat. Geoscience and Remote Sensing Symposium IGARSS Zürich, Vol. III, pp. 1571-1579, ESA Publications Division, Noordwijk.

LICHTENTHALER HK, 1987a. Chlorophyll fluorescence signatures of leaves during the autumnal chlorophyll breakdown. J. Plant Physiol. 131: 101-110.

LICHTENTHALER HK, 1987b. Chlorophylls and carotenoids: Pigments of photosynthetic biomembranes. Methods Enzymol. 148: 350-382.

LICHTENTHALER HK and PFISTER K, 1978. Praktikum der Photosynthese, Quelle & Meyer Verlag, Heidelberg.

LICHTENTHALER HK, BUSCHMANN C, DÖLL M, FIETZ H-J, BACH T, KOZEL U, MEIER D and RAHMSDORF U, 1981. Photosynthetic activity, chloroplast ultrastructure and leaf characteristics of high-light and low-light plants and of sun and shade leaves. Photosynth. Res. 2: 115-141.

LICHTENTHALER HK, BUSCHMANN C, RINDERLE U and SCHMUCK G, 1986. Application of chlorophyll fluorescence in ecophysiology. Radiation and Environmental Biophysics 25: 297-308.

LICHTENTHALER HK and RINDERLE U, 1988. Role of chlorophyll fluorescence in the detection of stress conditions in plants. CRC Critical Reviews in Analytical Chemistry 19, Suppl. I: S29-S85.

STRASSER R, SCHWARZ B and BUCHER J, 1987. Simultane Messung der Chlorophyll Fluoreszenz-Kinetik bei verschiedenen Wellenlängen als rasches Verfahren zur Frühdiagnose von Immissionsbelastungen an Waldbäumen. Ozoneinwirkung auf Buchen und Pappeln. Europ. J. Forest Pathology 17: 149-157.

ZIMMERMANN R and GÜNTHER KP, 1986. Laser-induced chlorophyll a fluorescence of terrestrial plants, In: Proc. Internat Geoscience and Remote Sensing Symposium IGARSS Zürich, Vol. III, pp. 1609-1613, ESA Publications Division, Noordwijk.

FLUORESCENCE CHANGES OF A DRYING MAPLE LEAF OBSERVED IN THE VISIBLE AND NEAR-INFRARED

Arnold F. Theisen, U.S. Geological Survey, DFC MS-964, Denver, CO, 80225

Key words: chlorophyll, fluorescence, Fraunhofer, FLD, plant stress.

ABSTRACT

When the source of nutrients for a leaf are disrupted, the mechanisms for energy transfer, that are part of the photosynthetic process, undergo changes that alter the absorbtion and emission characteristics of the leaf. A series of fluorescence measurements made of a maple leaf as it dried, permit some of these changes to be observed. The emission peaks at approximately 690 nm and 740 nm are present throughout the measurements, although the intensity of the peaks vary. Additionally, a feature with an emission maximum at approximately 440 nm and an excitation maximum at approximately 330 nm appeared as the leaf dried and may provide an alternate indication of stress. Removing a leaf from the tree accelerates changes that occur *in vivo* when a plant is under stress. These changes are not visually apparent; however, analysis of the data indicate the strong probability that they would be easily detected with a remote-sensing instrument such as the Fraunhofer luminescence detector (FLD).

INTRODUCTION

Previous studies of plant stress using the Fraunhofer luminescence detector (FLD) (Watson et al, 1973; Hemphill et al, 1977; and McFarlane et al, 1980), concentrated on the hydrogen alpha Fraunhofer line at 656.3 nm because of its proximity to the chlorophyll emission near 690 nm. Other investigators using laser-induced fluorescence (Lichtenthaler, 1986; and Lichtenthaler et al, 1986) measured the emission near 690 nm and 740 nm and used the F690/F740 ratio to monitor plant status. To gain a better understanding of the nature of chlorophyll emissions and other fluorescence phenomena, spectroscopic measurements, at this laboratory, are now made so that excitation and emission spectra can be observed simultaneously.

SPECTROSCOPIC METHODS

A Perkin-Elmer MPF-44B Fluorescence Spectrophotometer*, with a 150 watt Xenon excitation source and a pair of R-928 photomultipliers for sample and reference detection is used for collection of excitation-emission-matrix (EEM) data. Czerny-Turner monochromators, with a holographic grating for the excitation, are gear driven at 480 nanometers per minute. Excitation and emission slit widths are variable and adjusted to provide an effective width of 10 nm to match the sampling rate. The spectrophotometer is

*/ Use of trade names in this paper is for discriptive purposes only and does not constitute an endorsement by the U.S. Geological Survey.

H. K. Lichtenthaler (ed.) Applications of Chlorophyll Fluorescence, 197–201.

operated in a ratio mode; therefore, excitation and absorption spectra can be considered to be identical (Hamilton et al, 1978). For data collection, excitation is begun at 300 nm and stepped at 10 nm intervals to 600 nm. For each excitation wavelength, the emission is scanned from 410 nm, or 60 nm beyond the excitation wavelength, to 830 nm with the relative fluorescence intensity recorded at 10 nm intervals.

FLUORESCENCE MEASUREMENTS

To simulate a plant under stress, a leaf was removed from a maple tree, disrupting the leaf's source of nutrients. This apparently healthy maple tree was in full morning sun and should have been into its daytime mode, although no means of measuring photosynthesis was available. Periodic EEM measurements were made as the leaf dried to observe the absorption and emission characteristics related to the changing mechanisms of photosynthetic energy transfer. The first measurement was made within five minutes of leaf detachment and subsequent measurements were made approximately one hour apart until after 17:00 of the first day. An additional measurement was made at 12:55 the following day. Only measurements of the top of the leaf are included in this series. A spectral resolution of 10 nm was used. Although the emission peak near 690 nm would actually warrant a 5 nm interval, a full EEM at 5 nm resolution would take over two hours, too long for this initial test.

RESULTS

A maximum intensity for the series of measurements occured at 470 nm excitation and 740 nm emission with the 13:31 measurement. All data are normalized to this maximum. Figures 1 through 3 show 3-D perspective plots of the EEM data taken at 09:33, 13:31, and 17:06 respectively and figure 4 shows the 12:55 measurement of the next day. The emission resulting from the excitation of chlorophyll <u>b</u> at 470 nm, is the more prominent until the 14:30 measurement (not shown) where it is roughly equal to the emission from the 430 nm excitation of chlorophyll <u>a</u>. Thereafter, the emission from excitation at 430 nm is the more prominent. For a quantitative appraisal of the changes that occured during the full set of measurements, see Table I below. The ratios for all three selected wavelengths rise steadily with drying. This increase can be attributed to a decrease in either chlorophyll content or physiological activity (Lichtenthaler and Rinderle, 1988). The 450 and 470 nm excitations level off at 14:30 and the 600 nm excitation

TABLE 1. F690/F735 ratios at selected excitation wavelengths.

Time	450 nm	470 nm	600 nm
09:33	0.777	0.888	0.467
10:30	0.795	0.906	0.494
11:41	0.889	0.966	0.529
12:19	0.916	0.981	0.548
13:31	0.990	1.013	0.590
14:30	1.065	1.060	0.701
15:27	1.112	1.109	0.734
16:27	1.095	1.102	0.750
17:06	1.096	1.097	0.737
12:55	1.089	1.069	0.721

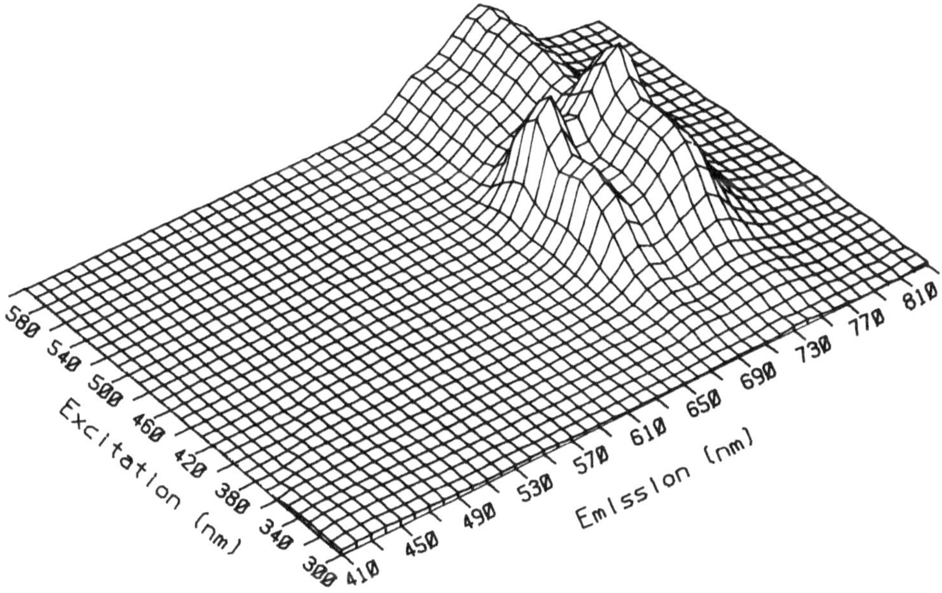

FIGURE 1. Maple leaf 3-D perspective plot. Measurement #1 at 09:33, 8/5/87.

peaks at 16:27 with a slight decline to the last measurement. In figure 2 the first definite signs of emission at 440 nm, in the violet, with a 330 nm excitation can be seen. The intensity of this emission doubles by 15:27, and more than doubles again by 17:06 (figure 3). The last measurement seen in figure 4 shows some increase in this emission. Table 2 is a listing of

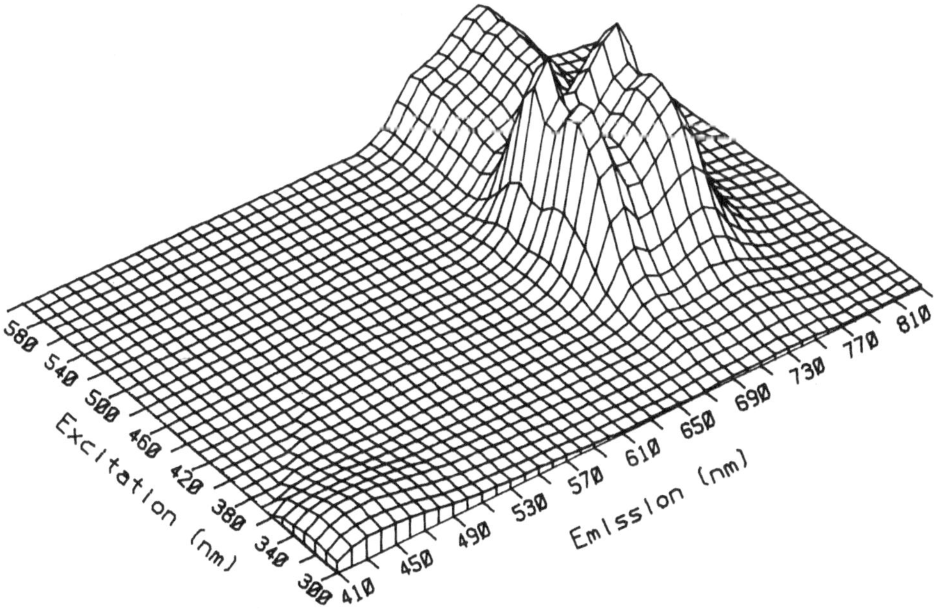

FIGURE 2. Maple leaf 3-D perspective plot. Measurement #5 at 13:31, 8/5/87.

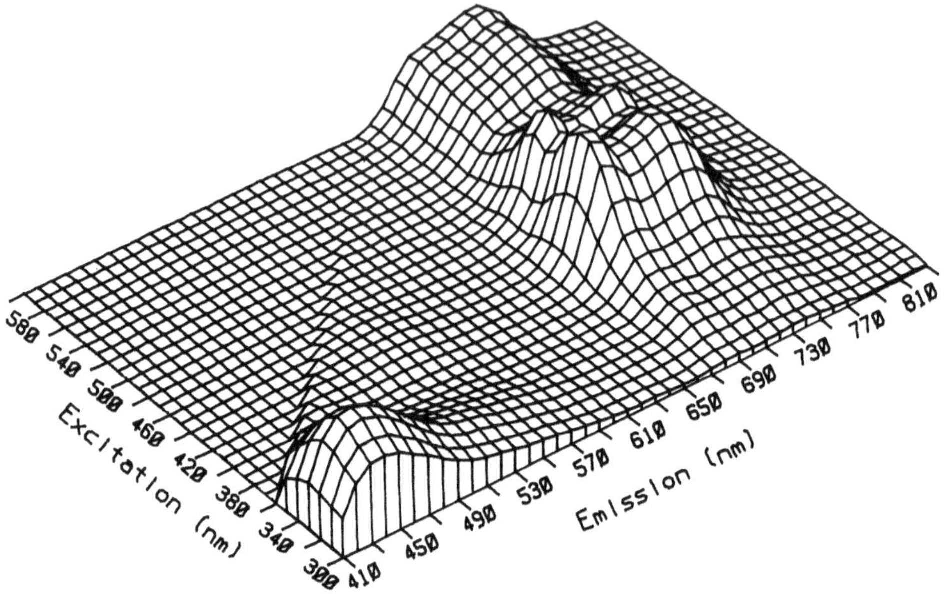

FIGURE 3. Maple leaf 3-D perspective plot. Measurement #9 at 17:06, 8/5/87.

the F440/F690 ratio for the entire series. These data reveal the violet emission as soon as 2 hours after picking. The apparent color of the maple leaf did not change during or immediatley after these measurements were taken. The humidity varied between 10 and 30 percent over the span of the measurements and the leaf was brittle when the measurements were completed.

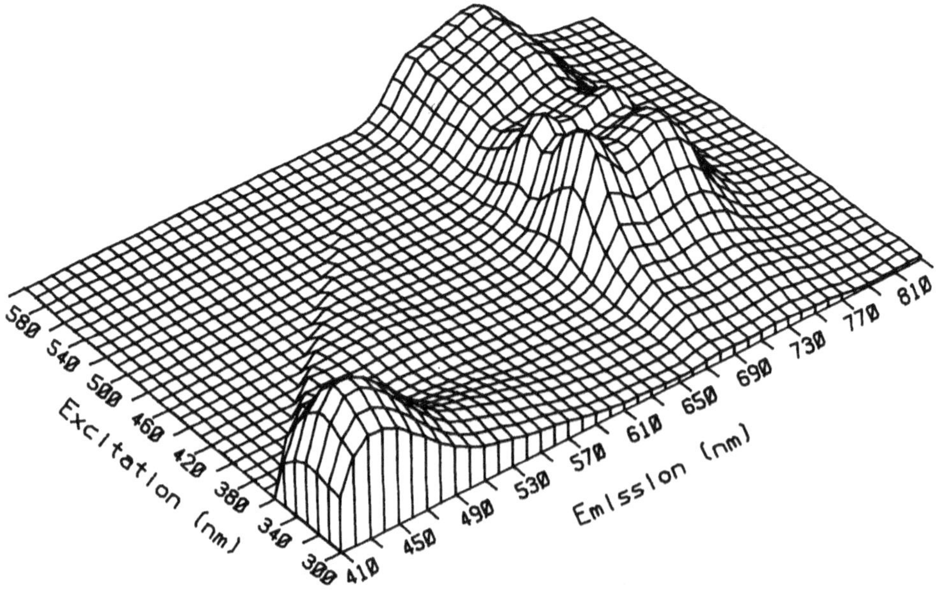

FIGURE 4. Maple leaf 3-D perspective plot. Measurement #10 at 12:55, 8/6/87.

TABLE 2. Ratio of 440 nm fluorescence emission vs 690 nm emission.

Time	09:33	10:30	11:41	12:19	13:31	14:30	15:27	16:27	17:06	12:55
	0.897	0.829	0.976	1.780	2.927	2.926	2.878	4.086	6.103	7.235

DISCUSSION

EEM fluorescence provides a means by which the complete visible and near-infrared absorption and emission characteristics of a leaf can be monitored to observe its response to stress. Further study is needed to determine if the results of this simulated stress can be repeated for plants stressed non-destructively. A passive instrument, such as the Fraunhofer luminescence detector (FLD), capable of monitoring the 438.4 nm Fraunhofer line in addition to the 656.3 nm line may be appropriate to observe two dynamic aspects of the fluorescence response of <u>in vivo</u> plants, and thus the state of their health.

ACKNOWLEDGEMENTS

I would to thank Prof. Dr. H.K. Lichtenthaler and Dr. Barry Rock for their generous assistance on this paper and help in attending these proceedings.

REFERENCES

Hamilton TDS, Munro IH and Walker G, 1978. Luminescence instrumentation. Luminescence Spectroscopy, Michael D. Lumb ed., Academic Press: 149-238.

Hemphill WR, Watson RD, Bigelow RC, and Hessen TD, 1977. Measurement of luminescence of geochemically stressed trees and other materials. USGS Professional Paper 1015: 93-112.

Lichtenthaler HK, 1986. Laser-induced chlorophyll fluorescence of living plants. Proceedings of the Remote Sensing Symposium (IGARSS) Zürich, vol. III, ESA Publication Division Noordwijk: 1571-1579.

Lichtenthaler HK, Buschmann C, Rinderle U and Schmuck G, 1986. Application of chlorophyll fluorescence in ecophysiology. Radiation and Enviromental Biophysics 25: 297-308.

Lichtenthaler HK and Rinderle U, 1988. Chlorophyll fluorescence spectra of leaves as induced by blue light and red laser light. Proceed. 4th International Colloquium on Spectral Signatures of Objects in Remote Sensing, Aussois, ESA Publication Division Noorwijk:

McFarlane JC, Watson RD, Theisen AF, Jackson R, Ehrler WL, Pinter PJ, Idso SB and Reginato RJ, 1980. Plant stress detection by remote measurement of fluorescence. Applied Optics, v. 19, no. 19: 3287-3289.

Watson RD, Hemphill WR and Hessen TD, 1973. Quantification of the luminescence intensity of natural materials. Symposium for the Management and Utilization of Remote Sensing Data, Sioux Falls, S.D. Proc.: 364-376.

THE EFFECT OF LONG TERM EXPOSITION OF POPLARS TO LOW CONCENTRATIONS OF SO_2 AND NH_3.

Olaf van Kooten, Bert van Hove and Klaas Jan van Wijk
Agricultural University Wageningen, Gen. Foulkesweg 72,
NL-6703 BW Wageningen, the Netherlands.

Key words: *air pollution, chlorophyll fluorescence, poplar, quantum yield, photosynthesis, 'non' photochemical quenching.*

Abstract

Young cuts of poplar (*P. euramericana* L.) were exposed to concentrations of SO_2 and NH_3 which are considered normal for certain areas in the Netherlands. The effect of 7 weeks of this treatment was monitored with the modulated fluorescence technique and with gas exchange measurements. Photosynthetic performance during the development of the leaves was also followed with the fluorescence technique. The effect of the fumigation on these stress-'free' plants was relatively small in comparison to the developmental changes during the 7 weeks of observation and treatment. The photosynthesis measurements revealed a slight increase in light efficiency for the higher concentrations of NH_3 and SO_2. However this increase was absent when the two gasses were combined. NH_3 also revealed an increased Pmax. Fluorescence measurements only revealed an enhanced non-photochemical quenching component under steady state conditions for the NH_3 treatment. The SO_2 treatment revealed enhanced activation kinetics of the CO_2 fixation process (Calvin cycle). When the 2 measurements are combined it becomes evident that the higher SO_2 concentration (37.2 ppb) and the NH_3 treatment (129 ppb) enhance the quantum yield of photosystem 2 for photochemistry under the condition of an 'energized' thylakoid membrane. It is therefore concluded that low concentrations of these air pollutants are able to induce a more pronounced state of sun adaptation (Björkman 1981) than their counter parts, which had received either filtered air, a low SO_2 concentration (18.8 ppb) or a combination of SO_2 and NH_3.

Introduction

The effect of air pollution on the physiology of plants is under intensive investigation (Heath 1980). We chose to study the effect of SO_2 and NH_3 concentrations comparable to actual

---

Abbreviations: F_0= minimal fluorescence, F_m= maximal fluorescence, Fv= variable fluorescence, (Fv)s=saturated variable fluorescence, LHC= Light Harvesting Complex, PAM= Pulsed Amplitude Modulated fluorometer, NO_X= NO + NO_2, PAR= Photosynthetic Active Radiation, Pmax= maximum photosynthesis rate, Pn= net photosynthesis rate, PSI= photosystem 1, PSII= photosystem 2, q_Q= photochemical quenching, q_E= non-photochemical quenching, Rd= dark respiration, Φ_{po}= app. quantum yield in the non-energized state (q_E=0 and all traps open), Φ_{pe}= app. quantum yield in the fully energized state (q_E=1 and all traps open), σ= light efficiency.

H. K. Lichtenthaler (ed.) Applications of Chlorophyll Fluorescence, 203–209.

values as measured in our country (Bell 1987). In contrast most studies have been performed at relatively high concentrations and short time intervals (e.g. Schmidt et al. 1988). The PAM fluorescence and photosynthetic gas exchange were applied as techniques to measure the impact of these gasses on poplar trees. Both techniques did register small but significant effects of fumigation. In combination the results reveal the unique possibilities of the simultaneous application of the techniques to probe the photosynthetic processes in the leaf.

Materials and methods

Fumigation of SO_2 and NH_3 was performed on poplars (*Populus euramericana* L. cv. 'Flevo'). The poplars were grown from 10 cm cuts yielding a genetically homogenous set of plants. Cuts were kept in an aerated nutrient solution consisting of 10% Hoagland solution for the first two weeks, 20% in the third week and 30% in the forth week when the plants were transferred from the phytotron to the fumigation chambers. After the 4th week the solution was kept at 35% Hoagland. The temperature was kept at 20 °C and relative humidity varied between 60 and 80%. A daily light period of 17 hours varied in intensity depending on the height of the plant. At the level of the cut the light intensity was 40 $W \cdot m^{-2}$ and near the roof of the chamber the light intensity was 80 $W \cdot m^{-2}$ (PAR).

Five fumigation chambers made of glass and an internal volume of 1 m^3 were situated in a phytotron. External air passed through active charcoal filters, a moistening device and finally through the chambers. Before entering the chambers a controlled flow of NH_3 and/or SO_2 was added to the air. The air entered the chambers through a perforated floor and left the chamber near the roof. The 5 fumigation treatments are described as follows:

1. Control (i.e. filtered air still containing ≈12 ppb NO).
2. SO2lo= 50 $\mu g \cdot m^{-3}$ SO_2 (i.e. 18.8 ppb at 20 °C).
3. SO2hi= 100 $\mu g \cdot m^{-3}$ SO_2 (i.e. 37.5 ppb at 20 °C).
4. SO2NH3= 50 $\mu g \cdot m^{-3}$ SO_2 + 75 $\mu g \cdot m^{-3}$ NH_3
5. NH3= 75 $\mu g \cdot m^{-3}$ NH_3 (i.e. 129 ppb at 20 °C).

For continuous measurement of the NH_3 concentration the gas is converted in a heated stainless steel pipe (850 °C) into NO (Van Hove et al. 1987) and then measured in a NO_X monitor (Mons Labs 8810). Photosynthesis measurements were performed in a specially designed leaf cuvette, made out of teflon and glass and described elsewhere (Van Hove et al. 1988). Differential CO_2 and H_2O measurements were performed by IRGA-monitors (ADC-225-MK3). The absolute SO_2 concentration was determined with a SO_2 monitor (Mons Labs 8850). The wind speed in the leaf cuvette was kept at 1 $m \cdot s^{-1}$ and the temperature at 20 °C. The cuvette was illuminated by 2 HPI lamps (Philips 400W). The light intensity could be varied allowing the measurement of a photosynthesis vs. light curve. The Pmax, Rd and σ were determined by least squares fitting these curves to a formula of Goudriaan (1982):

$$Pn(I) = (Pmax - Rd) \cdot \{1 - \exp[-I \cdot \sigma/(Pmax - Rd)]\} + Rd \quad (1)$$

After 7 weeks of fumigation the leaf of the 9th internodium was measured in the presence of the same pollutant concentration as during the treatment and 340 ppm CO_2. After this measurement the fresh and dry weight of the leaves, stem and roots of the whole plant were determined.

During the fumigation period each plant was measured every week with the PAM fluorometer (Schreiber et al. 1986). The plants were dark adapted for exactly 30 minutes. The leaf of the 8th internodium (in the first week the 6th internodium) was gently fixed at 3 mm from the end of a quadruforcated light fiber. Below the area of the fiber optics moistened gas is passed at 80 $ml \cdot min^{-1}$, i.e. 2% O_2 + 340 ppm CO_2 + 98% N_2 as in Schreiber and Bilger (1987). The measurement is performed at room temperature (≈ 22 °C). The measurement consists of an illumination period of 6 minutes interspersed with saturating light pulses (700 ms, 2000 $W \cdot m^{-2}$ PAR) followed by 1 minute darkness. The process is controlled by a personal computer and the data are stored for further analysis (Van Kooten and Van Hove 1988). The photochemical and the non-photochemical quenching, q_Q and q_E respectively, are determined with the following equations from Schreiber and Bilger (1987):

$$q_Q = ((Fv)s - Fv)/(Fv)s \quad (2)$$

$$q_E = ((Fv)m - (Fv)s)/(Fv)m \quad (3)$$

The values of q_Q and q_E are correct as long as F_0 is not notably quenched (Bilger and Schreiber 1986). At the level of q_E found in our results F_0 quenching is unlikely to occur.

The value Fv/F_m (Björkman and Demmig 1987) is taken as the maximum relative efficiency of photochemistry in the dark adapted state, i.e.

$$Fv/F_m = (F_m - F_0)/F_m = k_p/(k_p + k_t + k_f + k_d) \quad (4)$$

the k_i values represent the different rate constants for the exciton dissipation processes in the model of Kitajima and Butler (1975).

The equation which relates the quenching components to the photosynthesis rate is taken from Weis and Berry (1987)

$$J/(I \cdot q_Q) = \Phi_{po} - q_E \cdot (\Phi_{po} - \Phi_{pe}) \quad (5)$$

J is the electron transport rate in $\mu mol \cdot m^{-2} \cdot s^{-1}$ and I the incident light flux density also expressed in $\mu mol \cdot m^{-2} \cdot s^{-1}$ (unlike in eq. 1). Φ_{po} is the apparent quantum yield of PSII in a non-energized membrane (i.e. dark adapted or q_E=0) with all acceptors oxidized. After close scrutiny of eqs. 5 and 4 it becomes evident that $\Phi_{po} \equiv Fv/F_m$. Φ_{pe} is the apparent quantum yield of PSII in a fully energized membrane (i.e. q_E=1) but also with all acceptors oxidized.

Results

In fig. 1 the photosynthesis curves of poplars are shown after 7 weeks of treatment. The curves are averages of 5 to 10 measurements except for the control plants which are of 15 measurements. Both SO2hi and NH3 treated plants reveal an enhanced σ (see also table 1), but only the NH3 treatment has stimulated Pmax. In fig. 2A we see the changes in F_0 during the fumigation period. F_0 decreases in the first three weeks as the leaf develops and the exciton transfer between LHC and PSII becomes more efficient. This effect is also evident in fig. 2B where the quantum yield for photochemistry of PSII is shown. Apart from the effect of development on these parameters no clear effect of treatment can be discerned. The average standard deviation in F_0 is 7%; in F_v/F_m it is 1%. If we look at the fluorescence at the end of the 6 minute illumination period, i.e. in the steady state, we only see a slight effect of the fumigation treatment on the quenching components. In fig. 3 the effect of the NH3 treatment on the q_E in the steady state is evident and becomes already apparent in the second week of fumigation. The

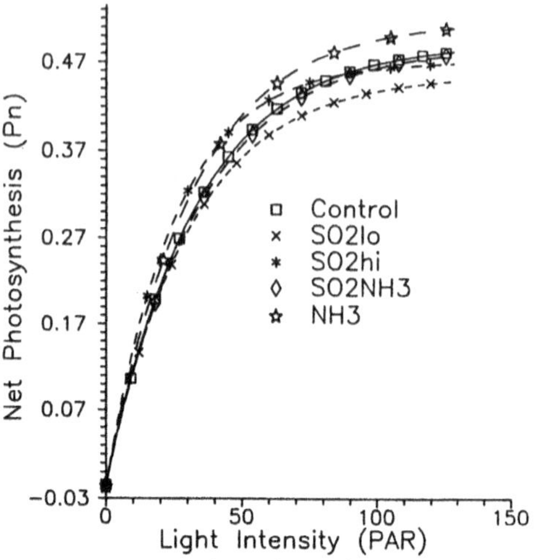

Figure 1: Light dependent photosynthesis rates from attached poplar leaves were measured after a seven week fumigation period. The results were fitted to eq. 1 resulting in the determination of σ and Pmax, Rd was measured directly in the dark. These values were then averaged for each treatment (Control=15, SO2lo=7, SO2hi=5, SO2NH3=10, NH3=6 measurements) and the resulting curves calculated with eq. 1 are shown here. The light intensity is expressed as $W \cdot m^{-2}$ and Pn as mg $CO_2 \cdot m^{-2} \cdot s^{-1}$.

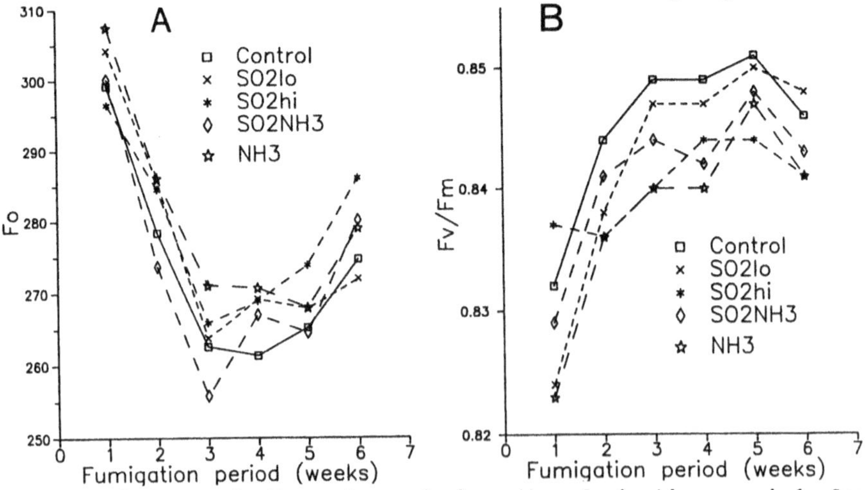

Figure 2: A) Averaged F_0 values during the fumigation period for the different treatments. The standard deviations of these averaged values vary between 2 and 15%. B) The averaged quantum yield for photochemistry of PSII calculated with eq. 4. These values were calculated from each experiment and then averaged. The standard deviation of these values varies between 0.7 and 1.9%. The differences between the treatments are not significant.

development effect in all the other treatments revealing first a rise in photosynthesis rate in the first 3 weeks and then a gradual decrease, is absent in the NH3 treatment. From fig. 4 it is clear that the difference in q_E in the steady state in comparison with the other treatments is small. The major changes occur in the first 2 minutes after the light is turned on. Here the SO2hi treatment discerns itself and reveals faster kinetics, which could imply a faster turning on of the Calvin cycle. When we combine the results of fig. 1 with the results of fig. 2B we can infer the value of Φ_{pe} for the different treatments. This is presented in table 1. It is evident that a slight rise in σ coincides with strong rise in Φ_{pe}.

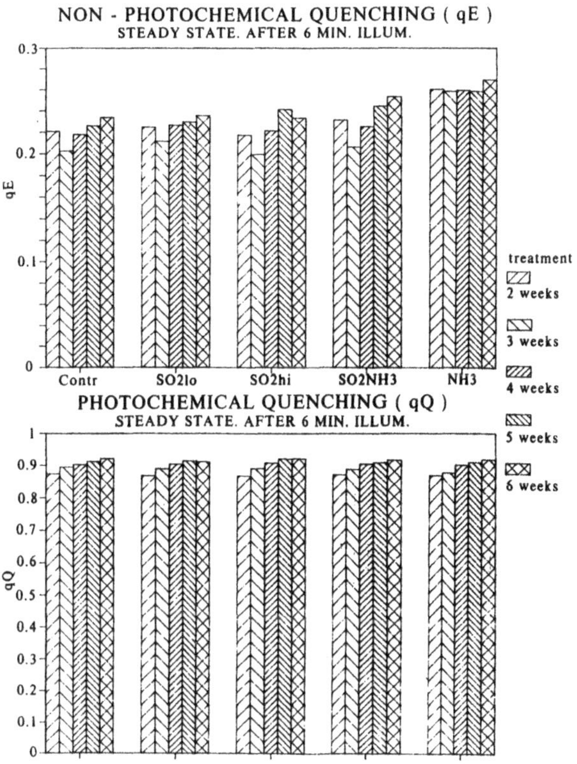

Figure 3: The averaged values of the photochemical and non-photochemical quenching components measured at the end of a 6 minute light period and calculated with eqs. 2 and 3. The effect of the fumigation treatment was negligible on q_Q as was the case with Fv and (Fv)s (data not shown). On q_E a barely significant effect can be seen for the NH3 treatment only.

Table 1: Rd, σ and Pmax are averaged measurements obtains with eq. 1. The actual light intensity in the fluorescence measurement is 25 $W \cdot m^{-2}$ (i.e. 100 $\mu mol \cdot m^{-2} \cdot s^{-1}$), but Pn was calculated at 30 $W \cdot m^{-2}$ to compensate for the low O_2 concentration in the fluorescence experiments. The electron transport rate was calculated from Pn by assuming 10 electrons per CO_2 fixed. For Φ_{po} the averaged value of 0.844 was taken from fig. 2B.

	contr	SO2lo	SO2hi	SO2NH3	NH3
Rd ($mgCO_2 \cdot m^{-2} \cdot s^{-1}$)	-0.015	-0.014	-0.019	-0.015	-0.016
σ ($mgCO_2 \cdot J^{-1}$)	0.0154	0.0151	0.0196	0.0150	0.0170
P_{max} ($mgCO_2 \cdot m^{-2} \cdot s^{-1}$)	0.493	0.457	0.472	0.490	0.519
Pn($30W \cdot m^{-2}$) (")	0.288	0.277	0.324	0.283	0.313
J ($\mu mol\ e^{-} \cdot m^{-2} \cdot s^{-1}$)	65.5	62.9	73.6	64.3	71.1
q_Q	0.93	0.91	0.93	0.93	0.93
q_E	0.235	0.238	0.235	0.255	0.270
Φ_{pe}	0.252	0.204	0.619	0.245	0.549

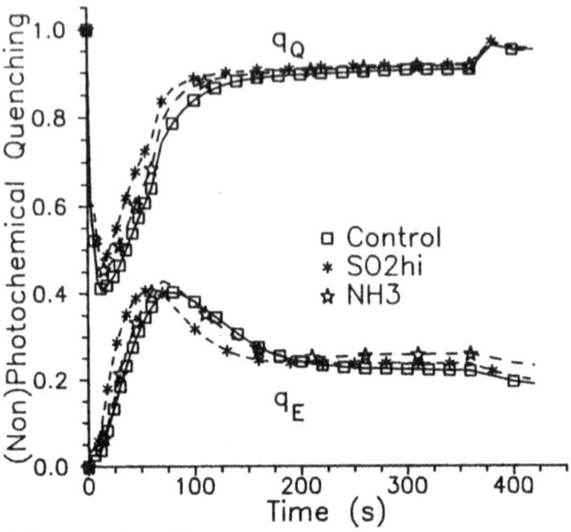

Figure 4: The averaged effect of a 6 weeks fumigation treatment (Control=11, SO2hi=11, NH3=10 measurements) on the changes in quenching. The measurement is described in the materials and methods. The effect of SO_2 is only evident in the first 2 minutes, while NH_3 differs from the control only in the steady state.

Discussion

The fact that the fumigation did neither affect F_0 nor F_v/F_m precludes the possibility that any of the treatments damaged the photosynthetic apparatus near PSII (Kyle 1987). Thus damage to the water splitting complex as proposed by Shimazaki and Sugahara (1980) does not occur at these SO_2 concentrations. The rise in light efficiency is explained by a decrease in the regulating mechanism known as 'energy'-dependent quenching (Weis and Berry 1987). The rise in σ is accompanied by a rise in Pmax for the NH3 treatment. With SO2hi Pmax is even a little lower than the control. Wether this is the result of inhibition of the Calvin cycle (Kropff 1987) has to be examined. From fig. 4 it is evident that the effects of these two gasses on the electron transport rate are on a different level. While NH_3 stimulates the steady state electron transport rate, SO_2 also enhances the rate of activation of the CO_2 fixation reactions. The combined fumigation SO2NH3 had no effect on σ, Pmax, Rd or Φ_{pe}, the gasses seem to counteract each other. When we try to compare these results to outdoor measurements, one should keep in mind that these results were obtained from healthy plants under optimal conditions. Combination of these fumigation treatments with other stress conditions, such as drought and low or high temperatures, could induce contrasting results from those obtained here.

Acknowledgements
We wish to thank dr. W.J. Vredenberg for fruitful discussions and for comments on the manuscript. This research was made possible by grants from the National Institute of Public Health and Environmental Hygiene and is part of the Dutch Priority Programme on Acidification (project nrs. 77 and 78).

References

Bell J.N.B. (1987) Effects of air pollution on plants and their interactions with other environmental factors. In: **Acid Rain: Scientific and Technical Advances** (Perry R. et al. eds.) pp. 535-545, Selper Ltd., London.

Bilger W. and Schreiber U. (1986) Energy-Dependent Quenching of Dark-Level Chlorophyll Fluorescence in Intact Leaves. **Photosynth. Res.** 10, 303-308.

Björkman O. (1981) Response to Different Quantum Flux Densities. In: Encyclopedia of Plant Physiology, vol. 12A (Lange O.L. et al. eds.) pp. 57-107, Springer-Verlag, Berlin

Björkman O. and Demmig B. (1987) Photon yield of O_2 evolution and chlorophyll fluorescence characteristics at 77 K among vascular plants of diverse origins. **Planta** 170, 489-504.

Goudriaan J. (1982) In: Potential production processes (Penning de Vries F.W.T. and Van Laar H.H. eds.) Simulation of plant growth and crop production. pp. 98-103, Pudoc, Wageningen

Heath R.L. (1980) Initial events in injury to plants by air pollutants. **Ann. Rev. Plant Physiol.** 31, pp. 395-431.

Kitajima M. and Butler W.L. (1975) Quenching of chlorophyll fluorescence and primary photochemistry in chloroplasts by dibromothymoquinone. **Biochim. Biophys. Acta** 376, 105-115.

Kropff M.J. (1987) Physiological effects of sulphur Dioxide. **Plant Cell Environ.** 10, 9.

Kyle D.J. (1987) The Biochemical Basis for Photoinhibition of Photosystem II. In: Photoinhibition (Kyle D.J. et al. eds.), pp. 197-226, Elsevier Science Pub. B.V.

Shimazaki, K. and Sugahara K. (1980) Inhibition Site of the Electron Transport System in Lettuce Chloroplasts by Fumigation of Leaves with SO_2. **Plant Cell Physiol.** 21(1), p.125-135.

Schmidt W., Schreiber U. and Urbach W. (1988) SO_2 Injury in Intact Leaves, as Detected by Chlorophyll Fluorescence. **Z. Naturforsch.** 43c, 269-274.

Schreiber U., Schliwa U. and Bilger W. (1986) Continuous recording of photochemical and non-photochemical chlorophyll fluorescence quenching with a new type of modulation fluorometer. **Photosynth. Res.** 10, 51-62.

Schreiber U. and Bilger W. (1987) Rapid Assessment of Stress Effects on Plant Leaves by Chlorophyll Fluorescence Measurements. In: Plant Response to Stress (Tenhunen et al. eds.), pp. 27-53, Springer Verlag, Berlin.

Van Hove L.W.A., Koops A.J., Adema E.H., Vredenberg W.J. and Pieters G.A. (1987) Analysis of the Uptake of Atmospheric Ammonia by Leaves of *Phaseolus vulgaris* L. **Atmosph. Environ.** 21, 1759-1763.

Van Hove L.W.A., Tonk W.J.M., Pieters G.A., Adema E.H. and W.J. Vredenberg (1988) A Leaf Chamber for Measuring the Uptake of Pollutant Gases at Low Concentrations by Leaves, Transpiration and Carbon Dioxide Assimilation. **Atmosph. Environ.**, in the press.

Van Kooten and van Hove L.W.A. (1988) Fluorescence as a Means of Diagnosing the Effect of Pollutant Induced Stress in Plants. In: Air Pollution and Ecosystems. (Mathy P. ed.) pp. 596-601 D. Reidel Pub. Co.

Weis E. and Berry J.A. (1987) Quantum Efficiency of Photosystem II in Relation to 'Energy'-Dependent Quenching of Chlorophyll Fluorescence. **Biochim. Biophys. Acta** 894, 198-208.

LUMINESCENCE OF ORGANIC MOLECULES
THEORY AND ANALYTICAL APPLICATIONS IN PHOTOSYNTHESIS

Werner Schmidt

University of Konstanz, D-7750 Konstanz, F.R.G.

Key words: Luminescence, delayed luminescence, potential energy surface, fluorescence, lifetime, quantum efficiency, chlorophyll, photosynthesis

ABSTRACT

This article surveys some essential features of luminescence spectroscopy as subdisciplin of optical spectroscopy, particularly focusing on photosynthesis. Theory and origin of various types of luminescence of molecules and complex biological systems are explored in terms of potential energy surfaces. Depending on particular electronic species and a great diversity of mechanisms involved, numerous biochemical, biophysical and photobiological mechanisms are accompanied by emission of light quanta, including prompt fluorescence and phosphorescence as well as a great variety of delayed light emission. It is exemplified how these phenomena may be experimentally adopted for assaying parameters not easily accessible by other, non-invasive methods.

INTRODUCTION

Optical spectroscopy as one of the most powerful tools in (photo-)biology (1) is operationally subdivided into absorption-, scattering-, reflexion- and luminescence spectroscopy as surveyed by fig. 1.

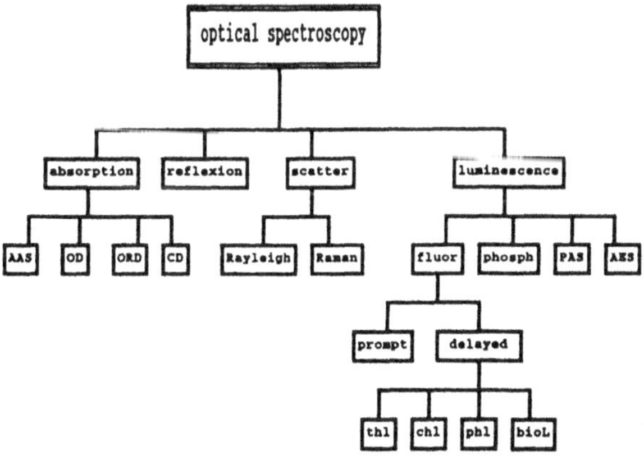

Figure 1. Classification of optical spectroscopy. AAS: atomic absorption, OD: (molecular) absorption, ORD: optical rotational dichroism, CD: circular dichroism, fluor: fluorescence, phosph: phosporescence, PAS: photoacoustic spectroscopy, AES: atomic emission spectroscopy, thl: thermoluminescence, chl: chemoluminescence, phl: photoluminescence, bioL: bioluminescence.

Here we will focus our attention on the latter topic. Any phenomenon of lightemission by electronically excited molecules upon their transition to the ground state is termed "luminescence" (4)(in the present article we will ignore AES, atomic emission spectroscopy). Luminescence, in turn, is further subdivided in fluorescence and phosphorescence, emanating from the first excited singlet or triplet state, respectively. Emission spectroscopy in a wider sence also includes photoacoustic spectroscopy (PAS),

H. K. Lichtenthaler (ed.) Applications of Chlorophyll Fluorescence, 211–216.

monitoring heat evolving from excited states. Depending on the mechanisms involved, we observe prompt or delayed fluorescence and phosphorescence (collectively termed "luminescence"). "Delayed" emission is defined by a lifetime τ lasting significantly longer than the natural ("intrinsic") lifetime τ_o of the emitting species (2, 6). Excitation of electronic states can be accomplished by several means such as heat, light, electromagnetic fields or chemical reactions giving rise to the extraordinary width of analytical applications of luminescence.

Compared to absorption spectroscopy, the advantage of luminescence spectroscopy is twofold: (i) While absorption spectroscopy essentially determines only one single parameter, i.e. the absorption coefficient as a function of the wavelength, $\varepsilon(\lambda)$, luminescence spectroscopy allows the measurement of a great number of parameters, each bearing independent information about the fluorophore, such as quantum efficiency Φ of fluorescence, range "R_c" of (radiationless) energy transfer, lifetime τ, intrinsic (p_o) and apparent polarisation p, rotational motility ρ, and viscosity η or polarity z of its microenvironment. (ii) Luminescence light is directly emitted by molecules and measured as absolute entity, whereas absorption is a dimensionless, relative measure: It is defined as a difference of logarithms ($A = \log I_o/I$), i.e. only transmitted light (I) is taken into account, which actually did not interact with the molecule and related to the intensity I_o of the impinging "measuring light". For this reason luminescence offers a spectral assay many orders more sensitive than absorption, presenting a highly versatile tool in photosynthesis research. However, due to the single beam character of fluorescence measurements, determining corrected fluorescence spectra particularly at wavelengths exceeding 600 nm remains a serious problem, even on the basis of computerized fluorimeters (5).

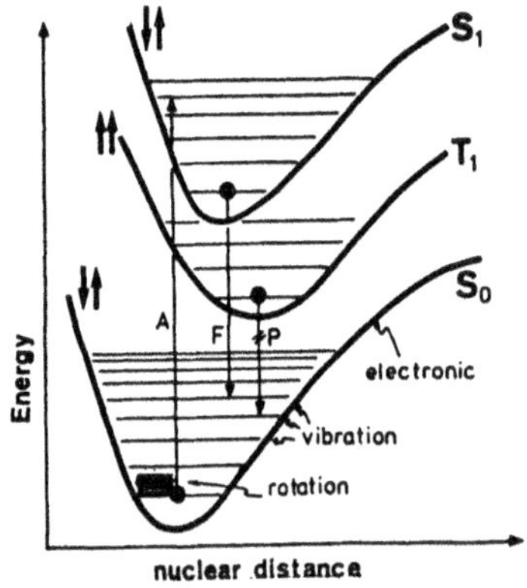

Figure 2. Potential energy diagram of a diatomic molecule as function of inter-atomic distance, for the ground and the first excited singlet (S) and triplet (T) states. The Franck-Condon principle is based on "vertical transitions" while nuclear distance remains constant. Three types of transition according to absorption (A), fluorescence (F) and ("forbidden") phosphorescence (P) are shown.

ORIGIN AND TYPES OF LUMINESCENCE

Atoms exhibit simple "line-spectra", identical for absorption and fluorescence (Einstein's relation), whereas spectroscopical properties of large molecules and molecular systems are much more complex. Nevertheless, essential features are readibly accessible by means of potential diagrams. For a hyphothetical biatomic molecule, for each electronic state the potential energy is plotted as a function of distance. Such a potential trough is distinctly "filled up" with vibrational states, each vibritional state, in turn, bearing on top a series of rotational states resulting in typical "band-spectra" bearly resolving rotational and vibrational sub-states (fig. 2). For multiatomic molecules such a diagram is readily extended to higher -even if less perceptive- dimensions.

Commonly, the various excitation levels of biological molecules are more comprehensively depicted by a Jablonski-diagram (i.e. omitting the potenial envelopes) such as for chlorophyll a in fig. 3, right, which is paralleled with the corresponding spectra on the left. Fluorescence exlusively takes place from the first excited singlet state and is inversely related to the probability of "radiationless" deexcitation: Photosynthetic activity, e.g., is can be traced by fluorescence: A decreasing fluorescence signal indicates an increased photosynthetic activity, and vice versa.

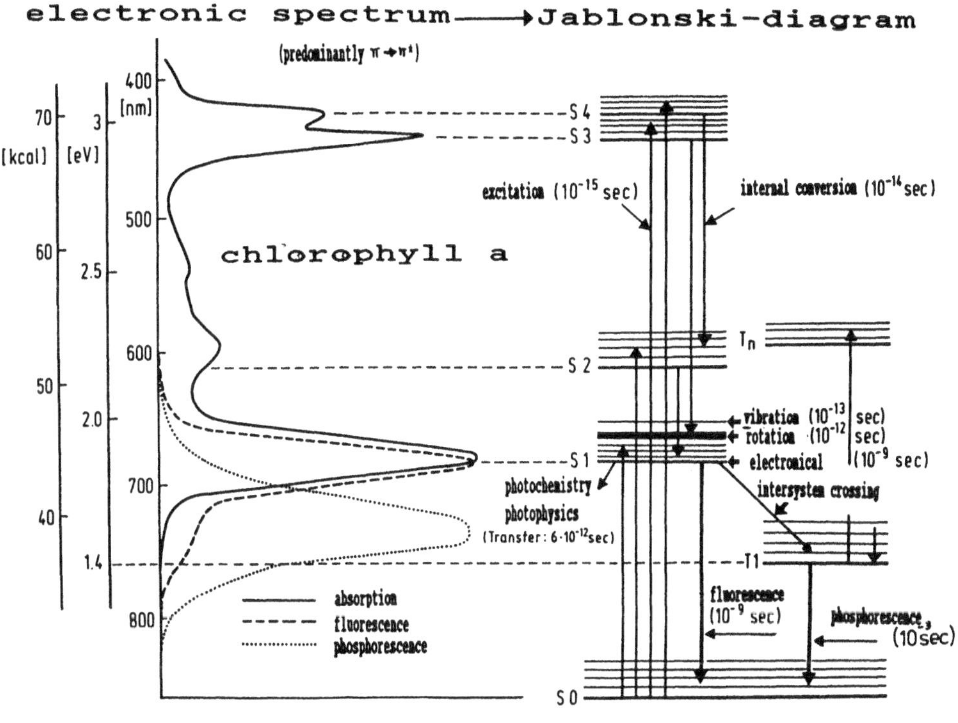

Figure 3. Right: Schemed Jablonski-diagram of chlorophyll a including crude reaction times. Left: Corresponding spectra elucidated on the common wavelength-scale.

All biologically relevant molecules except O_2 are singlet in their ground state, spins of valence electrons being paired (diamagnetic). In the excited states spins are either paired (S = 1/2 - 1/2 = 0) or alligned (S = 1/2 + 1/2 = 1, paramagnetic), giving rise to singlet- and triplet-ladders. For physical reason transitions between singlet- and triplet states ("intersystem crossing") are largely "forbidden", favoring transitions without "change in multiplicity". As a rule of thumb, triplet→singlet transitions (and vice versa) are about 1000 times less favorable, giving rise to extremely weak spectral bands when observed at room-temperature.

For physical reasons transitions between various electronic states always take place from their vibrational groundstates and end up in vibrationally excited states (Franck-Condon principle). If energy is being absorbed, the transition occurs "upwards", upon "downward" transitions (i.e. to lower electronically excited states), energy is set free, occasionally in form of light ("luminescence"). Assuming a similar order of vibrational states for various potential curves (cf. fig.2), the typical mirror-symmetry of absorption- and luminescence spectra is explained. It also becomes obvious that emission-spectra are "bathochromically" shifted (and phosphorescence spectra even more), peaking at longer wavelengths than the absorption spectra ("Stokes′ shift"). The Stokes′ shift, e.g., can be adopted for monitoring hydrogen-bonding. Another application: after normalization, (at first approach) the intersection of absorption- and fluorescence spectra marks the only common transition-wavelengths, the so-called "0-0-transition", which only takes place between vibrational ground states. It represents a sensitive function of temperature and polarity of the microenvironment offering itself as a "probe".

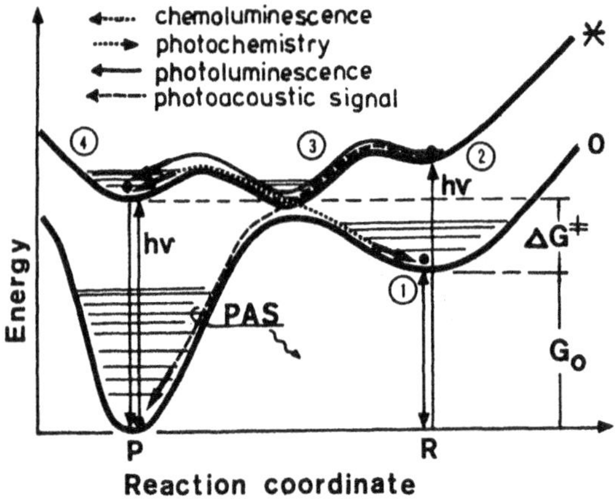

Figure 4. Generalized potential energy diagram for a more complex system of interacting molecules, but for clearity reduced to two dimensions. Various types of light reactions are sketched.

In photobiology, particularly photosynthesis, many different pigment systems such as flavins, carotinoids or chlorophylls of photosystem II and I are intimately correlated within (tylakoid-) membranes. Therefore, a generalization of the above mentioned potential diagram for a single molecule (fig. 2) to higher dimensional, multimolecular potential energy surfaces readily allows a principle understanding of (bio-) photochemistry and (delayed) bio-, photo-, chemo- and thermoluminescence (fig. 4). Basically there are two relevant surfaces, the ground and first electronically excited state. The abscissa of fig. 4 represents the "reaction coordinate" tracing the "progress" of a reaction, e.g., in terms of time or separation of the components, where "R" stands for reactants and "P" for "products", the ordinate the corresponding potential energy for all intermediate states in between. If the system in state "R" is irradiated with light, it is excited to an intermediate "1", from where it might proceed via the intermediate "3" to the excited state "4", provided sufficient activation energy can be taken up (photoluminescence). However, it is also feasible to supply the free energy ΔG_o by a chemical reaction thereby inducing a transition to "2", from where the system -again after uptaking activation energy $G^{\#}$- might pass over to the excited state "4" (chemoluminescence).

There are some prerequisites for various kinds of luminescence to occur: (i) For photoluminescence to occur the wavelength of the exciting light has to be sufficiently small ($\lambda < hc/G_o$). (ii) If the intrinsic lifetime of state "4" (τ_o) is not much smaller than some nanoseconds and the thermal deexcitation of state "4" is not likely to occur, than the fluorescence quantum efficiency Φ and the apparent fluorescence liftetime τ are sufficiently high. If this is not the case (e.g., for carotenoids as light-harvesting pigments in photosynthesis which do not fluoresce), an efficient "Förster-energy-transfer" towards another, ternary fluorophore ("enhancer") might nevertheless lead to appriciable fluorescence. (iii) There must be sufficient activation energy $G^{\#}$ available. At low temperatures this might not be the case. If such a system is warmed up, according to Arrhenius' equation ($k = \exp-(G^{\#}/RT)$, with reaction constant k, gas constant R and absolute temperature T), the route towards "4" might open again and (delayed) light emission occur as function of temperature (thermoluminescence). (iV) The adiabatic passage (i.e. without external heat exchange) from "R" to "P" on the excited state surface (cf. fig. 4) is only terminated by a "radiative jump" if the gap between both surfaces is not too small. Photochemistry based on the photoinduced, reverse transition from "P" to "R" with a free energy ΔG_o, however, is favored by a smaller gap between the potential energy surfaces.

APPLICATIONS OF LUMINESCENCE

We will confine ourselves to some repesentative examples. For reasons given above prompt fluorescence is the widestly utilized phenomenon in photosynthesis research, on the basis of spectral or kinetical analysis. A (properly corrected) excitation spectrum reflects exactly the absorption spectrum of a molecule, often used for identification. Emission spectra must not change with the excitation wavelength, which can be taken as criterion for purity. Prompt fluorescence as a function of time or concentration of external components allows to monitor the physiological status of photosynthetic organisms, e.g. in herbicide or pollution research. At physiological temperatures fluorescence emanates mostly from chlorophyll a of photosystem II: In organisms treated with the herbicide DCMU (3-(3,4-dichlorophenyl)-1,1-dimethylurea), as another example, the electron flow between photosystem II and I is blocked. Consequently, the excitation energy of photosystem II cannot be transferred to photosystem I but is rather wasted partly in form of fluorescence of various components, which increases within a few seconds after onset of light exhibiting complex kinetics (Kautsky-effect). The Kautsky-effect has been extensively utilized for studying photosynthesis.

Phosphorescence of chlorophylls so far has not been utilized for any photophysiological assay, but a detailed photophysical analysis is available (3).

Delayed luminescence (DL, 10^{-16} E/cm^2 s and below) is by several orders weaker than prompt fluorescence (10^{-10} E/cm^2 s) and not detectable by the unaided eye. Its measurement requires spectrophotometers not necessarily available in biochemical laboratories. Using highly sophisticated equipment only a few quanta of light per second are detectable. Kinetics of DL in photosynthetic systems range from µs up to several minutes; spectral analyses indicate complex underlying mechanisms. Three basic mechanisms can be distinguished:

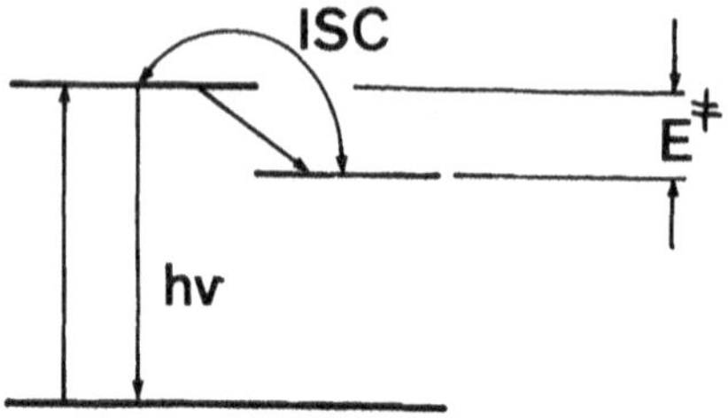

E-type DL: Decay time like phosphorescence, monomolecular process, signal proportional to intensity of exciting light, temperature-dependent, applies to pigments with $E^{\#}$ of 20-90 kJ/mol (chlorophyll a: 30 kJ/mol).

$$T_1^* \rightarrow S_1^* \rightarrow S_0,$$

P-type DL: Triplet-triplet annihilation, decay constant twice that of phosphorescence, bimolecular reaction, hence signal proportional to I^2, dependent on viscosity, temperature and concentration.

$$T_1^* + T_1^* \rightarrow (2{*}T)^* \rightarrow S_1^* + S_1$$

Electron transfer: Reaction taking place in complex biological systems by electron and energy-transfer according to the diagram of potential energy surfaces in fig. **4** . DL of photosynthetic systems as a representative example (common terminology) are further explored in a accompanying paper (Schmidt):

H_2O-OEC-P_{680}-Phe-Q's-PQ-RFeS-cyt f-PC-P_{700}-Fd-NADP.

PS II ⇝ (H₂O-OEC-P680-Phe-Q's) PS I ⇝ (f-PC-P700-Fd-NADP)

The term "bioluminescence" commonly includes only phenomena of relatively strong, visible light emission by organisms like bugs, fish or bacteria used for communication, regardless of its physical/biochemical basis, however, does not include weak and ultraweak light emission observed in numerous biological systems. Bioluminescence is a kind of chemoluminescence which molecular basis is in principle known: In essence it is a consequence of biochemical oxidation processes, yielding free energy -partly in form of light-, a phenomenon widely adopted for various clinical and biochemical luminescence assays.

REFERENCES

1. Brown, S.B. (ed.) An introduction to spectroscopy for biochemists, Academic Press Inc., 403 pages, ISBN 0-12-137080-1, 1980
2. Jursinic, P.A. Delayed Fluorescence: Current concepts and status. In: Light emission by plants and bacteria, pp. 291-382, eds. Govindjee, J. Amesz and D. C. Fork, Academic Press Inc., 1986
3. Krasnovsky, A.A., Jr, Romanyuk, V.A., and Litvin, F.F. Dok. Akad. Nauk SSSR 209, 965-973, 1973
4. Lakowicz, J.R. Principles of fluorescence spectroscopy, Plenum Press, New York and London, 496 pages, ISBN 0-306-41285-3, 1983
5. Schmidt, W. On-line-computer capability in fluorescence spectroscopy. Optical engeneering 22, 576-582, 1983
6. Turro, J.N. Modern molecular photochemistry, Benjamin/Cummings Publishing Co., Inc, 628 pages, ISBN 0-8053-9354-4, 1978

LONG TERM DELAYED LUMINESCENCE IN GREEN ORGANISMS

Werner Schmidt

University of Konstanz, D-7750 Konstanz, F.R.G.

KEY WORDS: Delayed luminescence, forest decline (Waldsterben), pollution control, photosynthesis, chlorophyll fluorescence

ABSTRACT

Long term delayed luminescence of green organisms in the time range between 0.3 s and several minutes (LDL) has been investigated on a kinetical and a spectral basis. A specific influence of various exogeneous parameters such as herbicides or essential elements, different pH values, temperature, preillumination and diurnal rhythms is found. LDL is particularly dependent on nutrinional deficiencies and depletion of essential elements and environmental pollutive components, probably involving both photosynthetic reaction centers PSI and PSII. Therefore, LDL offers itself as a convenient, highly sensitive, specific and non-invasive assay for a number of stress factors in photosynthesizing plants in the field.

INTRODUCTION

Delayed luminescence (DL) has been accidently discovered in 1951 by Strehler and Arnold (11). As a valuable guide Lavorel (5) introduced the so-called "τ-scale", which separates DL in terms of three distinct regions of decay kinetics with halflives ranging from 10^{-5} seconds (short term delayed luminescence, SDL) up to several minutes (LDL). The domain larger than $t_{1/2} = 5$ s comprises very little data in literature even though valuable information may be extracted from this type of experiments (1).

Emission- and excitation-spectra of SDL often resemble those of prompt fluorescence (1). Consequently, SDL has been assumed to involve the same molecular species as prompt fluorescence: the lowest excited singlet state of chlorophyll a of PS II. Various possible mechanisms generating DL have been suggested (6). The most accepted explanation at present favors a thermally induced back-reaction of the transmembraneously accumulated charges as built up by the photosynthetic electron transport chain (cf. Schmidt, 3rd mechanism, accompanying paper). Nevertheless, there have been claims that some types of DL originate from PS I (6), which are strongly supported by our own spectral data. Since screening, selfabsorption and scattering as indispensable obstacles inherent in optical spectroscopy of opaque materials such as leaves of higher plants will truncate both spectra and kinetics in a badly controlled manner (8), in our later experiments we adopted the green alga *Scenedesmus obliquus* (4, 9, 10). Samples are easier to handle and allow a better control of various optical and physiological parameters. In addition, there are several specific mutants available with well-defined defects offering further analyis of LDL.

H. K. Lichtenthaler (ed.) Applications of Chlorophyll Fluorescence, 217–221.

MATERIALS AND METHODS

Cultures of the unicellular algae Scenedesmus obliquus strain D3 were grown at a temperature of 30° C in liquid inorganic medium and aerated with 3% CO_2 in air. In depleted media salts of Ca, Mg, N, K or Fe were substituted by salts of other elements of the medium. For some experiments the medium was supplemented with Cl-salts of Cu (2.5 mg element/l = 39.4 μM), Cd (2 mg/l = 17.8 μM), Ni (2.5 mg/l = 42.6 μM), Hg (1 mg/l = 5μM) or Pb (2 mg/l = 9.3 μM). NO_2^- (80 mg/l = 1.74 μM) was given as Na-salt. The concentrations were chosen according to (7). These are highest concentrations tolerated in German rivers. Further datails are given elsewhere (4).

Measurement of LDL with typical fluence rates smaller than 10^8 quanta/ml s requires custom made set-ups described in detail previously (8). Absorption was measured with a homemade, computerized single-beam spectrophotometer, and corrected fluorescence spectra with the Shimadzu fluorimeter, model RF-505.

RESULTS AND DISCUSSION

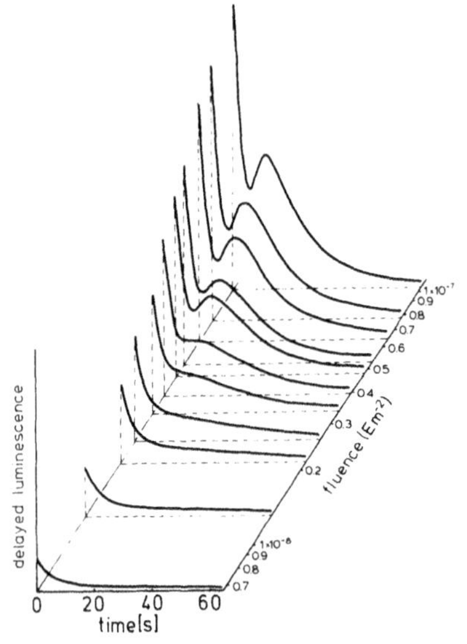

Fig.1

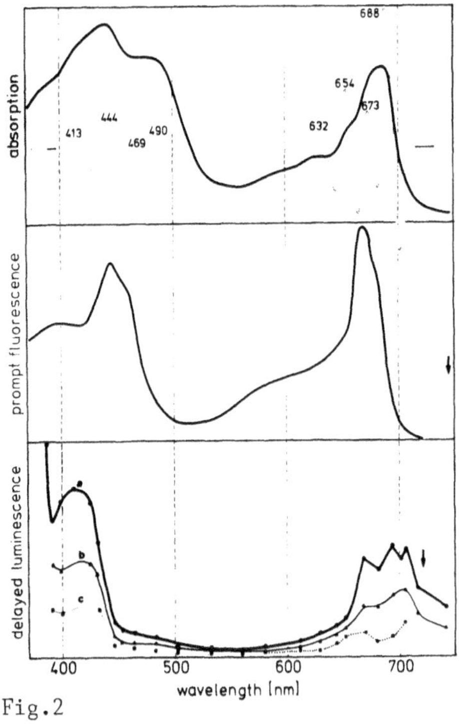

Fig.2

Figure 1, Kinetics of LDL as function of the fluence of a 1 s exciting light pulse of 721 nm. Light emission was monitored at 712 nm; after (9), with permission.

Figure 2, Top Absorption spectrum of wild type Scenedesmus. Total extinction is approximately A = 1. Dotted line: fourth derivative. Middle Corrected excitation spectrum of prompt fluorescence. Excitation wavelength: slit width 5 nm; emission wavelength: 750 nm, slit width 15 nm. Bottom Action spectra derived from fluence response curves at three different responses "a","b","c"; after (9), with permission.

The dependency of the decay kinetics of LDL of Scenedesmus on the fluence rate of a 1 s exciting light pulse is shown in fig. 1.

At fluence rates exceeding $0.3*10^{-7}$ E/m²s and with light emission monitored at 712 nm and excitation at 721 nm, after about 10 s a long term luminescent intermediate is observed but fades out with smaller fluence rates of the inducing light. This intermediate only develops upon excitation with light above 700 nm, indicating the involvement of PS I. Similar results are obtained with the emission-wavelength: The intermediate bump exhibits its maximum at 720 nm.

Following standard procedure, we determined the corrected excitation spectra for prompt and delayed fluorescence (fig. 2). Compared with absorption and prompt fluorescence, LDL clearly turns out to be far more complex than representing a plain reversal of primary charge separation by PS II, particularly involving pigments of PS I.

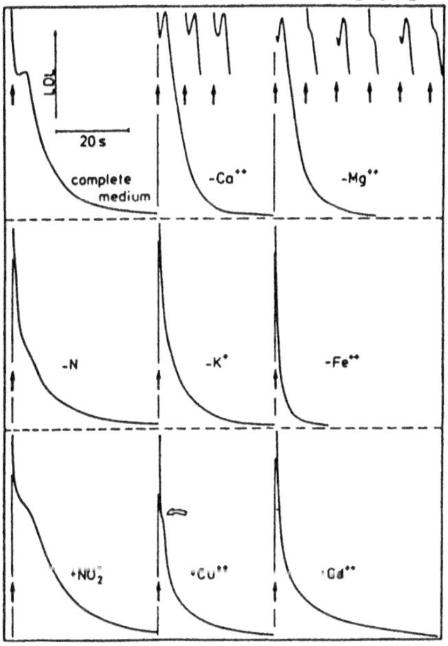

Figure 3 LDL-kinetics of Scenedesmus obliquus grown under various nutritional conditions (from left to right, top to bottom): Complete medium (similar to +187 μM Pb), under depletion of calcium, magnesium, nitrogen, potassium, iron, and under supplementation with 1.74 mM nitrite, 788 μM copper (similar to +5 μM Hg) and 17.8 μM cadmium (similar to +852 μM Ni), all applied as chlorine-salts. Kinetics were induced by "flashes" of 1 s light of 10^{-7} E m^{-2} (solid arrows) and monitored at 712 nm (AL-interference filter, Schott). For clarity, the empty arrow indicates a "residual hump" in the kinetics for copper (and mercury-) deficiency (after (4), with permission).

Figure 3 exemplifies some of the most significant effects of nutrition deficiencies and environmental stress factors on the kinetics of LDL. In some cases the intermediate hump is completely lost ($-K^{+}$, $-Fe^{++}$, $+Cd^{++}$), sometimes it is emphasized reaching the same height as the initial component of LDL ($-Ca^{++}$) and in one particular instance ($-Mg^{++}$) we observed even a repetetive behaviour: every second kinetic (largely independent of light intensity, however dependent on the duration of darkness between 1 s-"flashes") exhibits a strongly amplified intermediate, with kinetics in between showing only minute shoulders rather than a distinctive hump. Both nitrogen deficiency and nitrite supplementation result in relatively small shoulders but prolonged kinetic decay. In the case of supplementation of the growth medium with copper only a minute bump remains.

Copper is known to be toxic in higher concentrations. Chlorotic effects originate obviously from displacing other metal-ions, particularly Fe, at physiologically sensitive sites (3). Only a small shoulder in

LDL-kinetics indicates similarities with the kinetics observed after K^+- and Fe^{++}-depletion. Cadmium degrades photosynthesis in maize by inhibition of PS-II-activity (2).

By far the fastest kinetics observed is that obtained under conditions of Fe-deficiency, the slowest that with nitrite-supplementation ($+NO_2^-$). The effect of iron is probably related to its indispensible requirement by compounds such as ferredoxin or various cytochromes representing essential components of the photosynthetic electron transport chain. Calcium, potassium and magnesium ions known to reveal important roles in membrane-correlated mechanisms also might be expected to be reflected by modification of LDL-kinetics.

Summarizing, the effects of environmental stress factors on LDL present a potential assay of mineral deficiencies and toxifications of plants and photosynthesizing microorganisms. This is further exemplified in the current example of fig. 4, comparing LDL-kinetics taken from a healthy and a partly damaged pine tree.

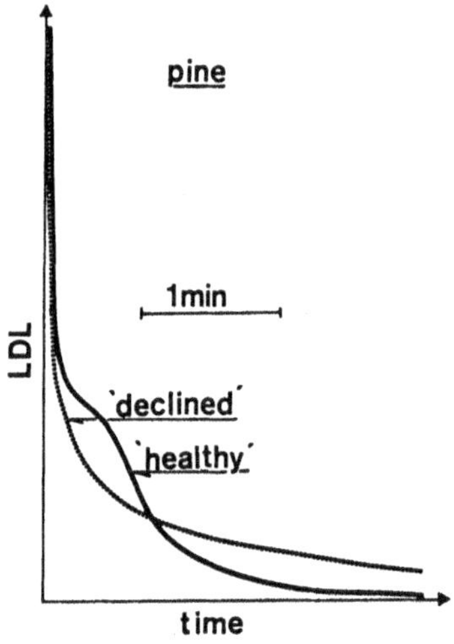

Figure 4 LDL-kinetics of needles of two pine trees of different status of decline selected from the forest (Marburg/Germany) by eye.

A microcomputer-based library of all kinds of LDL-kinetics could help to further cope with our environmental problems. It has been argued that the light signal of LDL evolves either from the reaction centers of PS I or PS II themselves or it close vicinity, whereas prompt fluorescence originates in the bulk-antenna pigments. Therefore the former is supposed to be a more immediate probe of the state of vitality of photosynthetic organisms. Finally, preliminary measurements promise an assay by remote field measurements of LDL-kinetics employing laser excitation and tele-spectroscopy.

ACKNOWLEDGEMENT

The experiments could be performed with the help of Prof. Senger in Marburg.

REFERENCES

1. Amesz, J. and Gorkom, H. J. v. (1978): Delayed fluorescence in photosynthesis. Ann. Rev. Plant Physiol. 29, 47-66

2. Bazzas, F.A., Rolfe, G.L. and Carlson, R.W. (1974): Effect of Cd on photosynthesis and transpiration of excised leaves of corn and sunflower. Plant Physiol. 32, 373-376

3. Butjas, C. and Cseh, E. (1981): Effect of heavy metals and chelating agents on potassium uptake of cereal roots. Plant and Soil 63, 97-100

4. Bürger, J. and Schmidt, W. (1988): Long term delayed luminescence: A fast and convenient assay for nutritional deficiencies and environmental pollution damages in plants. Plant and Soil, in press

5. Lavorel, J. (1975): Luminescence. In: Bioenergetics of Photosynthesis (Govindjee, R. ed.), pp. 223-317, Academic Press, New York, San Francisco, London

6. Malkin, S. (1977): Delayed Luminescence. In Primary Processes of Photosynthesis., (Barber, J., ed.), pp. 349-431, North Holland Biomedical Press

7. Ökotest Magazin (1986): Kommunale Abwasser-Satzungen, Grenzwerte für Schadstoffeinleitungen der Länderarbeitsgemeinschaft Wasser (LAWA), 9, 36-41

8. Schmidt, W. (1985): Prompt and delayed fluorescence in mature tabacco leaves: a spectral comparison. Photobiochem. Photobiophys. 9, 89-97

9. Schmidt, W. and Senger, H. (1987a): Long-term delayed luminescence in Scenedesmus obliquus I. Spectral and kinetic properties. Biochim. Biophys. Acta 890, 15-22

10. Schmidt, W. and Senger, H. (1987 b): Long-term delayed luminescence in Scenedesmus obliquus II. Influence of exogeneous factors. Biochim. Biophys. Acta. 891, 22-27

11. Strehler, B. L. and Arnold, W. (1951): Light production by green plants. J. Gen. Physiol. 34, 809-820

EARLY DETECTION OF DAMAGE CONDITIONS IN PLANTS BY DELAYED CHLOROPHYLL FLUORESCENCE

Rolf Blaich, Bundesforschungsanstalt für Rebenzüchtung, Geilweilerhof, D-6741 Siebeldingen, West-Germany

KEY WORDS

delayed fluorescence, luminescence, herbizides, frost damage, heat damage

ABSTRACT

A device is described which allows the analysis of delayed chlorophyll fluorescence (luminescence) recorded continuously 20 msec after 50 msec pulses of exciting light. Both qualitative analyses of luminescence images produced by an image intensifier and quantitative measurements using a photomultiplier are possible. Luminescence kinetics are recorded by means of a microcomputer which allows saving of data files, automatic statistic treatment and printout of single or combined curves. The device has been applied to analyse damages caused by herbizides, heat, and freezing.

INTRODUCTION

Delayed luminescence phenomena in green parts of plants, often referred to as "luminescence", may be considered as a special case of the Kautsky effect of fast fluorescence which has been discussed in detail in the previous papers of this seminar (for literature see Blaich et al., 1982). It should be pointed out, though, that an intact photosynthetic system must be present for these effects to occur: fast fluorescence of isolated chlorophyll is well known, but the Kautsky effect and related phenomena are restricted to chloroplasts or at least to complete chlorophyll-protein complexes.

The analysis of luminescence is more difficult than that of fast fluorescence because is it very weak and hence needs complete darkness which cannot be provided by filters but affords a shutter mechanism which may lead to a bulky device. We feel, however, that luminography has some advantages which sometimes outweigh those drawbacks and which will be discussed in the following paper. A direct comparison of both methods is given by Stein et al. (1986).

MATERIAL AND METHODS

Kinetics of delayed fluorescence. The device we use to measure luminescence is based on the apparatus described by Björn and Forsberg (1979). We published a first version 6 years ago (Blaich et al., 1982). A much larger one was described at the same time by Ellenson and Amundson (1982). In the meantime the equipment has been enhanced, its size is now reduced and a microcomputer controls the phosphoroscope and processes data including some statistic treatment.

H. K. Lichtenthaler (ed.) Applications of Chlorophyll Fluorescence, 223–228.

The specimen (normally a leaf) is mounted on a stage by means of a variable metal mask which allows to expose 0.5, 1, 2, 4 and 8 square centimeters of the leaf surface. The form of the diaphragm is either a square which is used for normal leaves or a long rectangle for grass leaves or needles. If a leaf is large enough, subsequent measurements on different locations may be carried out without exposing the rest of the leaf.

A rotating slot disc allows alternating illumination of the specimen by means of a fiber optic. Quality and intensity of the exciting light are kept constant at 3800 ^{0}K and 15 W/m^2, repectively.

Basing on preliminary experiences the turning rate of the slot disk has been set to 8 revs/sec. The specimen is thus illuminated for 50 msec, the measurement of the afterglow takes place 20 msec after closing the path of the exciting light, when the diaphragm has opened the light path to the photomultiplier. At this point an impulse is transmitted to the microcomputer which samples the value of luminescence by means of an AD-converter. Depending on the recording speed, 2 or more values are integrated to yield one point of a curve which is displayed on the screen of the microcomputer.

The grafics resolution of the APPLE II microcomputer used in this case is limited to 191 x 280 pixels. The program which records and displays the luminescence kinetics had to be adapted to these restrictions.

The program (for a detailed listing see Blaich, 1985) was written in Applesoft Basic to allow easy changes during the development period. Time critical parts of the AD-converting subroutines are written in machine code which is poked in at the beginning. The program starts with the display of a menue offering the following possibilities:

1) Recording of luminescence kinetics

In this case some parameters are requested:

a) After the run, the curve and its parameters are saved on a floppy disc: a name for the file has to be entered. Statistical evaluations afford repeated measurements. If this is to be done automatically by the computer (up to 9 repeats), the filenames have to be identical with the exception of a single cipher (1 through 9) at the end.

b) the date of the run and some informations (up to 255 characters) may be entered (optionally).

c) the recording speed has to be entered which may be about 30 sec for the fastest run, 33 min for the slowest. The standard record takes 3 min which is usually sufficient to reach the steady state of the kinetics, although this needs more time than in fluorescence measurements (see Stein et al., 1987) due to the pulse illumination. The speed of the record depends (1) on the rotation speed of the slot disk, kept constant by an electronic control and (2) on the number of measurements integrated to obtain a single dot of the curve. The values of two rotations are integrated for the fastest run, the points of a standard 3 min kinetic correspond to the mean value of 12 measurements, and of 128 for a 33 min curve.

At the beginning of the run the program plots a grid depending on the record speed on the screen. Time zero of the diagram is set after some seconds of recording which starts with the light turned off. When the plot reaches the time zero line a sound signal reminds the operator to switch on the light. The run may be stopped at any time by one of several options: type "SPACE" to clear the screen and start again with identical parameters; type "S" to save the (even uncomplete) curve on disk before turning back to the main menue; type "N" to go directly to the menue to change parameters, type "P" to get a printout of the curve as a screendump on a printer.

2) Automatically calculate a mean curve from repeated measurements, and the standard deviation of every 20th point and save it onto diskette.

3) Plot saved curves from the diskette. For the sake of comparison several curves may be combined within one diagram. The standard deviation of a mean curve obtained by option (2) may be plotted optionally. Typing a "P" causes a printout of the screen.

4) A fast plot option allows a continuos scanning of the luminescence for tests and demonstration purposes

Luminescence imaging. To obtain photographs, the photomultiplier was replaced by an image intensifier (Proxitronic type BV 2542 QG 35) which allowed to get undistorted high resolution images over the whole screen area. The relatively low amplification (7500) of this type was compensated by its fiber glass screen, which allowed contact prints of the screen without loss of light. The gain in sensitivity was about 50 as compared with normal macrophotography of the screen.

RESULTS AND DISCUSSION

Although traces of luminescence may still be demonstrated more than 20 sec after the end of excitement, the following results base on kinetics obtained by measurements after a constant delay of about 20 msec.

Controls. Green parts of plants pre-incubated under light conditions (at least 3 min at 100 W/m^2) show a faint luminescence which hardly changes during observation (figure 1 L). Leaf areas kept in the dark for 30 min, show a strong luminescence which, due to the influence of the exciting (pulse) light reaches the level of the pre-illuminated area (Fig. 1 D).

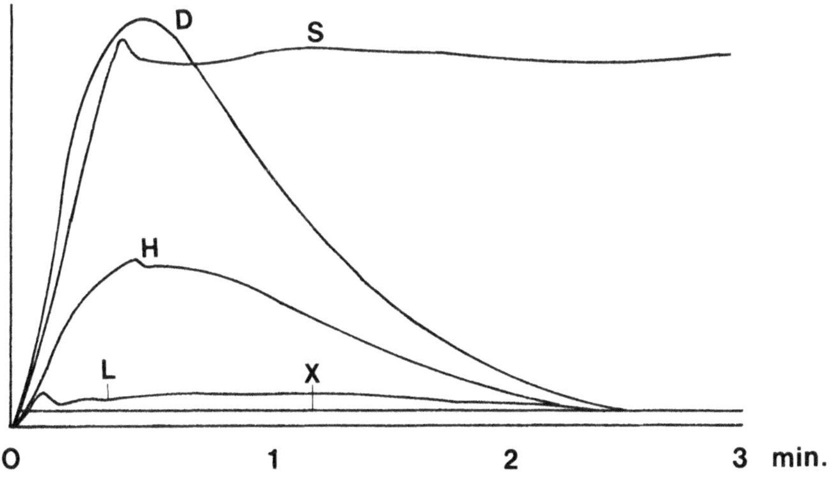

FIGURE 1. Luminescence kinetics of leaf areas after different pretreatments. D control (normal leaf, kept in dark for 30 min); L normal leaf, pre-illuminated; H heat treated (42 °C), kept in dark; X killed by heat (50 °C), the line is drawn too high, actually its value is zero; S photosynthesis blocked by herbicide.

Herbicide treatments. Inhibitors of the electron transport (like Diuron, Simazin, Atrazin) lead to a strong luminescence in damaged areas indicating a complete lack of CO_2 fixation (Blaich et al., 1982). This luminescence has the same intensity as that of controls kept in the dark but it does not disappear after illumination (Fig. 1 S). The penetration of photosynthesis inhibitors may be followed immediately in one single leaf without the necessity to use radioactive markers and great number of leaves which have to be killed before autoradiography (Fig. 2).

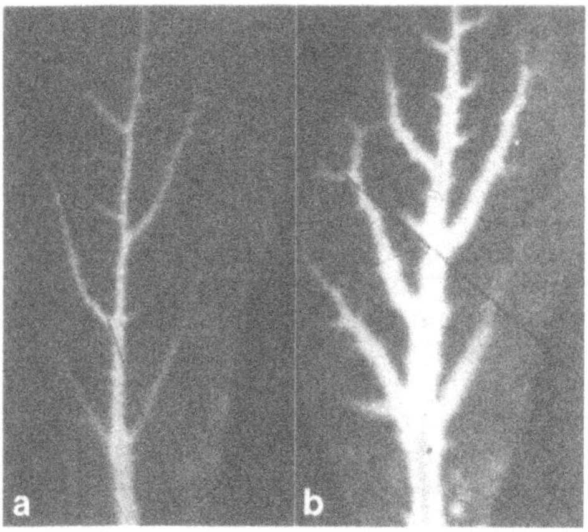

FIGURE 2. Luminograph of a primary leaf of *Raphanus sativus* ; a) after the uptake of simazine via the petiole for 5 min; b) after 10 min.

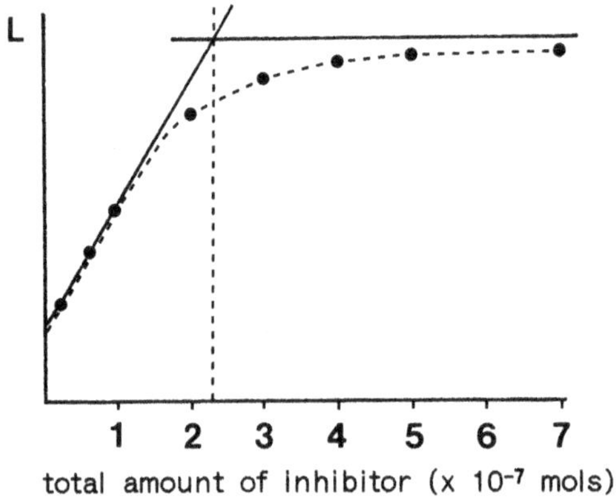

FIGURE 3. Delayed fluorescence of *Elodea* sprouts (2 g wet weight, containing about 2 mg chlorophyll) submersed in a DCMU solution, the amount of which was increased every 30 min during a continuous pulse illumination. This time was sufficient for the luminescence to reach a constant level (plotted as a dot), which depended on the amount of inhibitor added.

The application of herbizides to terrestrial plants leads to unknown local concentrations within the leaf and it is not possible to remove the substances. In contrast, the submersed growth of *Elodea canadense* allows the application of defined amounts of inhibitors which penetrate quickly (waterplants lack a cuticula) and may be removed again. Fig. 3 shows the effect of different amounts of DCMU on the luminescence of *Elodea* sprouts. The linear correlation at the beginning between luminescence and the amount of inhibitor added indicates a high affinity for its binding sites and allowed to estimate their maximum number (4-8 per 100 molecules of chlorophyll). After removal of the DCMU by rincing with water, the experiment could be repeated several times.

Heat treatment. Strong heating destroys luminescence, but not fluorescence (fig. 4). Kinetik measurements show that slight heating yields intermediate curves (between C and X, however with normal kinetics (Fig. 1 H).

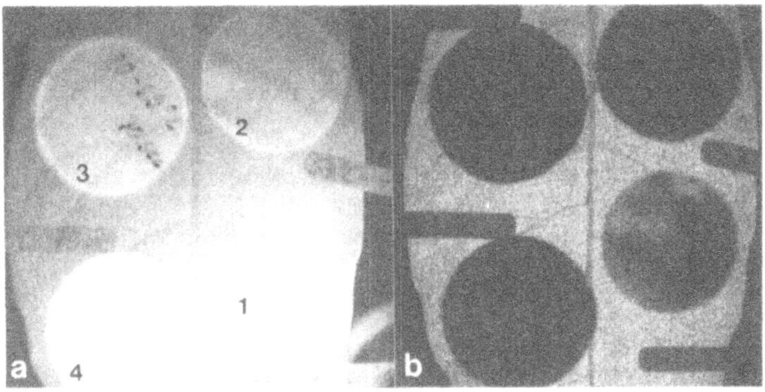

FIGURE 4. Heating of circular areas on a leaf of *Tetrastigma voineriana*. Hot brass cylinders (1 46 °C, 2 54 °C, 3 70 °C, 4 96 °C) were set on the spots for 70 sec.
a) Fast fluorescence image of the heated leaf: heating, even to complete killing of the tissue, enhances fluorescence.
b) Delayed fluorescence image: luminescence is partly (spot 1) or totally destroyed in heated areas.

Cold treatment. Freezing below 6-9 °C (depending on duration and plant species) of the leaves led to an immediate disappearance of the luminescence - the curves were identical with the zero line. Scanning electron microscopy showed that in these cases the cell walls were pierced by growing ice cristals. On the other hand cautious chilling either had no effect, or - at certain temperatures - an intermediate stage of damage could be obtained

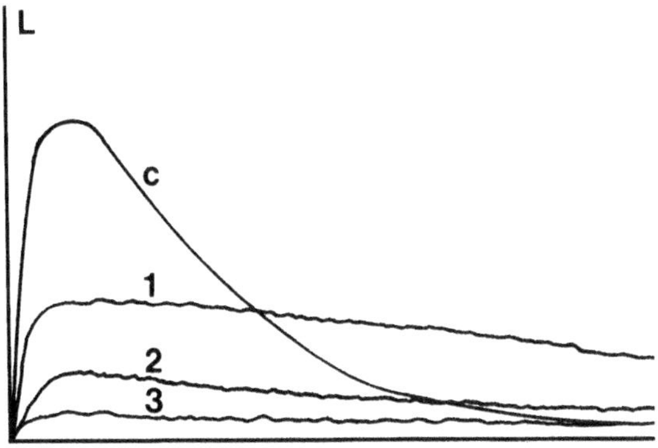

FIGURE 5. Effects of chilling a *Ficus elastica* leaf to irreversible damage: c control which was then chilled to -7 °C for 15 min and brought to room temperature within 3 min. 1, 2, 3 luminescence kinetics after 3, 7, and 11 min, respectively. Evidently the leaf has been dying during the subsequent measurements.

where luminescence was not quenched by the exciting light, but disappeared after some minutes due to the gradual dying of the leaf. This effect could be followed by subsequent measurements (Fig. 5). The transition temperature where this stage did occur seems to correspond to the general frost resistance of the plant.

CONCLUSIONS

As compared with fast fluorescence measurements, luminescence offers two major advantages:

1) The intensity of luminescence in photosynthically active vs. blocked areas is normally around 1:20 but only about 1:3 for fluorescence. On the one hand, if the specimen is photographed by means of a light amplifier (Fig. 1) this means (a) high contrast images of damaged leaves which have not been obtained by fluorescence and (b) distinct differences of kinetic curves which facilitate their interpretation.

2) Fluorescence of the chlorophyll is also present in dead parts of green plants, and sometimes even enhanced (Fig. 4), Therefore the distinction between parts of a leaf where photosynthesis is blocked reversibly and parts where the system is disintegrated irreversibly due to non-specific damage is sometimes difficult by this technique whereas those two levels of damage are easily differentiated by phytoluminography:

a) Healthy areas of leaves pre-incubated in the dark show a luminescence which disappears within some minutes due to the exciting pulse light. A pre-illuminated leaf shows only a faint luminescence from the beginning.

b) Leaf areas where only the energy flow is blocked or otherwise disturbed always show a strong luminescence which does not diminish even during prolonged observation.

c) Areas where the photosynthetic structures are disintegrated show no luminescence at all, whether the chlorophyll is present or not.

REFERENCES

Björn LO, Forsberg AS, 1979. Imaging by delayed light emission (phytoluminography) as a method for detecting damage to the photosynthetic system. Physiol. Plant. 47: 215-212.

Blaich R, 1985. Erfassung und grafische Auswertung von Meßdaten mit dem Apple II Mikrocomputer. Computer Anwendung im Labor 1: 160-167.

Blaich R, Bachmann O and Baumberger I, 1982. Studies of photosynthesis inhibition by phytoluminography. Z. Naturforsch. 37c: 452-457.

Ellenson JL and Amundson RG, 1982: Delayed light imaging for the early detection of plant stress. Science 215: 1004-1006

Stein U, Buschmann C and Blaich R, 1986. Fluorescence kinetics of chloroplasts as indicators of disorders in the photosynthetic system I. Comparative studies with greening leaves of *Vitis* and *Hordeum*. Vitis 25: 129-141.

Strehler BL and Arnold W, 1951. Light production by green plants. J. Gen. Physiol. 34: 809-820.

ASSESSMENT OF UV-B STRESS BY CHLOROPHYLL FLUORESCENCE ANALYSIS

TEVINI, M., GRUSEMANN, P. and FIESER, G.

Botanisches Institut II, Universität Karlsruhe (TH), Kaiserstr. 12, D-7500 Karlsruhe, F.R.G.

KEY WORDS: Ozone depletion, UV-B radiation stress, sunflower, chlorophyll fluorescence, q-analysis, photosystem II

ABSTRACT

UV-B stress was evaluated by measuring chlorophyll fluorescence induction kinetics with the PAM-fluorometer. Commonly used parameters such as Fo, Fm, Fv, Fp, qQ and qE are compared in sunflower seedlings grown under enhanced artificial or solar UV B radiation. It was shown that Fo increased whereas Fm and Fv decreased under enhanced artificial UV-B. Photochemical phase Fo-I1 decreased at a similar degree as "thermal" phase I1-I2. The half life of fluorescence dark decay after a single turnover flash was shorter under UV-B stress. In the steady state photochemical quenching qQ decreased with UV-B exposure whereas energy dependent quenching qE increased. Similar results were obtained under solar radiation at a Mediterranian latitude.

INTRODUCTION

Damaging effects of UV radiation, mainly UV-C (200-280 nm) on photosynthesis of algae and isolated chloroplasts are well documented (Renger et al. 1986). However, the highly deleterious UV-C is not present in daylight due to its absorption by stratospheric ozone. Enhanced levels of UV-B radiation (280-320 nm) constituting the shortwave part of the global spectrum are also known to inhibit growth and photosynthesis of UV-sensitive plants (Iwanzik 1983, Tevini and Iwanzik 1983). These UV-B induced phenomena have attained great importance since the documentation of stratospheric ozone depletion by photoactive gases particularly by chlorofluoromethanes (CFM) (Molina and Rowland 1974, Fabian 1986). CFMs which are used worldwide as refrigerants, propellents, solvents and foames are decomposed by solar UV-C radiation. The resulting chlorine radicals catalyze ozone destruction and thereby reduce UV-absorption capacity of the stratosphere. Therefore stratospheric ozone depletion increases UV-B intensity and shifts UV-B radiation to shorter wavelengths, than normally found in global radiation. Since vital biological molecules like DNA, proteins, some phytohormones and also photosynthetic quinones absorb in the UV range, an enhancement of the solar UV-B radiation may lead to damage of plants in different ways. As far as

H. K. Lichtenthaler (ed.) Applications of Chlorophyll Fluorescence, 229–238.

photosynthesis is concerned, it was shown that photosystem II (PS II) activity monitored as variable fluorescence is reduced by UV-B in dependence of fluence rate and wavelength in intact radish plants (Tevini and Iwanzik 1983). The decrease of variable fluorescence due to decreased maximal fluorescence has been explained by a transformation of PS II-reaction centres into dissipative sinks since the QB-protein remains intact at moderate UV-B levels which otherwise would lead to increased maximal fluorescence (Tevini and Pfister 1985).

In stress physiology, measurement of fluorescence parameters like Fo, Fm, Fv are suitable to characterize stress potency. In addition, the effects of UV-B radiation on photochemical (qQ) and energy dependent (qE) quenching were measured in this study. These parameters may also give insights into primary and secondary photosynthetic reactions under UV-B stress (Schreiber et al. 1986).

MATERIALS AND METHODS

Sunflower seedlings (*Helianthus annuus, cv. Polstar*) were grown at 20° C on standard greenhouse soil TKS 1 (Floratorf, FRG). During the experiments the plants were irradiated continously with two white light lamps (Philips TL 40/29) and two UV-B lamps (Philips TL 40/12). Different UV-irradiances were supplied by a series of cut off filters (WG-series, Schott, FRG) transmitting photosynthetic active radiation (PAR, 400-750 nm) and delimiting UV radiation at different degrees . If not otherwise specified photosynthetic active radiation (PAR, 400-750 nm) was 5.6 W/m^2 and UV-B irradiance was 0 mW/m^2 (WG 360), 300 mW/m^2 (WG 320), 695 mW/m^2 (WG 305), 830 mW/m^2 (WG 295) and 1240 mW/m^2 (WG 280), respectively.

Further experiments were performed to assess the effects of enhanced natural UV-B radiation on fluorescence characteristics. Sunflower seedlings were grown under natural daylight by passing light through transparent plexiglass cuvettes on top of two growth chambers. These were placed at a mediterranian station (Caparica, Portugal, 38° N) where higher UV-B-irradiances represent enhanced values relative to Middle European conditions. For a reference group of plants, UV-B was reduced by ozone streaming through the plexiglass cuvette leading to a reduced UV-B irradiance simulating natural radiation of more northern latitudes.

The transparent cuvettes on top of the growth chambers were made of plexiglass (GS 2458, Röhm & Haas, FRG) which had a total transmission of 92 % (280-750 nm). The ozone was provided by an Ozone generator (Technomed, FRG) and streamed through the cuvette in a closed circuit. A constant ozone concentration of 45 µg/ml corresponded to 42 Dobson Units (D.U., i.e. 0.042 cm atm) which add to the average natural ozone layer of 300 D.U. above Portugal in summer. The resulting value of 342 D.U. is similar to the ozone layer above middle europe in summer. UV-B-irradiances under the air filled cuvette simulated a destruction of the ozone layer of about 12%. The average UV-B-irradiance in August at the suns zenith was 1.30 W/m^2 under the air filled cuvette. The ozone streaming through the other cuvette led to a relative UV-B diminuation of about 25% corresponding to 0.99 W/m^2. Calculated daily UV-B fluences were 20.1 and 16.2 kJ/m^2, respectively. Measurements of spectral energy distribution were carried out by a spectroradiometer equipped with a double monochromator optic (Optronic 742, Optronics, Florida). Global radiation was collected by flexible fiberoptics. A computer controlled mechanism adjusted the optics of the radiometer to the solar zenith angle.

Chlorophyll fluorescence was measured with the PAM fluorometer (Walz, FRG) at 21°C. Instantaneous fluorescence (Fo) was induced by a weak modulated measuring light of 1-7 mW/m^2. Actinic light of 9 W/m^2 provided photosynthesis and induced the fluorescence kinetics. The maximum fluorescence yield (Fm) was obtained by application of saturating flashes of white light (Osram Xenophot HLX) with an intensity of 400 W/m^2 which also served for the determination of saturated variable fluorescence ((Fv)s) during the recording of fluorescence kinetics. Light intensity of the saturating flashes was enhanced to 800 W/m^2 for plants grown under natural radiation. Fast fluorescence rise resolving two intermediate maxima I1 and I2 was induced by white light of approximately 1500 W/m^2. The opening time of the shutter used in these experiments was < 1 ms. Dark decay kinetics of 5 ms duration followed a saturating single turnover flash (XST 103, Walz, FRG). Irradiance of the Xenon flash probably was > 10^4 W/m^2. Maximum intensity was reached after 8 µs and relaxation to half intensity after 12 µs. Signals of fluorescence yield were recorded on an analog recorder or digitized and recorded by a strip chart recorder (asystant+, Keithley, FRG). The fast fluorescence intermediates were recorded by a fast transient recorder and a high speed recorder, respectively. The factors of photochemical fluorescence quenching (qQ) and of non-photochemical fluorescence quenching (qE) were determined as

$$qQ = \frac{(Fv)s - Fv}{(Fv)s} \quad \text{and} \quad qE = \frac{(Fv)m - (Fv)s}{(Fv)m}$$

according to definitions by Schreiber et al. (1986). Before measuring, plants were darkened for 60 min. Darkening time of seedlings grown under global radiation was 30 min.

RESULTS

The effect of different UV-B-irradiations on sunflower cotyledons was investigated after 3, 6 and 10 days exposure. Tab. 1 summarizes the effects on Fo, Fm and Fv and figure 1 gives examples of the parameters at the extreme radiation conditions with 1240 mW/m^2 and without UV-B. Fo was slightly raised with increasing UV irradiance up to 695 mW/m^2. After 3 and 6 days these differences were levelled whereas Fo was significantly raised under the highest UV-B intensity of 1240 mW/m^2. Variable (Fv) and maximum fluorescence yield (Fm) were not significantly altered by UV-B treatment up to 695 mW/m^2 compared to the reference receiving no UV-B. Cotyledons grown under the highest UV-B fluence rate displayed decreasing levels of Fm as well as of Fv. This effect was fluence dependent as shown after six and ten days of UV-B-exposure.

Table 1: Fluorescence parameters Fo, Fm and Fv of sunflower cotyledons irradiated with different UV-B fluence rates

UV-B (W/m^2)		0	%	0.300	0.695	1.24	%
	3d	1.36	100	1.43	1.52	1.58	116
Fo	6d	1.48	100	1.51	1.50	2.03	137
	10d	1.65	100	1.71	1.70	2.39	145
	3d	7.93	100	8.91	8.81	7.75	98
Fm	6d	8.34	100	8.10	8.15	6.92	82
	10d	8.67	100	8.73	8.78	6.54	75
	3d	4.59	100	4.71	5.06	4.01	87
Fv	6d	4.51	100	4.39	4.34	3.04	67
	10d	4.75	100	4.64	4.79	2.89	61

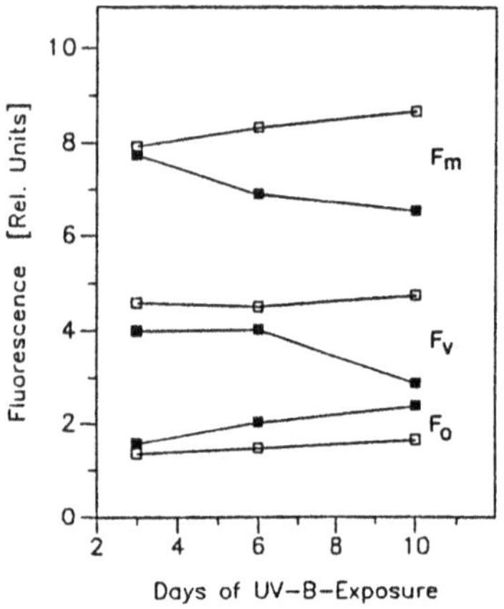

Fig.1: Fluorescence parameters Fm, Fv and Fo of sunflower cotyledons irradiated without and with 1024 mW/m^2 UV-B in dependence of time of exposure. Fv determines the fluorescence peak induced by actinic light ($9W/m^2$) minus Fo.

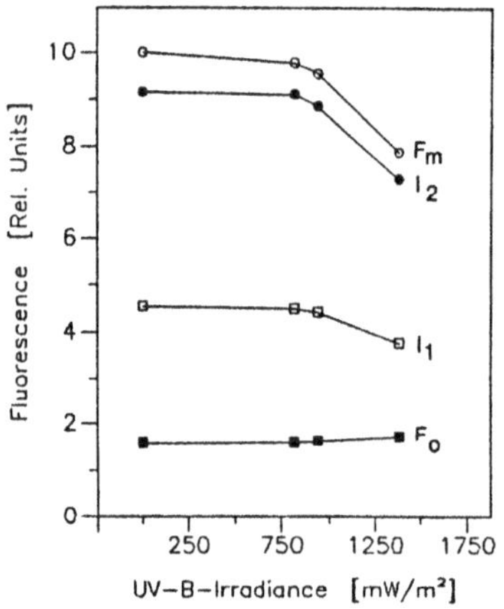

Fig.2: Fluorescence parameters Fm, Fo, I1 and I2 in dependence of UV-B irradiance. PAR in this experiment was 90 W/m^2.

For a more differentiated evaluation of the UV-B effect on the primary reactions of photosynthesis seedlings of sunflower were exposed to different UV-B-irradiances. PAR (400-750 nm) in this experiment was 90 W/m^2. Primary leaves were darkened for 2 hours and irradiated by a flash with an intensity of 1500 W/m^2. As decribed by Neubauer and Schreiber (1987) two intermediate maxima I1 and I2 can be resolved in the fast fluorescence rise.

Figure 2 depicts the fluorescence yield of Fo, I1, I2 and Fm. Primary leaves of sunflower showed the typical decrease of Fm, as found in cotyledons before. The transients I1 and I2 were also diminished with increasing UV-B. A slight increase of Fo was observed. The phase I2-Fm was not much affected by these irradiances.

Time courses of the quenching coefficients qQ and qE after different times of UV-B-irradiation are shown in figure 3. Seedlings exposed to the highest UV-B-irradiance of 1240 mW/m^2 for 3 days showed a remarkably high Q-quenching in the beginning of the kinetics. A more oxidized state of the electron transport chain could be explained by a UV induced damage of PS II. Approaching the steady state, however, the qQ values of the UV-B-exposed seedlings were only a little higher than those of the reference plants with no UV-B. After 3 days, the initial qE-rise was more pronounced in seedlings exposed to UV-B and they displayed some higher energy-quenching in the steady state. With prolonged time of UV-B exposure, the predominant effect in the steady state was the significantlyy reduced photochemical quenching while the thylakoids of the UV-B-stressed plants developed a relative high energy state which is indicative for an inhibited turnover of the reduction and energy equivalents in the Calvin cycle.

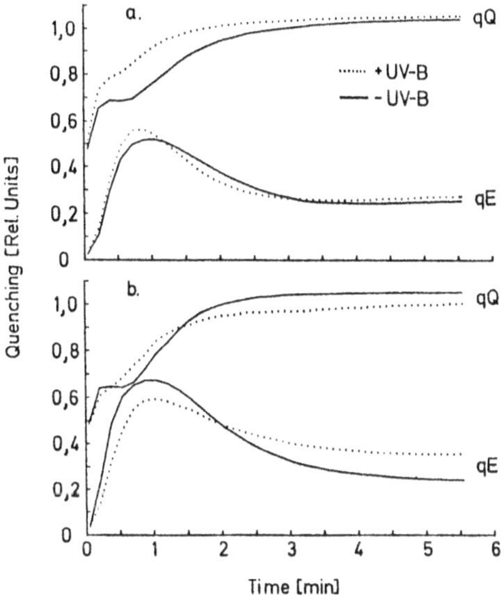

Fig.3:Quenching coefficients qQ and qE of sunflower cotyledons irradiated without (solid line) and with 1.24 W/m^2 UV-B (dotted line). Time of irradiation was 3 d (3a) and 10 d (3b), respectively. Values of qQ are enhanced by 0.2 in the graphic.

Fluorescence dark decay kinetics of sunflower primary leaves following a single turnover flash are depicted in Fig. 4 and Tab. 2. Instantaneous fluorescence is not included in the values. The fluorescence peak induced by a saturating flash of less than a ms is not the real maximum value (Schreiber 1986). Therefore the peak was designated "Fm1". Its value should correspond to the transient I1 (Fig. 2) at saturating light intensity. The fluorescence level after a 5 ms dark decay was designated "Fd". The fluorescence kinetics following the flash, displayed a biphasic decay which is seen more clearly in UV-B irradiated plants. A fast phase < 0.5 ms was interpreted as the reoxidation of $QA^- QB$ --> $QA\ QB^-$ (Crofts and Wraight 1983). Since the decay up to 5 ms is largely caused by the fast component the reoxidation of QA^- was estimated by the half life between the peak and the fluorescence level Fd. Increasing UV-B irradiances successively diminished Fm1 and Fd while UV-B irradiances above 0.83 W/m^2 did not further decrease the decay rate and the fluorescence decay relative to Fm1 (i.e. (Fm1-Fd)/Fm1)). Hence, the half life vs. the level of Fm1 increased upon UV-B irradiances greater than 0.83 W/m^2 indicating a slowed down reoxidation of the first stable electron acceptor of PS II under the highest UV-B irradiance.

Table 2: Fluorescence dark decay parameters of sunflower primary leaves grown with different UV-B fluence rates.

UV-B (W/m^2)	0	0.695	0.830	1.24
Fm_1 - Fo	7.82	6.62	5.39	4.54
Fd - Fo	1.59	1.21	.88	.76
(Fm_1 - Fd) / (Fm_1 - Fo)	.797	.817	.837	.834
$t_{1/2}$ (ms)	.326	.261	.183	.184
$t_{1/2}$/Fm1	.042	.039	.034	.040

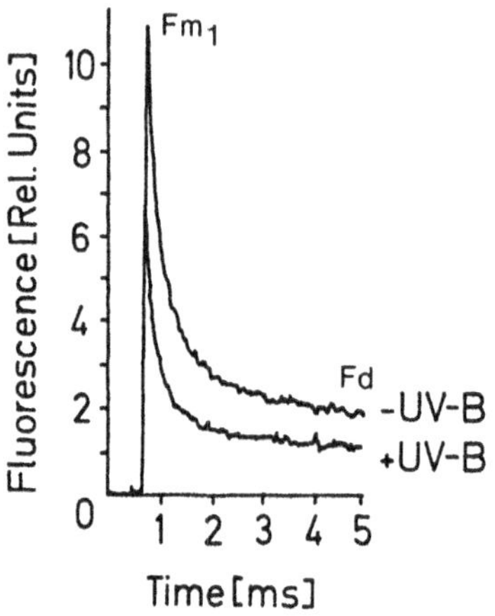

Fig.4: Dark decay kinetics of sunflower primary leaves irradiated without UV-B and with 1.24 W/m^2 UV-B. PAR was 90 W/m^2.

Sunflower seedlings grown with natural radiation displayed characteristic diurnal courses of fluorescence parameters (Fig. 5). Maximum fluorescence Fm sank to a minimum value shortly after the sun's zenith and subsequently rose again in the afternoon. Instantaneous fluorescence Fo went through a slight reversible increase at noon.

Seedlings grown with reduced or enhanced levels of UV-B dis-played nearly the same decrease of Fm in the morning. While the minimum at noon reached a lower level in plants grown with enhanced UV-B, Fm under both con-ditions rose to similar levels again in the afternoon. The midday rise of Fo was insignificantly more pronounced in seedlings grown with enhanced UV-B.

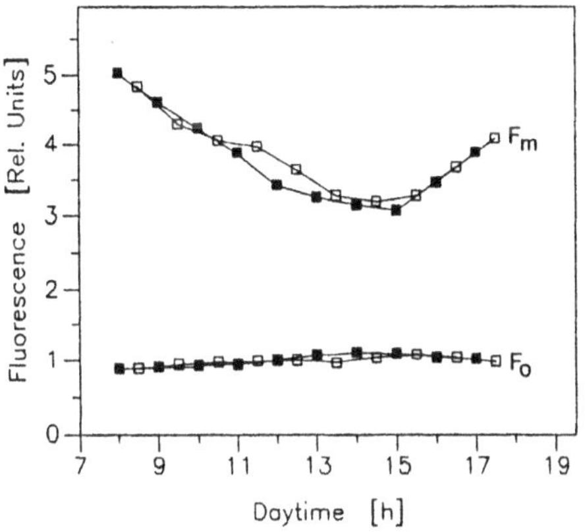

Fig.5:Diurnal course of Fm and Fo of sunflower cotyledons irradiated with enhanced (solid squares) and reduced global UV-B (open squares), respectively.

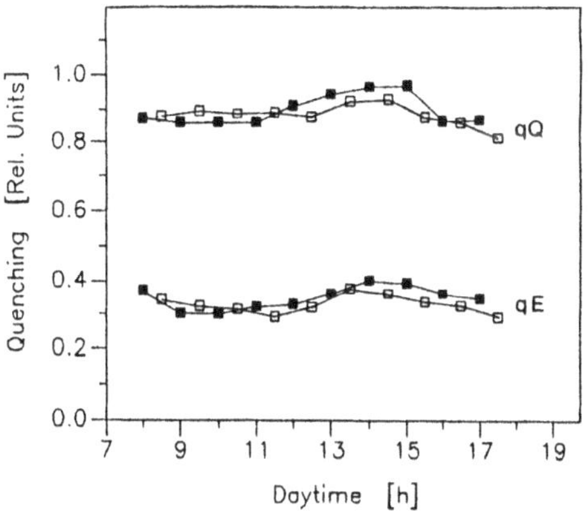

Fig.6: Diurnal course of qQ and qE in the steady state (10 min) of sunflowe cotyledons irradiated with enhanced (solid squares) and reduced (open squares) global UV-B, respectively.

The diurnal courses of the quenching factors qQ and qE in the steady state showed that in the morning photochemical quenching was almost constant and a little lower in seedlings grown under enhanced UV-B while values of qQ in these plants reached a higher maximum at late noon (Fig. 6). In the afternoon qQ declined under both radiation conditions. The course of energy quenching displayed a pattern similar to qQ at noon. However, in the morning there was a slight decrease of qE. The maximum of energy quenching at noon was reached about 1 hour earlier than that of photochemical quenching and qE-values of plants grown with enhanced UV-B-irradiance rose earlier and the maximum was even higher.

DISCUSSION

The experiments performed with sunflower leaves reveal manyfold effects of UV-B radiation on fluorescence parameters. The decrease of Fm confirms earlier studies with different species (Tevini and Iwanzik 1983). From the fact that artificial electron donors did not restore either Hill activity reduced by UV-B (Tevini and Pfister 1985) or the lowered 320 nm-absorption change (Iwanzik et al. 1983) it was concluded that PS II reaction centres are transformed into dissipative sinks for excitation energy upon UV-B treatment. Tevini und Pfister (1985) showed that fluorescence characteristics of PS II α-centres were lost during UV irradiation in isolated chloroplasts. A greater rate constant for dissipation solely does not explain a rise of Fo. Another explanation for a raised Fo might be a reduced probability for exciton transfer in the antenna or an impaired rate constant of PS II photochemistry.

A decrease of the phases Fo-I1 of varying degree and still more of I1-I2 was established upon treatment with substances affecting the PS II donor site or after heating stress (Schreiber and Neubauer 1987). Application of DCMU which inhibits electron flow at the PS II-acceptor site caused a weak decrease of Fm and I2 at low concentrations (Neubauer und Schreiber 1987). The phases I1-I2-Fm were totally eliminated when a fast increase of I1 was induced with increasing concentrations of the inhibitor. The analysis of the fast fluorescence rise, showing a decrease of I1 together with reduced values of I2 and Fm, leads to the conclusion that PS II acceptors are not the primary target of UV-B. A decrease of Fm along with an enhancement of Fo progressed with prolonged time of UV-B exposure and was more pronounced under high UV-B irradiances suggesting a UV-B fluence dependency of photosynthetic damage.

An impairment of UV absorbing quinones or of the HBP-32 under enhanced levels of UV-B (1.24 W/m^2) is concluded by the data of the half life of fluorescence decay versus the fluorescence peak. Decreasing values of $t_{1/2}$/Fm1 with UV-B irradiances increasing up to 830 mW/m^2 are probably due to an impaired PS II. However, Fm1 still decreases with further enhancement of UV-B up to 1.24 W/m^2 while the half life fluorescence decay versus Fm1 is slowed down again with 1.24 W/m^2 suggesting a secondary target of UV-B at the acceptor site of PS II. This also would explain the lower values of photochemical quenching and energy quenching at the beginning of fluorescence induction in cotyledons irradiated for 10 days compared with plants irradiated for only 3 days (Fig. 3a and 3b). The effect of a possible destruction at the acceptor site on the rise kinetics under high UV-B irradiances might be superimposed by a greater damage at or near PS II leading to a decrease of the fluorescence intermediates in the ascending part of the induction.

A possible mechanism enhancing qQ and qE at the same time (Fig. 3a) could be a state I/state II-shift supporting cyclic electron flow around PS I which is known to be less sensitive to UV irradiation. However, it is not yet known if further quenching mechanisms are induced by UV-B or if altered membrane compositions may lead to different quenching characteristics. The predominant effect in the steady state after 10 days of UV-B treatment (Fig. 3b) is the drastically enhanced energy quenching coinciding with a more reduced state of the electron transport chain indicating a severely reduced capacity for linear electron transport. A reduced activity of Ribulose-1,5-bisphosphate carboxylase upon high doses of UV-B was observed by Vu et al. (1984). A slight impairment of UV-absorbing enzymes or a lack of metabolites at earlier times might not be recognized in the quenching kinetics induced by a relatively moderate actinic light intensity of 9 W/m^2 if processes of CO_2-fixation are not limiting photosynthesis under these conditions. Sunflower is known to display a very high light saturation of CO_2-fixation.

The ecological relevance of prognoses about the

consequences of a reduced ozone layer, however, is restricted by the transferability of effects caused by artificial UV-sources which display different optical qualities than global UV-radiation. For a comparison of the effects of artificial UV-B radiation to global UV-B radiation, sunflower seedlings were grown with natural daylight representing enhanced UV-B levels relative to Middle European conditions. The decrease of Fm together with the diurnal increase of global irradiance (Fig.5) is probably an adaptation to high light intensity. A photoinhibitory damage does not seem to exist because Fm recovers soon after the sun's zenith and plants were adapted to the high light regime. The decrease of Fm around noon was slightly more pronounced in seedlings grown with enhanced UV-B but Fm rose to levels similar to those of plants under reduced UV-B in the afternoon. Effects of UV-B damage were shown to be irreversible whereas a fluorescence decrease caused by moderate photoinhibitory blue light treatments were recovered by the stressed plants (Björn et al. 1986). Hence, one can suppose an effect of enhanced global UV-B in terms of intensified reactions caused by strong sunlight. The reversible Fo-rise in the opposite course of the decrease of Fm was insignificantly more pronounced in seedlings under enhanced UV-B. A rise of Fo was also observed upon photoinhibitory treatments and a decrease of Fm relative to Fo was correleted with a reduced quantum yield of the water splitting process (Demmig et al. 1987).

In seedlings grown with enhanced UV-B, the rising values of qQ and qE (Fig. 6) around noon resemble steady state conditions of cotyledons irradiated with artificial UV-B for three days (Fig. 3a). A state 1/state 2-shift might be a cause of totally higher quenching but it is hard to explain in regard to a lower Fm suggesting a reduced quantum of excitons reaching the reaction centre of PS II and thereby leading to a more oxidized state of the electron transport chain.

ACKNOWLEDGEMENTS

This work was supported by the German Bundesministerium für Forschung und Technologie (Project KBF 51).

REFERENCES

BJÖRN LO, BORNMAN JF and OLSSON E, 1986. Effects of ultraviolet radiation on fluorescence induction kinetics in isolated thylakoids and intact leaves. In: Stratospheric Ozone Reduction, Solar Ultraviolet Radiation and Plant Life (Worrest AC and Caldwell MM., ed.). NATO ASI Series G: Ecological Sciences, Vol. 8, Springer Berlin

CROFTS AR and WRAIGHT CA, 1983. The electrochemical domain of photosynthesis. Biochim. Biophys. Acta 726, 149-185

DEMMIG B, WINTER K, KRÜGER A and CZYGAN FC, 1987. Photoinhibition and zeaxanthin formation in intact leaves. Plant Physiol., 84, 218-224

FABIAN P, 1988. Antarktisches Ozonloch: Indizien weisen auf

Umweltverschmutzung, Phys. Bl. 44, Nr. 1, 2-7

IWANZIK W, 1983. Wirkung von UV-Strahlung auf das Wachstum von Keimlingen und die Zusammensetzung und photosynthetische Aktivität von Blättern und isolierten Chloroplasten. Karlsruher Beiträge zur Entwicklungsphysiologie der Pflanzen (Hrsg.: M. Tevini), Nr.5, Universität Karlsruhe

IWANZIK W, TEVINI M, DOHNT G, VOSS M, WEISS W, GRÄBER P and RENGER G., 1983. Action of UV-B radiation on pho-tosynthetic primary reactions in spinach chloroplasts. Physiol. Plant. 58, 401-407

MOLINA MJ and ROWLAND FS, 1974. Stratospheric sink for chlorofluoromethanes: Chlorine atom catalysed destruction of ozone. Nature 249, 810-812

NEUBAUER C and SCHREIBER U, 1987. The polyphasic rise of chlorophyll fluorescence upon onset of strong continuous illumination: I. Saturation characteristics and partial control by the photosystem II acceptor side. Z. Naturforsch. 42 c, 1246-1254

RENGER G, VOSS M, GRÄBER P and SCHULZ A, 1986. Effect of UV irradiation on different partial reactions of the primary processes of photosynthesis. In: Stratospheric Ozone Reduction, Solar Ultraviolet Radiation and Plant Life (Worrest AC and Caldwell MM., ed.). NATO ASI Series G: Ecological Sciences, Vol. 8, Springer Berlin

SCHREIBER U, 1986. Detection of rapid induction kinetics with a new type of high-frequency modulated chlorophyll fluorometer. Photosynth. Res. 9, 261-272

SCHREIBER U and NEUBAUER C, 1987. The Polyphasic Rise of Chlorophyll Fluorescence upon Onset of Strong Continuous Illumination: I. Partial Control by the Photosystem II Donor Side and Possible Ways of Interpretation. Z. Naturforsch. 42 c, 1255-1264

SCHREIBER U, SCHLIWA U and BILGER W, 1986. Continous recording of photochemical and non-photochemical chlorophyll fluorescence quenching with a new type of modulation fluorometer. Photosynth. Res. 10, 51-62

TEVINI M and IWANZIK W, 1983. Inhibition of photosynthetic activity by UV-B radiation in radish seedlings. Physiol. Plant. 58, 395-400

TEVINI M and PFISTER K, 1985. Inhibition of photosystem II by UV-B radiation. Z. Naturforsch. 40c, 129-133

VU CV, ALLEN LH and GARRARD LA, 1984. Effects of Enhanced UV-B Radiation (280-320 nm) on Ribulose-1,5-Bisphosphat Carboxylase in Pea and Soybean. Environm. Experim. Bot. 24, No 2, 131-143

COMPARISON OF PHOTOACOUSTIC AND CHLOROPHYLL FLUORESCENCE SIGNATURES OF GREEN LEAVES

Eckehard M. NAGEL and Hartmut K. LICHTENTHALER

Botanisches Institut II (Plant Physiology and Plant Biochemistry)
University of Karlsruhe, Kaiserstr. 12, D-7500 Karlsruhe, F.R.G.

KEY WORDS: aurea mutant, chlorophyll fluorescence, CO_2-assimilation P_N, heat, fluorescence induction kinetics, photoacoustic signals, photosynthesis, sun and shade leaves.

ABSTRACT

The photoacoustic (PA) signals of tobacco and beech leaves are compared with measurements of the chlorophyll fluorescence, the CO_2-assimilation and the pigment content of the leaves to show that the PA-method is a valuable tool to obtain additional information about the photosynthetic apparatus. The PA-signals are shown as excitation spectra in the visible range and as induction kinetics excited by a He/Ne-laser.

The PA-signals are determined by the light-induced heat but are superimposed on the photosynthetically produced oxygen pulses at low chopping frequencies (150 Hz).

During the autumnal chlorophyll breakdown the PA-signal declines in the red-light part of the spectrum due to the loss of chlorophyll. In the blue-light part the PA-signal remains at a constant height, indicating that the light-energy absorbed by the carotenoids can no longer be used in photosynthesis and is dissipated as heat.

During the light-induced induction kinetics of the green leaves the heat signal (PA-signal at 238 Hz) and the chlorophyll-fluorescence signal decrease after a very fast initial increase, whereas the PA-signal at 22 Hz increases furthermore after the fast increase at the onset of illumination. The PA-signals at 22 and 238 Hz and the net CO_2-assimilation reach the steady state after 20 min, whereas the chlorophyll fluorescence is much faster and already reaches the steady state after 4 min. The increase of the CO_2-assimilation seems to be related more closely to the decrease of the heat signal than to the decrease of the fluorescence signal.

If continuous saturating white light is added to the chopped excitation laser light (238 Hz) at the steady state of the kinetic, the heat signal increases and remains at a constant height during the illumination, whereas the fluorescence signal increases and then declines. This may indicate that the heat signal arises from both photosystems, whereas the chlorophyll-fluorescence emission at room temperature primarily emanates from photosystem II. The height of the increase of the PA-signal reflects the photosynthetic activity of the leaves.

INTRODUCTION

In this investigation the photoacoustic (PA) method was used to determine the light-induced heat production from non-radiative deexcitation processes in leaves of tobacco and beech with different photosynthetic activity

H. K. Lichtenthaler (ed.) Applications of Chlorophyll Fluorescence, 239–248.

and different chlorophyll content. The light-energy absorbed by the photosynthetic pigments (chlorophylls and carotenoids) can either be used for photosynthesis or is emitted as fluorescence and phosphorescence or dissipated as heat from non-radiative deexcitation. The measurements of the chlorophyll fluorescence induction kinetics are a valuable tool in photosynthesis research (Butler, 1977; Karukstis and Sauer, 1983; Lichtenthaler et al., 1986). The height of the fluorescence signal during the induction kinetic depends on the redox state of the quencher Q (Duysens and Sweers, 1963) and on several non-photochemical quenching mechanisms.

The heat from non-radiative deexcitation processes in plant leaves can be measured with the photoacoustic method (Bults et al., 1982; Buschmann and Prehn, 1981; Buschmann et al., 1984; Malkin et al., 1981; Canaani et al., 1985; Nagel et al., 1987; Nagel and Lichtenthaler, 1988).

The photoacoustic method is based on the photoacoustic effect (Bell, 1880). A light-absorbing sample loses the absorbed energy at least partially via non-radiative processes. These non-radiative processes lead to an increase of the temperature of the sample and a rise of the temperature of the air above the sample. The sample, which is enclosed in an airtight PA-cell, is illuminated with chopped light. Inside the PA-cell pressure changes are created and can be measured with a microphone. The PA-signals recorded at different wavelengths yield heat-excitation spectra. If the PA-signal is recorded during the illumination period, one obtains heat-induction kinetics.

Photosynthetically active leaves produce oxygen in the light. When the illumination proceeds with chopped light the rhythmically produced oxygen can contribute to the PA-signal (Bults et al., 1982), but only when measured at low chopping frequencies (e.g. 22 Hz). At high chopping frequencies (above 150 Hz) the PA-signals of the leaves are only determined by the heat production of the leaves, since the modulation of the oxygen evolution is damped under these conditions (Bults et al., 1982; Poulet et al., 1983).

The aim of this work was to determine whether the PA-spectra and the PA-kinetics depend on the photosynthetic activity of the leaves. Another question which was to be solved was how the heat signal changes during the induction kinetics of leaves as compared to the changes of the chlorophyll fluorescence signal.

MATERIALS AND METHODS

Tobacco leaves (Nicotiana tabacum L., green form su/su and aurea form Su/su) of 4 month old plants (4th to 6th leaves from the top) were used for the measurements. The leaves of the beech (Fagus sylvatica L.) were collected from a 20 m high tree from the Schloßgarten in Karlsruhe both from a sun-exposed branch (sun leaves) and from inside the tree (shade leaves).

The PA-spectra were measured with a system which has been described in detail earlier (Nagel et al., 1987). As light source a 450 W Xenon high-pressure lamp was used. The light was chopped by a mechanical chopper and focused onto the entrance slit of a monochromator (grating: 1200 lines/mm, blazed at 500 nm). The monochromatic light (about 0-54 $\mu E \cdot m^{-2} \cdot s^{-1}$) was focused onto the PA-cell (EG&G 6003). A part of the monochromatic light was used as reference light which was measured with an pyroelectrical detector. Both the signal of the PA-cell and the signal of the pyroelectrical detector were fed into a lock-in amplifier and then processed in a microcomputer.

The PA-kinetics were measured using a 5 mW He/Ne-laser (λ_{max}=632,8 nm, Spectra Physics). The light was chopped with a mechanical chopper and guided inside an arm of a glasfiber on to the PA-cell. The intensity of the measuring light was chopped from 0 to 290 $\mu E\ m^{-2}\ s^{-1}$. The second arm of the glasfiber served for guiding the fluorescence on to the photodiode. In front of the photodiode a cut-off filter (RG 665, Schott Mainz) and an interference filter (DAL 685) were mounted. The signal of the photodiode was fed into a one-channel lock-in amplifier (5101 EG & G) and the signal of the PA-cell was fed into a two-channel lock-in amplifier . Both signals were recorded with a yt-recorder. For the saturating of the photosynthesis, continuous white light (2600 $\mu E \cdot m^{-2} \cdot s^{-1}$) was added by means of two additional glassfibers.

The pigments of the leaves were measured after extraction with 100% acetone and the pigment content was calculated after Lichtenthaler (1987). The absorption spectra of leaves were measured with a photometer (MPS 5000, Shimadzu, Kyoto, Japan). The photosynthetic net CO_2-fixation was measured using a CO_2/H_2O-porometer (Walz, Effeltrich, FRG).

RESULTS

PA-spectra (22 Hz) and absorption spectra of tobacco leaves

The photoacoustic spectra of green tobacco (su/su) leaves have a maximum in the red-light region with a peak at about 675 nm (Fig. 1A). From the minimum at about 550 nm the PA-signal increases towards shorter wavelengths via two shoulders (475 nm, 440 nm) to the highest level at 380 nm. The PA-spectrum of the aurea mutant is lower than the PA-spectrum of the green form except for the maximum at 675 nm which is higher in the aurea form (Fig. 1A).

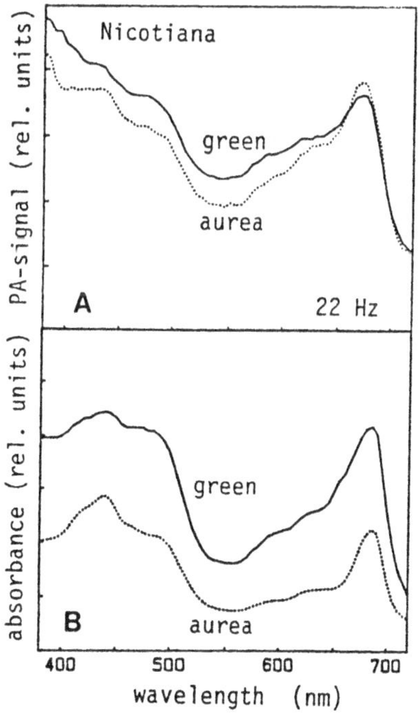

Figure 1. (**A**) Photoacoustic spectra measured at a chopping frequency of the excitation light of 22 Hz and (**B**) absorption spectra of a leaves of a green form and an aurea mutant of Nicotiana tabacum. The PA-spectra are the mean of 20 (green) and 10 (aurea mutant) single spectra.

The absorption spectra (Fig. 1B) of the green tobacco leaves have maxima in the blue-light part of the spectrum (440 nm) and the red-light part of the spectrum (680 nm). From the minimum at 550 nm the signal increases to shorter wavelengths via a shoulder (475 nm) to the maximum at 440 nm. From this maximum the signal declines towards shorter wavelengths. The absorption spectra of the leaves of the aurea mutant are lower than these spectra of the green leaves, but have nearly the same characteristics.

Photoacoustic spectra (22 Hz) and pigment content of shade and sun leaves from a beech tree during the autumnal chlorophyll breakdown.
The photoacoustic spectra of the green shade and sun leaves of beech trees have a maximum in the red-light part of the spectrum and a broad plateau in the blue-light part of the spectrum (Fig. 2A). With decreasing chlorophyll content (Tab. 1) the signal in the red-light part decreases, whereas it remains in the blue-light region. The PA-spectra of the sun leaves are lower than the spectra of the shade leaves. The spectra of the shade leaves remain more or less at a constant height between 500 and 380 nm, whereas the signal of the spectra of the sun leaves rises very strongly below 420 nm.

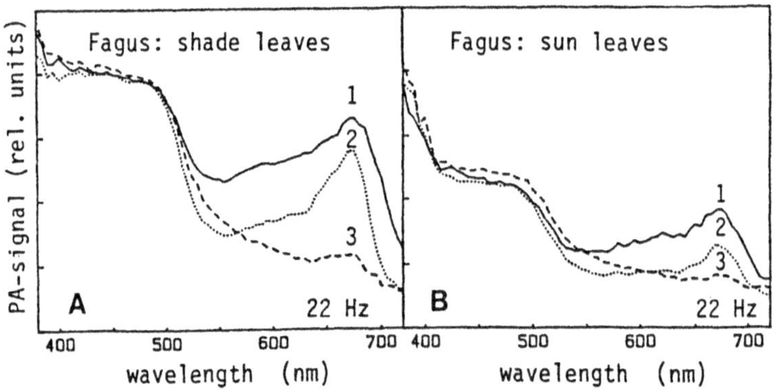

Figure 2. Photoacoustic spectra of shade (**A**) and of sun (**B**) leaves of a beech (Fagus sylvatica L.) during the autumnal chlorophyll breakdown. The spectra were measured with a chopping frequency of 22 Hz. Mean values of 4 measurements. 1) green leaves, 2) yellowish-green leaves, 3) yellow leaves. For chlorophyll content see Tab. 1.

Table 1: Chlorophyll (a+b) and carotenoid content (x+c) as well as pigment ratios of sun and shade leaves of the beech (Fagus sylvatica L.) during the autumnal chlorophyll breakdown. Mean values with standard deviation (n = 14 in green, n = 8 in yellowish-green, n = 6 in yellow leaves).

	a+b	x+c	a/b	a+b/x+c
shade leaves:				
green	**25.7** ± 2.1	**6.6** ± 0.4	**2.3** ± 0.1	**3.9** ± 0.2
yellowish-green	**5.4** ± 0.4	**4.6** ± 0.2	**1.9** ± 0.1	**1.2** ± 0.1
yellow	**0.9** ± 0.2	**4.2** ± 0.3	**1.1** ± 0.1	**0.2** ± 0.04
sun leaves:				
green	**29.7** ± 2.2	**9.0** ± 0.6	**2.9** ± 0.2	**3.3** ± 0.2
yellowish-green	**12.5** ± 1.6	**9.0** ± 0.5	**2.5** ± 0.1	**1.4** ± 0.2
yellow	**2.4** ± 1.0	**7.9** ± 0.9	**1.6** ± 0.4	**0.3** ± 0.1

PA-kinetics of tobacco leaves at 22 and 238 Hz and simultaneously measured fluorescence kinetics.

At the onset of the illumination of a tobacco leaf the 22 Hz PA-signal rises very fast, declines somewhat and then slowly increases to a maximum within about 20 minutes (Fig. 3A). When measured at a chopping frequency of 238 Hz, the PA-signal rises very fast to a maximum and then slowly declines.

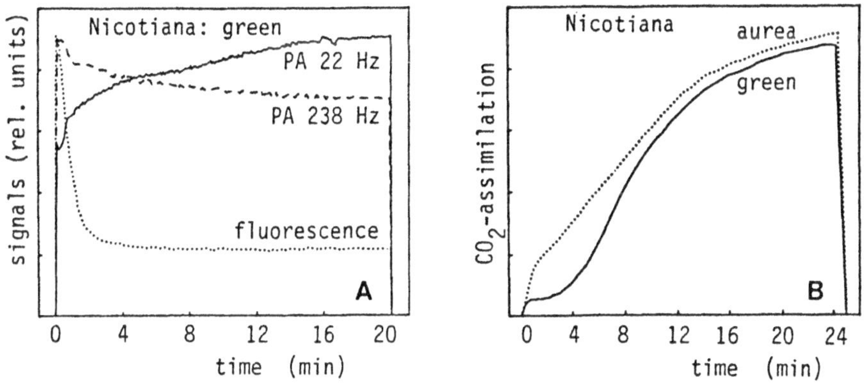

Figure 3. A. Kinetics of the chlorophyll fluorescence and the photoacoustic signals measured at a chopping frequency of the excitation light of 22 and 238 Hz (PA 22 Hz, PA 238 Hz) of a green tobacco leaf during a 20 min illumination period. **B.** Photosynthetic CO_2-uptake in freshly illuminated leaves of a green and an aurea form of tobacco during a 24 min illumination period.

The chlorophyll-fluorescence induction kinetics measured at 22 Hz and 238 Hz are similar and therefore given only for 238 Hz. The fluorescence rises very fast to a maximum and then declines rather fast to a steady-state level (Fig. 3A).

The PA-kinetics of the aurea mutant are similar to that of the green form and are therefore not shown here. Differences between aurea and green leaves occur in the height of the decline of the PA-kinetics measured at 238 Hz. The signal of the green form declines to 85 % ($\pm$ 7 %; n = 16) compared to the signal at the onset of the illumination, whereas the signal of the aurea mutant declines to 73 % ($\pm$ 3%, n = 6). These differences between both leaves are significant (p=0.001).

CO_2 uptake of the tobacco leaves during the illumination

If dark-adapted tobacco leaves are illuminated, they need about 20 minutes to reach the full CO_2-fixation rate (Fig. 3B), when illuminated with a light intensity of 2600 $\mu E \cdot m^{-2} \cdot s^{-1}$. The CO_2 fixation rate of the aurea mutant is slightly higher than the CO_2-fixation rate of the green form (see Tab. 3).

PA and fluorescence kinetics of sun leaves of a beech tree.

At the onset of the illumination of the PA-kinetic measured with 22 Hz, the signal increases fast (Fig. 4A). Within 20 min the signal increases further and reaches its steady state.

The PA-signal in the induction kinetics measured with a chopping frequency of 238 Hz declines after its initial rise, within 20 min to a steady-state value. The fluorescence kinetics for 22 Hz and 238 Hz are similar. After the rise at the onset of the illumination the signal declines within for min to the steady-state level.

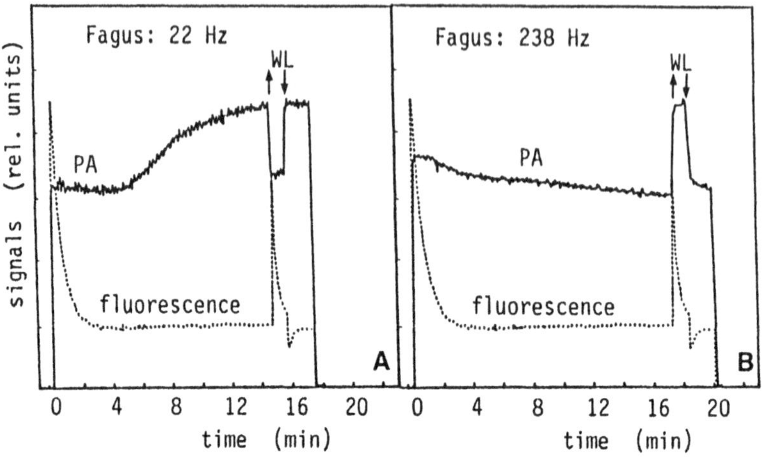

Figure 4. PA-kinetics and fluorescence-kinetics of a sun leaf of a beech tree at 22 Hz (**A**) and 238 Hz (**B**). The addition of continuous white light is indicated by arrows.

Changes of the PA-signal and the fluorescence-signal when continuous white light is added at the steady state

When continuous white light (2600 $\mu E \cdot m^{-2} \cdot s^{-1}$) is added at the steady state of the 22 Hz kinetics of the sun leaves the PA-signal declines to 78 % of the steady-state level (Tab. 2) and remains at a constant height during the illumination (Fig. 4A). In contrast, the signal of the shade leaves does not change when white light is added.

The signals also decrease in the 22 Hz PA-kinetics of tobacco leaves upon addition of continuous white light to 61% (green) and 46% (aurea mutant) of the steady-state level.

Table 2: Effect of continuous white light on the PA-signals at the steady state of the induction kinetics (22 and 238 Hz). The values are given as percentages of the steady-state signal without continuous white light. The differences between green and aurea mutant as well as shade and sun leaves are significant ($p \leqslant 0.001$).

	22 Hz		238 Hz	
Nicotiana tabacum				
green leaf su/su	**61** ± 8 %	23	**138** ± 18 %	15
aurea mutant Su/su	**46** ± 9 %	8	**174** ± 13 %	6
Fagus sylvatica				
shade leaf	**102** ± 6 %	9	**108** ± 6 %	5
sun leaf	**78** ± 11 %	8	**144** ± 6 %	3

The continuous white light given at the steady state of the 238 Hz kinetics leads to an increase of the signals of all leaves. It remains at a constant height during the illumination (Fig. 4B).

The PA-signal at 238 Hz increases to 138% in the green tobacco leaves, to 174% in the leaves of the aurea mutant, to 108% in the shade leaves of the beech and to 144% in the sun leaves of the beech (Tab. 2 and Fig. 5A).

Net CO_2-assimilation (P_N) and chlorophyll-content of tobacco and beech leaves

The CO_2-fixation rate per leaf area unit of the aurea mutant is slightly higher than the CO_2-fixation of the green form (Tab. 3), though the chlorophyll content of the aurea form only amounts 25 % of that of the green form. Consequently the CO_2-assimilation per chlorophyll content is even more than 5 times higher for the aurea mutant than for the green form (Fig. 5B).

Table 3: Net CO_2-assimilation rate (P_N) and chlorophyll content (a+b) of differently pigmented beech leaves. P_N was measured with a saturating white light (2600 $\mu E \cdot m^{-2} \cdot s^{-1}$). Mean of 9 leaves per each condition. The differences in the pigment content and P_N between green and aurea leaves as well as between shade and sun leaves are significant (p = 0.001), except for the P_N value per m^2 s for the tobacco leaves (P = 0.20).

	a+b ($\mu g \cdot cm^{-2}$)	P_N* (per m^2)	P_N** (per a+b)
Nicotiana tabacum			
green leaf su/su	**53.7** ± 6.1	**7.0** ± 2.1	**13.1**
aurea mutant su/su	**12.4** ± 2.6	**8.6** ± 0.9	**69.5**
Fagus sylvatica			
shade leaf	**44.1** ± 3.7	**3.1** ± 0.9	**6.9**
sun leaf	**54.1** ± 5.1	**11.1** ± 1.0	**20.4**

* P_N in $\mu mol\ CO_2 \cdot m^{-2} \cdot s^{-1}$ ** P_N in $mmol\ CO_2 \cdot mg(a+b)^{-1} \cdot s^{-1}$

The CO_2-assimilation of the sun leaves of the beech is more than 3 times higher than the CO_2-assimilation of the shade leaves, no matter whether referred to a total chlorophyll content or on a leaf area unit. The chlorophyll content of the sun leaves is only slighty higher than the chlorophyll content of the shade leaves.

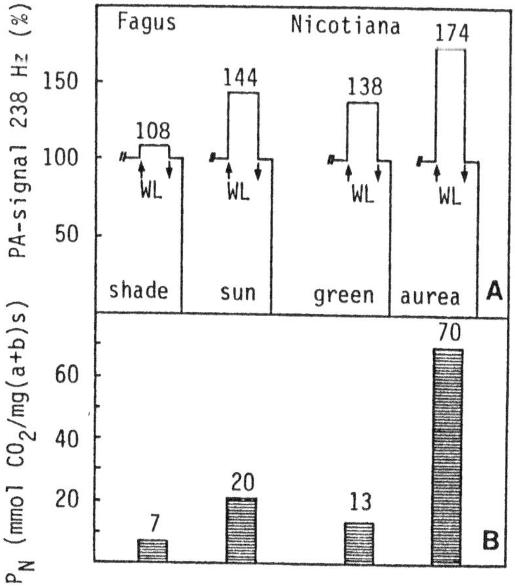

Figure 5. A. Effect of continuous white light on the PA-signal (238 Hz) at the steady state of the induction kinetics of tobacco and beech leaves. **B.** CO_2-assimilation per chlorophyll content of these leaves.

DISCUSSION

The photoacoustic spectra (22 Hz) of the tobacco leaves have about the same characteristics as the absorption spectra (Fig. 1). This similarity has also been shown before for other leaves (Buschmann and Prehn, 1981; Malkin et al., 1981; Pandey, 1983; Nagel et al., 1987; Nagel and Lichtenthaler, 1988). In the red-light part of the spectra the PA-signal arises from the chlorophylls, whereas in the blue-light part the PA-signal arises from chlorophylls and carotenoids. Because the heat production of the aurea leaf is lower than the heat production of the green leaf the PA-signal of the aurea leaf is lower between 380 and 600 nm. The red light maximum of the spectra of the aurea leaves is higher than this maximum in the spectra of the green leaves. This difference between the aurea form and the green form may be induced by the higher oxygen evolution of the leaves of the aurea form (Nagel, 1988). As shown by Bults (Bults, 1982) the photosynthetically produced oxygen can induce a PA-signal at a low chopping frequency of the excitation light.

Another difference between the PA-spectra and the absorption spectra is visible in the region below 420 nm. Here the signal of the absorption spectra decreases towards shorter wavelengths whereas the signal in the PA-spectra increases. The PA-spectra are determined not from the absorption characteristics of the whole leaf but mainly from outer cell layers because the heat waves from deeper layers are damped (Rosencwaig and Gersho, 1976). At a chopping frequency of the excitation light of 22 Hz the PA-signal is mainly determined from the heat of a layer of 46 µm (Nagel, 1988; calculated with the thermal properties of water according to the PA-theory of Rosencwaig, 1976). The heat production of the epidermal layer can contribute good to the PA-signal. The flavonoids inside the epidermis (Weissenböck et al., 1987; Schnabl et al., 1986) seem to be the cause for the increase of the signal below 420 nm.

This can also be seen in the PA-spectra of the beech leaves (Fig. 2). In the spectra of the sun leaves the signal below 420 nm increases very strongly whereas the signal of the shade leaves in this region remains constant. Sun leaves possess a five times higher flavonoid content as shade leaves (Nagel, 1988).

The PA-spectra of the shade leaves of the beech are higher than the spectra of the sun leaves (Fig. 2), although the pigment content and the photosynthetic CO_2-assimilation of the shade leaves are lower (Tab. 1 and 3). Therefore the heat production and the fluorescence emission (Nagel, 1988) of the sun leaves are lower than for the shade leaves because the absorbed light energy is used more efficiently for photosynthesis.

With decreasing chlorophyll content the red-light part of the PA-spectra becomes lower. This decrease is due to the lower absorption in this region. In the blue light region the signal remains at the same height. The reason is that the energy which is absorbed by the carotenoids cannot be used in photosynthesis and is lost as heat.

The PA-signals in the kinetics of the leaves measured at 22 Hz increase with time. Within about 20 minutes the signal reaches its maximum level (Fig. 3 and 4). The increase of the PA-signal at a low chopping frequency of the excitation light was explained by the increase of the photosynthetically produced oxygen (Bults et al., 1982). For tobacco leaves (Fig. 3B) and beech leaves (Nagel, 1988), the time which is necessary to reach their full CO_2-assimilation rate equals the time in which the PA-signal reaches its maximum.

At higher chopping frequencies (238 Hz) the PA-signal decreases after the initial increase within 20 minutes. Here only the heat production contributes to the PA-signal because the oxygen signal is damped at higher chopping frequencies of the excitation light (Bults et al., 1982; Poulet et al., 1983).

The fluorescence signal needs only 4 min to reach its steady-state level. This shows that the increase of CO_2-assimilation is related more closely to the decrease of the heat signal (PA-signal at high chopping frequencies) than to the decrease of the fluorescence signal.

If continuous saturating white light is added at the steady state the chopped measuring light can no longer be used for photosynthesis. Measured at a low chopping frequency (22 Hz) the adding of continuous white light leads to a decrease of the PA-signal (Tab. 2), because the chopped light cannot induce a chopped oxygen production. If this white light is added at the steady state of the 238 Hz kinetics, the signal increases because the chopped measuring light is transferred into heat. That means that the increase of the PA-signal by adding white light reflects that part of the energy which was used for photosynthesis. This increase of the heat signal (PA-signal at a high chopping frequency) was called the photochemical loss (Cahen et al., 1978). This photochemical loss should depend on the photosynthetic activity per absorbed quanta, because the heat signal with added white light depends on the totally absorbed energy (Malkin and Cahen, 1978). The photosynthetic CO_2-assimilation per chlorophyll content of the aurea form of tobacco is 5 times higher than that of the green form (Fig. 5). The increase of the 238 Hz PA- signal is 74 % in the aurea mutant, whereas the signal of the green form increases only by 38 %. In the PA-kinetic (238 Hz) of the shade leaves the signal increases by 8 %, whereas the signal in the kinetics of the sun leaves increases by 44 %. The CO_2-assimilation per chlorophyll content in the sun leaves is three times higher than that of the shade leaves. The increase of the PA-signals at a high chopping frequency by adding of continuous white light always depends on the photosynthetic activity of the leaves.

During the illumination with white light the PA-signal remains at a constant height, both in tobacco and beech leaves (Fig. 4). In contrast to this, the fluorescence signal decreases after the rise at the onset of the illumination with white light. The decrease of the fluorescence signal during the illumination may be interpreted in terms of a state1/state2 transition (Bonaventura and Myers, 1969). At the onset of the additional illumination the absorbed light energy is guided to PS II, which has a high fluorescence. During the illumination with white light the fluorescence decreases, which indicates that the light energy is guided more to the low-fluorescing PS I. During this state transition the heat signal (PA-signal at 238 Hz) remains at a constant height. This can be interpreted by the assumption that the heat signal arises from both photosystems.

CONCLUSION: It has been shown that the PA-method is a valuable tool in photosynthesis research. The heat signal is related more closely to the photosynthetic activity of the leaves than the fluorescence signal. In contrast to the fluorescence, the heat arises from both photosystems.

ACKNOWLEDGEMENTS

This work was sponsored in part by a grant from the PEF (Projekt Europäisches Forschungszentrum) Karlsruhe, which is gratefully acknowledged. We wish to thank Dr. C. Buschmann for valuable discussions and Mr. B. Volk for technical assistance.

REFERENCES

BELL AG, 1880. Upon the production and reproduction of sound by light. Journal of the Society of Telegraph Engineers 9: 404-426.

BULTS G, HORWITZ, BA, MALKIN, S and CAHEN D, 1982. Photoacoustic measurements of photosynthetic activities in whole leaves. Photochemistry and gas exchange. Biochim. Biophys. Acta 679: 452-465.

BUSCHMANN C and PREHN H, 1981. In vivo studies of radiative and non-radiative de-excitation process of pigments in Raphanus seedlings by photoacoustic spectroscxopy. Photobiochem. Photobiophys. 2: 209-215.

BUSCHMANN C, PREHN H and LICHTENTHALER H, 1984. Photoacoustic spectroscopy PAS and its application in photosynthesis research. Photosynth. Res. 5: 29-46.

BUTLER WL, 1977. Chlorophyll fluorescence: a probe for electron transfer and energy transfer. In: Encyclopedia of Plant Physiology. New Series, Vol. 5, pp. 149-167, Pirson A and Zimmermann MH eds., Springer Verlag, Berlin.

CAHEN D, Malkin S and LERNER EI, 1978. Photoacoustic spectroscopy of chloroplast membranes; listening to photosynthesis. FEBS LETTERS 91: 339-342.

CANAANI O, MOTZAN Z and MALKIN S, 1985. Comparison of photosynthetic parameters of an aurea mutant Su/su of tobacco and the wild-type by the photoacoustic method. Planta 164: 480-486.

DUYSENS LNM and SWEERS HE, 1963. Mechanism of two photochemical reactions in algae as studied by means of fluorescence. In: Microalgae and Photosynthetic Bacteria, pp. 353-372, Ashida. Jap. Soc. Plant Physiol., Univ. of Tokyo Press, Tokyo.

KARUKSTIS KK and SAUER K, 1983. Fluorescence decay kinetics of chlorophyll in photosynthetic membranes. Journal of Cellular Biochemistry 23: 131-158.

KRAUSE GH and WEIS E, 1984. Chlorophyll fluorescence as a tool in plant physiology. II. Interpretation of fluorescence signals. Photosynth. Res. 5: 138-157

LICHTENTHALER HK, 1987. Chlorophylls and carotenoids: pigments of photosynthetic biomembranes. Methods Enzymol. 148: 350-382.

LICHTENTHALER HK, BUSCHMANN C, RINDERLE U and SCHMUCK G, 1986. Application of chlorophyll fluorescence in ecophysiology. Radiat. Environ. Biophys. 25: 297-308.

MALKIN S and CAHEN D, 1978. Photoacoustic spectroscopy and radiant energy conversion: theory of the effect with special emphasis on photosynthesis. Photochem. Photobiol. 29: 803-813.

MALKIN S, LASSER-ROSS N, BULTS G and CAHEN D, 1981. Photoacoustic spectroscopy in photosynthesis. In: Proc. of the Fifth International Photosynthesis Congress, Photosynthesis III. Structure and Molecular organisation of the Photosynthetic Apparatus, pp. 1031-1042, Akoyunoglou G ed., Balaban International Science Services, Philadelphia.

NAGEL EM, 1988. Photoakustische Untersuchungen an Pflanzen. Dissertation am Botanischen Institut (Lehrstuhl für Pflanzenphysiologie und Pflanzenbiochemie) der Universität Karlsruhe.

NAGEL EM and LICHTENTHALER HK, 1988. Photoacoustic spectra of green leaves and of white leaves treated with the bleaching herbicide. In: Proc. of the 5th Internat. Top. Meeting on Photoacoustic and Photothermal Phenomena, pp. 568-569, Hess P and Pelzl J eds., Springer Series in Optical Sciences, Springer Berlin, Heidelberg.

NAGEL EM, BUSCHMANN C and LICHTENTHALER HK, 1987. Photoacoustic spectra of needles as an indicator of the activity of the photosynthetic apparatus of healthy and damaged conifers. Physiol. Plantarum 70: 427-437.

PANDEY GC, 1983. Photoacoustic spectroscopy in the study of metal toxicity. In: 3rd International Conference on Photoacoustic and Photothermic Spectroscopy, 5.12/pp. 1-3, Paris.

POULET P, CAHEN D and MALKIN S, 1983. Photoacoustic detection of photosynthetic oxygen evolution from leaves. Quantitative analysis by phase and amplitude measurements. Biochim. Biophys. Acta 724: 433-446.

ROSENCWAIG A and GERSHO A, 1976. Theory of the photoacoustic effect with solids. Applied Physics 47: 64-69.

SCHNABL H, WEISSENBÖCK G and SCHARF H, 1986. In vivo-microspectrophotometric characterization of flavonol glycosides in Vicia faba guard and epidermal cells. Experimental Botany 37: 61-72.

WEISSENBÖCK G, SCHNABL H, SCHARF H and SACHS G, 1987. On the properties of fluorescing compounds in guard and epidermal cells of Allium cepa L. Planta 171: 88-95.

THE MEASUREMENT OF RFD-VALUES AS PLANT VITALITY INDICES WITH THE PORTABLE FIELD CHLOROPHYLL FLUOROMETER AND THE PAM-FLUOROMETER

Michael HAITZ and Hartmut K. LICHTENTHALER

Botanisches Institut II (Plant Physiology and Plant Biochemistry), University of Karlsruhe, Kaiserstr. 12, D-7500 Karlsruhe 1, F.R.G.

KEY WORDS: Chlorophyll fluorescence induction kinetics at 690 and 730 nm, Kautsky effect, Rfd-values at 690 and 730 nm, stress-adaptation index A_p, vitality index.

ABSTRACT

The chlorophyll fluorescence vitality indices of green leaves (the Rfd-values) were comparatively determined from the same leaf samples using **1)** a self-constructed, red laser equipped, **portable field chlorophyll fluorometer** (Rfd-values in the 690 and 730 nm region) and **2)** the **PAM-fluorometer** (Rfd-values in the 730 nm region). The results indicate that the PAM-fluorometer, originally built for the measurement of Q- and E-quenching, can also be applied to determine the Rfd-values (in the 730 nm region) using its continuous white saturating light source for fluorescence induction. Since Rfd-values as a vitality index contain additional and complementary physiological information on the whole process of photosynthesis, it is strongly recommended to measure in all leaf samples - in addition to Q- and E-quenching - also the Rfd-values. Though the PAM-fluorometer can be successful used for the determination of Rfd-values, the portable field fluorometer has the advantage that it also permits calculation of the stress adaptation index A_p from the two Rfd-values at 690 and 730 nm.

INTRODUCTION

The measurement of the red-laser-induced induction kinetics of the in vivo chlorophyll fluorescence of pre-darkened intact leaves and needles proved to be a very suitable non-destructive method for the detection of stress and damage to the photosynthetic apparatus (Lichtenthaler, 1984 and 1986; Lichtenthaler et al. 1986; Lichtenthaler and Rinderle 1988). By application of a new portable field fluorometer, which was developed by us for outdoor measurements, one can screen the physiological state of agricultural plants in the field and of trees in the forests. By use of adequate filter systems this apparatus measures simultaneously the fluorescence-induction kinetics in the 690 and 730 nm regions, which correspond to the maxima (or shoulders) of the chlorophyll fluorescence-emission spectra of intact leaves (Figs. 1 and 2). From the slow component of the fluorescence-induction kinetics one can determine the ratio of the fluorescence decrease fd to the steady state fluorescence fs (Rfd = fd/fs; see Fig. 3). These ratios (Rfd-values) are a valuable indicator of the potential photosynthetic activity of leaves as has been repeatedly shown by parallel measurements of the net CO_2-assimilation P_N (Lichtenthaler, 1984 and 1986; Lichtenthaler et al. 1986, Lichtenthaler and Rinderle, 1988; Nagel et al. 1987). The **variable fluorescence ratio Rfd** is a vitality indicator of the leaf and plant and has been

H. K. Lichtenthaler (ed.) Applications of Chlorophyll Fluorescence, 249–254.

termed vitality index (Lichtenthaler 1986, Lichtenthaler and Rinderle 1988, Strasser et al. 1987). It permits a fast initial outdoor screening of the vitality of crop plants and trees, before other more time-consuming methods to further localize stress and damage are applied. Rfd-values above 3 (as determined in the 690 nm region) indicate very efficient photosynthesis and high photosynthetic rates per leaf area unit; at Rfd-values below 1.0 the leaves no longer exhibit a net CO_2-assimilation P_N.

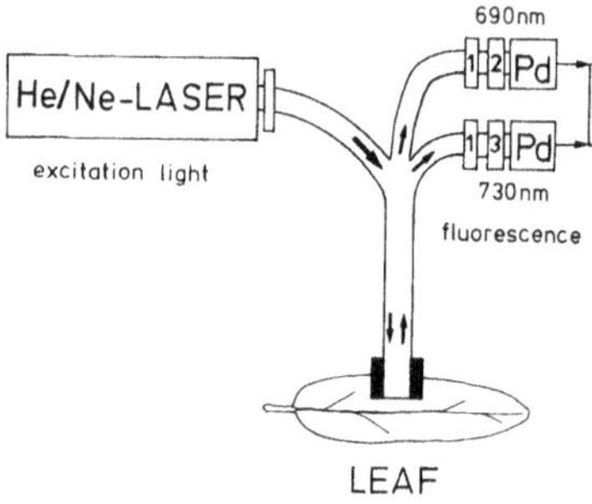

FIGURE 1. Laser-equipped **portable field chlorophyll fluorometer** for the simultaneous measurement of the fluorescence induction kinetics (Kautsky effect: slow component) of intact leaves in the 690 and 730 nm region. The red-laser light is sent through fiber optics to the predarkened leaf and the resulting chlorophyll fluorescence directed via fiberoptics to the photodiodes and the kinetics measured using a two-channel recorder.

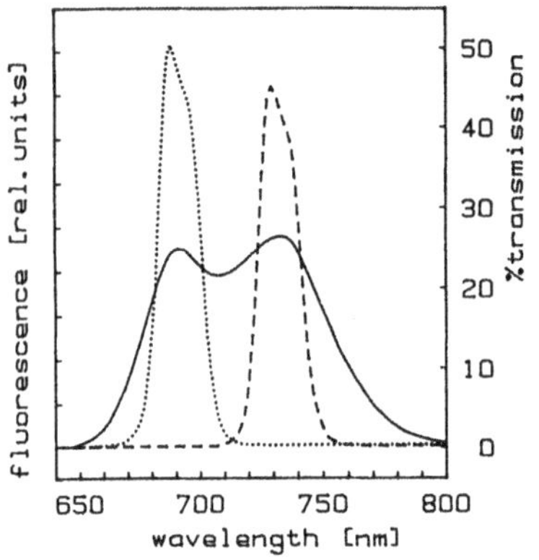

FIGURE 2. Transmission range of the filter systems used in the portable field fluorometer to sense the fluorescence induction kinetics in the 690 (·····) and 730 nm region (-----). The chlorophyll fluorescence emission spectrum (excitation light 470+30 nm) of a green leaf taken 5 min after onset of illumination, is indicated as a solid line.

Another newly developed chlorophyll fluorometer is the PAM-fluorometer (Schreiber, 1986; Schreiber et al. 1986) which permits determination of photochemical Q-quenching and the non-photochemical E-quenching of chlorophyll fluorescence. Due to the filter combinations applied the PAM-fluorometer only registers the fluorescence in the 730 nm region (Fig. 3), and it was originally not constructed to measure Rfd-values. Rfd-values, however, provide complementary and truly physiological information on the whole process of photosynthesis in addition to the quenching factors qQ and qE. It was therefore investigated whether the PAM-fluorometer - using its continuous saturating white-light source for fluorescence induction - can be applied for the determination of Rfd-values.

MATERIALS AND METHODS

The chlorophyll fluorescence of intact leaves or leaf disks was induced by a He/Ne laser (5mW; 632.8 nm; Spectra Physics) in a self-constructed portable apparatus as shown in Fig. 1. The chlorophyll fluorescence was simultaneously recorded in the 690 and 730 nm regions (Rfd 690 and Rfd 730) i.e.

in the chlorophyll-fluorescence maxima of intact leaves (Fig. 2). The leaves were predarkened for 20 min to obtain the real fluorescence maximum of the induction kinetics.

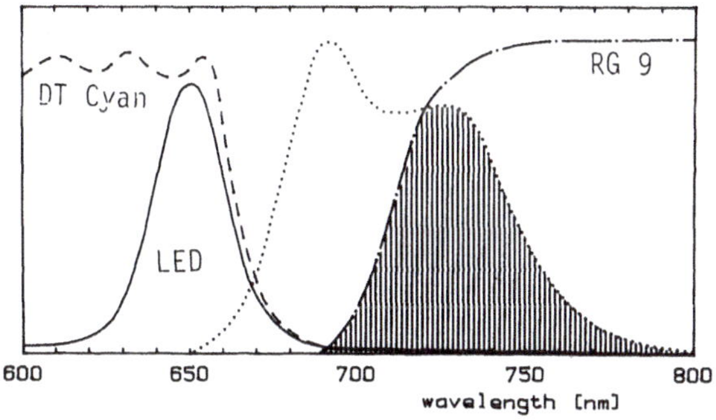

FIGURE 3. Transmission range of the filter system applied in the PAM-fluorometer which allows measurement of the induction kinetics only in the 730 nm region. The chlorophyll fluorescence emission spectrum of a green leaf as excited by 470±20nm light is indicated by a dotted line (LED = light source; RG 9 and DT Cyan = transmission filters).

The PAM-fluorometer, a pulse-amplitude modulation instrument, (Schreiber et al. 1986) was obtained from A. Walz, Effeltrich FRG. The ground fluorescence fo was measured in predarkened leaves (20 min) by excitation with 1 µs pulses of low intensity. The maximum level of the fluorescence was induced by a single saturating light pulse. By applying the white saturating light not as pulses but as continuous light, we measured the full fluorescence induction kinetics, from which the Rfd-values in the 730 nm region (Rfd 730) were determined.

RESULTS AND DISCUSSION

The red-laser light-induced chlorophyll fluorescence-induction kinetics of green leaves as measured with the portable field fluorometer are shown in Fig. 4. With the particular filter systems and the red excitation light (632.8 nm) applied, the chlorophyll fluorescence sensed in the 730 nm region is higher than that in the 690 nm region. The chlorophyll fluorescence kinetics at both wavelength regions go through a maximum (fmax), which is reached after ca. 250 to 400 ms, and then decline with the onset of photosynthesis and the energetization of the photosynthetic membrane. The relative fluorescence decrease fd with respect to the steady-state fluorescence fs is larger in the 690 nm than the 730 nm region. Correspondingly the Rfd-values determined from the induction kinetics at 690 nm (Rfd 690) are higher than those obtained from the 730 nm region kinetics (Rfd 730) as is indicated in Table 1. This is valid for intact healthy leaves and also for leaves which are under stress conditions. Only when the leaves are mechanically injured or when the autumnal or stress-induced senescence is much progressed, can Rfd 690 and Rfd 730 reach similar values (see review of Lichtenthaler and Rinderle, 1988).

From the Rfd values obtained in the 690 and 730 nm region one can calculate the stress-adaptation index A_p (Strasser et al. 1987) using the following equation:

$$A_p = 1 - \frac{\text{Rfd } 730 + 1}{\text{Rfd } 690 + 1}$$

The A_p-values found in different plants, using our portable field fluorometer, lie in the range of 0.17 to 0.33 for intact and photosynthetically active green plant tissue (see also Table 1). A_p-values decline with senescence, independently and later than the Rfd-values, and are usually lower in needles of damaged and partially bleached spruce trees (Lichtenthaler and Rinderle, 1988).

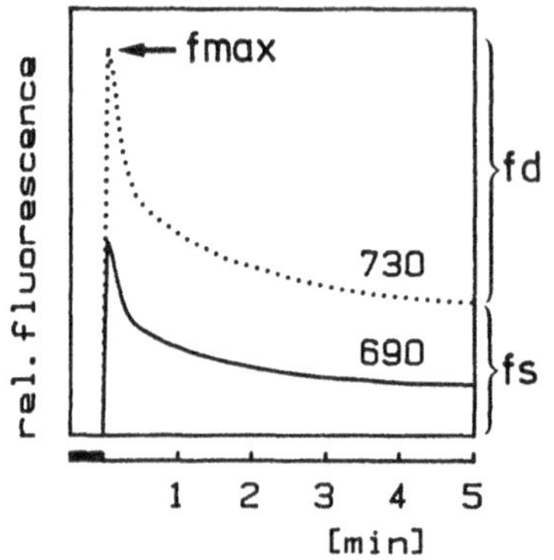

FIGURE 4. Chlorophyll fluorescence induction kinetics (Kautsky effect: slow component) measured separately in the 690 and 730 nm region of green tobacco leaves using the portable field fluorometer. The Rfd-values (Rfd = fd/fs) are higher for the 690 nm (ca. 2.9) than the 730 nm fluorescence (ca. 2.0). Fmax = maximum fluorescence; fd = fluorescence decrease; fs = steady state fluorescence.

With parallel samples of the same leaves tested in the portable field fluorometer, we also measured the chlorophyll fluorescence-induction kinetics in the PAM-fluorometer. The usual sequence in the registration of the fluorescence signals (Fo, Fm, fluorescence spikes induced by saturating light pulses) needed for the calculation of the photochemical Q-quenching and non-photochemical E-quenching indicated in Fig. 5. The intensity of the actinic light given at C to induce a small fluorescence-induction kinetic, is much too low and cannot be used for the determination of Rfd-values. Determination of Rfd-values as vitality index requires a much stronger light for the fluorescence induction, which must saturate the fluorescence emission and the process of photosynthesis. A light source, which can provide saturating light, is contained in the PAM-fluorometer. Instead of giving saturating light pulses every 10s to obtain the fluorescence spikes, one has to apply this saturating white light continuously to a predarkened leaf sample. The resulting fluorescence-induction kinetic (Fig. 6) shows the same characteristic as that obtained with the portable field fluorometer (Fig. 4). Due the particular filter combination applied in the PAM-fluorometer, the chlorophyll fluorescence is sensed in the 730 nm region (Fig. 3).

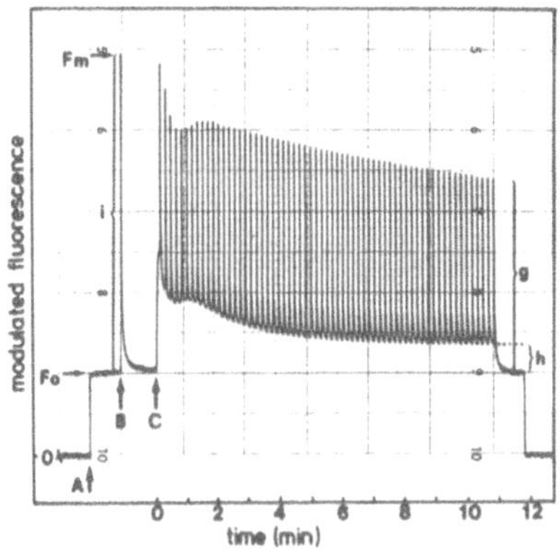

FIGURE 5. Original tracings of a modulated chlorophyll fluorescence-induction curve of a green leaf of Platanus hybrida used to determine Q- and E- quenching. **A.** measuring light pulses of low intensity (to detect the Fo level); **B.** 1s saturation pulse (to determine maximum fluorescence Fm); **C.** actinic light (+saturating light pulses every 10s).

The Rfd 730-values obtained with the PAM-fluorometer are very similar to the Rfd 730-values determined with the portable field fluorometer (Table 1). The PAM-fluorometer Rfd 730-values are practically identical for all green broadleaf plants investigated as well as for dark-green 2nd year spruce needles with the Rfd 730-values obtained using the field fluorometer. Only in the case of the first-year needles of spruce and fir were the PAM-Rfd 730-values somewhat higher (by 16 and 19 %, respectively) than the Rfd 730-values determined from the fluorescence kinetics measured with the field fluorometer. Yet in both cases these slightly higher PAM-Rfd 730-values ranged far beyond the significantly higher Rfd 690-values (Table 1).

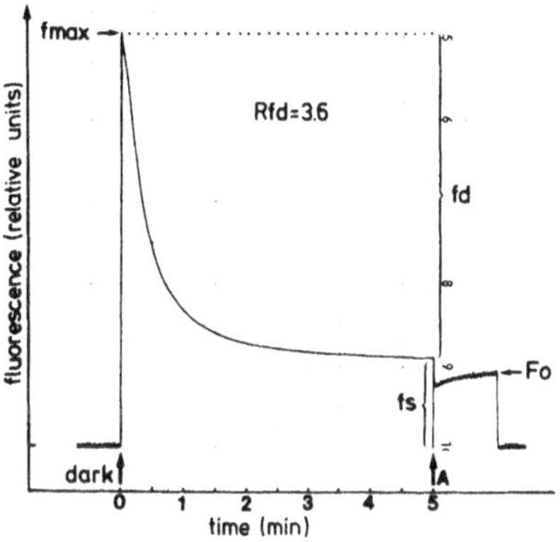

FIGURE 6. Chlorophyll fluorescence induction kinetic of a predarkened green leaf of Platanus hybrida (as in Fig. 5) measured with the PAM-fluorometer (in the 730 nm region). The fluorescence kinetic was induced by applying the saturating white light source continuously. After the continuous white light is cut off, the light pulses of low intensity are given at A to detect the Fo level.

Table 1. Stress-adaptation index A_p and comparison of the Rfd-values of green leaves measured in the 690 and 730 nm region with the portable field fluorometer and the Rfd-values obtained with parallel samples using the PAM-fluorometer (Rfd 730). The ratio fmax/fo is also indicated. Mean of 5 determinations per condition, with standard deviation.

	portable field fluorometer			PAM-fluorometer	
	A_p	Rfd 690	Rfd 730	Rfd 730	fmax/fo
broad leaves:					
Raphanus, cotyledon:	0.23 ±0.01	2.99 ±0.3	**1.99** ±0.2	**1.83** ±0.2	5.2 ±0.3
Raphanus, 1st leaf:	0,20 ±0.01	2.57 ±0.3	**1.85** ±0.3	**1.83** ±0.4	4.9 ±0.3
Monstera, deliciosa:	0,20 ±0.03	1.68 ±0.1	**1.15** ±0.1	**1.12** ±0.1	4.5 ±0.2
Zea mays, 5th leaf:	0.32 ±0.03	2.63 ±0.2	**1.46** ±0.1	**1.57** ±0.2	3.3 ±0.3
Nicotiana tabacum					
(aurea mutant Su/su):	0.19 ±0.01	2.75 ±0.1	**2.06** ±0.2	**2.18** ±0.2	4.5 ±0.2
conifer needles:					
Picea abies (spruce)					
dark-green, 2nd year:	0.27 ±0.06	5.43 ±0.6	**3.79** ±0.4	**3.86** ±0.5	3.6 ±0.6
green, 1st year:	0.26 ±0.03	5.40 ±0.3	**3.71** ±0.2	**4.31** ±0.4	4.0 ±0.2
Abies koreana (fir)					
green, 1st year:	0.26 ±0.03	6.05 ±0.4	**4.20** ±0.3	**4.98** ±0.5	4.0 ±0.3

CONCLUSION: The results of this comparative investigation indicate that the PAM-fluorometer with its saturating white light can be used to determine the Rfd 730-values of green leaf and needle material. Though the Rfd 730-values are lower than the Rfd 690-values determined from the 690 nm kinetics, they fully reflect the physiological activity and vitality of leaves.

The Rfd 730-values as well as the Rfd 690-values indicate the potential photosynthetic activity per leaf area unit. They decline with decreasing rate of photosynthesis and thus indicate stress and damage to the photosynthetic apparatus. Since the Rfd-values provide complementary and additional physiological data to the Q- and E-quenching information, which is routinely determined with the PAM-fluorometer, it is strongly recommended also to determine the PAM-Rfd-values of all leaf samples investigated in a PAM-fluorometer. This is particularly emphasized since PAM-fluorometers are now widely used in many laboratories. This would very much help to better define stress conditions in plants and would allow standardization, comparison and better understanding of the results obtained in different laboratories.

ACKNOWLEDGEMENTS

This work was sponsored by a grant from the European Joint Research Centre (JRC), Ispra and partly by the PEF (Projekt Europäisches Forschungszentrum für Maßnahmen der Luftreinhaltung), Karlsruhe, which is gratefully acknowledged. We wish to thank Dr. G. Rossi, Ispra for his interest in this research and Ms. B. Weigelt for technical assistance.

REFERENCES

LICHTENTHALER HK, 1984. Influence of enviromental factors on composition and function of the photosynthetic apparatus. In: Advances in Photosynthesis Research, C. Sybesma ed., Vol. IV, pp. 241-244, Martinus Nijhoff Publisher, The Hague (ISBN 90-247-2946-7).

LICHTENTHALER HK, 1986. Laser-induced chlorophyll fluorescence of living plants. In: Proc. Internat. Geoscience and Remote Sensing Symposium, IGARSS Zürich 1986, Vol. III, pp. 1571-1579, ESA Publications Division, Noordwijk.

LICHTENTHALER HK, BUSCHMANN C, RINDERLE U and SCHMUCK G, 1986. Application of chlorophyll fluorescence in ecophysiology. Radiation and Environmental Biophysics 25: 297-308.

LICHTENTHALER HK and RINDERLE U, 1988. Role of chlorophyll fluorescence in the detection of stress conditions in plants. CRC Critical Reviews in Analytical Chemistry 19, Suppl I: S 29-S 85.

NAGEL EM, BUSCHMANN C, and LICHTENTHALER HK, 1987. Photoacoustic spectra of needles as an indicator for the activity of the photosynthetic apparatus of healthy and damaged conifers. Physiol. Plantarum 70: 427-437.

STRASSER R, SCHWARZ B and BUCHER J, 1987. Simultane Messung der Chlorophyll Fluoreszenz-Kinetik bei verschiedenen Wellenlängen als rasches Verfahren zur Frühdiagnose von Immissionsbelastungen an Waldbäumen. Ozoneinwirkung auf Buchen und Pappeln. Europ. J. Forest Pathology 17: 149-157.

SCHREIBER U, 1986. Detection of rapid induction kinetics with a new type of high-frequency modulated fluorometer. Photosynth. Res. 9: 261-272.

SCHREIBER U, SCHLIWA U and BILGER W, 1986. Continuous recording of photochemical and non-photochemical chlorophyll fluorescence quenching with a new type of modulation fluorometer. Photosynth. Res. 10: 51-62.

FIBER - OPTIC DETECTION OF CHLOROPHYLL FLUORESCENCE

Herbert Schneckenburger[1,2] and Joachim Bader[1]

[1]Fachhochschule Aalen, Beethovenstr. 1, D-7080 Aalen, F.R.Germany

[2]Institut für Lasertechnologien in der Medizin an der Universität Ulm, Postfach 4066, D-7900 Ulm, F.R.Germany

Key words: Fiber-optic sensor, fluorescence detection

A fiber-optic system for detection of in vivo fluorescence was developed. Exciting argon laser light and fluorescence are transmitted through the same fiber using appropriate micro-optic devices (dichroic mirrors, filters, selfoc lenses), an optional spectral analyzer and an avalanche photodetector. This system allows for optimum flexibility, remote sensing and detection of small areas in the submillimeter range of biological samples. Preliminary measurements include the induction and decrease of chlorophyll fluorescence during light exposure of intact spruce needles and needles submitted to water stress.

INTRODUCTION

Different methods and instruments for detecting chlorophyll fluorescence of intact and damaged organisms have so far been reported. Those devices include portable fluorescence spectrometers (Lichtenthaler et al. 1986), pulse modulation fluorometers (Schulze et al. 1983, Schreiber 1986), special probes for detection of delayed fluorescence (Krause et al. 1984) and remote sensing. Fiber-optic devices were applied for measuring several fluorescent pigments or metabolites in analytical chemistry (Wolfbeis 1986) or biotechnology (Scheper and Schügerl 1986). Different fibers were therefore used for light excitation and fluorescence detection, thus limiting the sample volume which could be attained by both fibers.

A very sensitive and highly resolving technique is fluorescence microscopy (Schneckenburger et al. 1985), in particular since this method can be combined with measurements of emission spectra or picosecond decay times (Schneckenburger and Frenz 1986,Schneckenburger et al. 1987). The method is based on epi-illumination via a dichroic mirror, appropriate filters and a microscope objective lens. Sample fluorescence is collected by the same lens, passes the dichroic mirror and is measured by a highly sensitive detector (usually a photomultiplier or camera system). Since fluorescence microscopy is not flexible enough for many in vivo applications, we integrated some of its main components into a fiber-optic sensor system. This device allows for excitation and fluorescence detection of small samples in the submillimeter range via the same fiber, even at long distances from the light souce. Due to its flexibility it can be applied to chlorophyll measurements of various plants, as well as to many other fluorophores in living organisms.

H. K. Lichtenthaler (ed.) Applications of Chlorophyll Fluorescence, 255–258.

EXPERIMENTAL METHOD

An air-cooled argon ion laser with adjustable wavelenghths in the range between 457.9 nm and 514.5 nm is used as excitation source. For chlorophyll detection the 457.9 nm blue light is focused on the plain surface of a glass fiber with 100 µm of core diameter (Fig. 1). The light is guided to a micro-optic bench (size 3 x 3 cm²) which is the central element of the whole system. It contains a dichroic mirror (Zeiss, FT 580) oriented under 45° to the exciting light beam and additional holders for filters. The incident light passes an adjustable selfoc lens, which consists of a gradient fiber (as frequently used in communication techniques) with large diameter (1-2mm) and which is used to convert the light coming from a small spot (fiber tip) into an almost parellel beam. This beam is deflected by the dichroic mirror onto another selfoc lens, and via a measuring fiber of appropriate length reaches the sample. Fluorescence is detected by the same fiber, passes - due to its longer wavelength- the dichroic mirror and an additional long pass filter ($\lambda \geq$ 590 nm), and is measured by a Si avalanche photodiode. An optional spectral analyzer can be inserted into the detection fiber. Intensity losses of 53% for the excitation path and 58% for the detection path were calculated (Bader 1988). These values were confirmed in the experiment after optimum adjustment of the whole system. Additional losses of blue excitation light of about 12 dB/km - corresponding to a factor 2 for 250 m of fiber length - would occur for measurements at long distance.

Fig. 2 shows the geometry of illumination and fluorescence detection of the samples using plain fiber tips. In preliminary experiments a laser power of 1 mW at the end of the fiber and a distance l = 0.6 mm were adjusted, thus yielding a power density of 0.52 W/cm² on the sample surface. This was about 5 times as much as solar irradiance. The detection angle Ω was 0.022 sr corresponding to 1/576 of the whole spatial angle. Assuming complete absorption of incident light and a fluorescence quantum yield of 3 %, this means that about 50 nW could be taken up by the fiber. Taking into account the losses in detection path and the sensitivity of 1V/µW of the avalanche detector, signals of about 20 mV were expected. The experimental values were in the range of 5 - 20 mV.

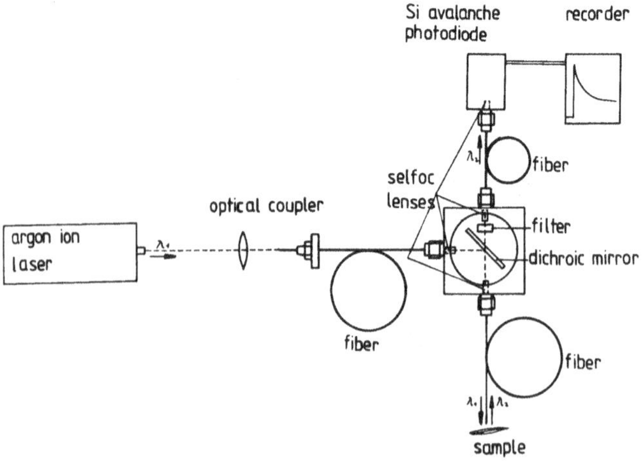

FIGURE 1. Fiber-optic fluorescence detection.

For measuring the induction and decrease of chlorophyll fluorescence (Kautsky curves) of spruce needles, individual needles or small branches were placed for at least 20 min in a black box prior to illumination by the fiber. One year old intact sun and shade needles (5 in each case) of an about 15 years old spruce were measured and compared with needles which had been plucked 4 days and 18 days before, and which had therefore been submitted to a certain water stress. Some needles of the previous age class were used for comparison.

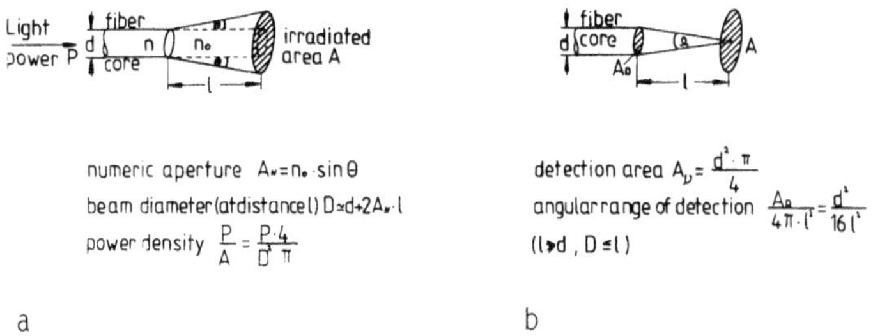

FIGURE 2. (a) Illumination and (b) light detection by a glass fiber.

RESULTS AND DISCUSSION

Fig. 3 shows the time course of fluorescence of an intact ("fresh") sun needle and needles which had been plucked 4 days and 18 days before. At the beginning of illumination the fluorescence increased within 1 s to a maximum value, and then during the onset of photosynthesis generally decreased, until after 5 - 8 min a constant level f_s was attained. The ratio of fluorescence decrease f_d and f_s (determined after 7.5 min) was found to depend on the time interval between plucking of the needle and measurement and was used as a measure for the efficiency of photosynthesis. This ratio varied between 2.5 and 4.5 for intact needles; a difference between sun and shade needles, as reported by Lichtenthaler et al. (1986) was not found, probably due to the limited number of needles so far investigated. 4 days after plucking a ratio f_d/f_s of 1.9-2.5 for sun needles and 1.2 - 1.8 for shade needles was detected. This indicates that the sun needles were better adapted to this kind of water stress. After 18 days ratios of 0.15 - 0.5 were measured for both types of needles, demonstrating a very low efficiency of photosynthesis.

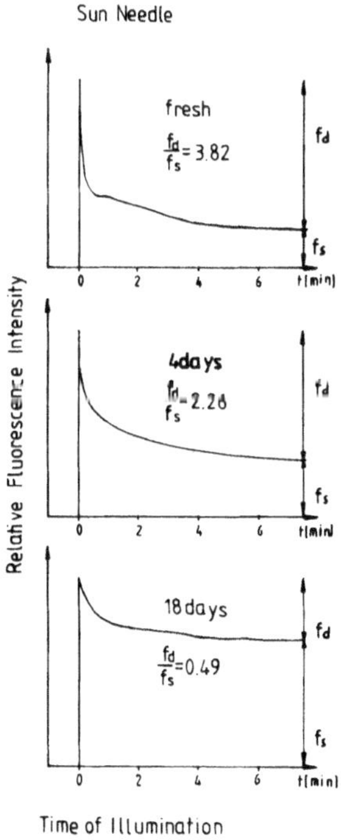

FIGURE 3. Kautsky curves of spruce needles at different times after plucking.

These preliminary results show that the experimental setup may well be appropriate for measuring damages to trees (e.g. by environmental pollutants) in vivo. Fibers of some hundreds of meters of length can be used for remote measurements. The whole apparatus can be easily removed from one measuring station to another, since all fiber-optic components are linked by fixed SMA plugs. Only the optical coupler between the argon laser and the fiber must be re-aligned. Additional components, such as beam modulators or attenuators can also be introduced, e.g. for application of pulse modulation techniques. The dichroic mirror and filters in the micro-optic bench can be exchanged, if another wavelength range is desired. The system is therefore quite universal and may be used as a fluorescence sensor in a broad field of biotechnology and medicine in the visible and near infrared range. Preliminary tests of measuring photosensitizing dyes in experimental tumors have been started.

A problem of several applications may be that at typical distances l between the fiber tip and the object of 0.5 mm - 1 mm the angle of fluorescence detection is rather low, whereas the power density of irradiation is so high that photochemical modifications occur. By reduction of the distance l, attenuation of the laser beam and use of a more sensitive detector this problem may be resolved.

REFERENCES

Bader J, 1988. Aufbau und Optimierung eines faseroptischen Sensors zur Messung schwacher Fluoreszenzsignale. Diplomarbeit, Fachhochschule Aalen.

Krause H, Gerhardt V and Gebhardt W, 1984. Verzögerte Fluoreszenz lebender Pflanzenzellen. Phys. unserer Zeit 15: 182-190.

Lichtenthaler HK, Buschmann C, Rinderle U and Schmuck G, 1986. Application of chlorophyll fluorescence in ecophysiology. Radiat. Environ. Biophys. 25: 297-308.

Scheper T and Schügerl K, 1986. Characterization of bioreactors by in-situ fluorometry. J. Biotechnol. 3: 224-229.

Schneckenburger H, Reuter BW and Schoberth SM, 1985. Fluorescence techniques in biotechnology. Trends Biotechnol. 3: 257-261.

Schneckenburger H and Frenz M, 1986. Time-resolved fluorescence of conifers exposed to environmental pollutants. Radiat. Environ. Biophys. 25: 289-295.

Schneckenburger H, Frenz M, Tsuchiya Y, Denzer U and Schleinkofer L, 1987. Picosecond fluorescence microscopy for measuring chlorophyll and porphyrin components in conifers and cultured cells. Lasers Life Sci. 1: 299-307.

Schreiber U, 1986. Differenzierung zwischen verschiedenen Typen der Fluoreszenzlöschung und Bestimmung der relativen Elektronen-Transportrate in vivo mit Hilfe der Sättigungspuls-Methode. Symp. Chlorophyll Fluorescence in Ecology, Hannover (FRG).

Schulze E, Pätzold G, Röther-Steinke M and Ernst D, 1983. Chlorophyllfluoreszenz zur Früherkennung von Photosythesebeeinträchtigungen an Coniferen. Forstarchiv 54: 229-232.

Wolfbeis OS, 1986. Analytical chemistry with optical sensors. Fresenius Z. Anal. Chem. 325: 387-392.

CHAPTER 3

Applications in hydrobiology, limnology and oceanography

APPLICATION OF CHLOROPHYLL FLUORESCENCE IN HYDROBIOLOGY

Dieter E.W. Ernst, Inst. f. Biophysik, Hannover

Key words:
Chlorophyll fluorescence, hydrobiology, limnology, eutrophication, poisioning, oceanography, O_2-depletion, monitoring of algae, chlorophyll content, electron transport chain blocker, chlorophyll extraction, primary production, quench effect, delayed fluorescence, colour classes.

ABSTRRACT

The most important classical limnological parameters are mentioned and the possibility to measure corresponding values by chlorophyll fluorescence are reviewed. These are mainly the chlorophyll content and the photosynthetic turnover which allows estimates of primary production. In contrast to forestry or agricultural applications hydrobiological work faces mixtures of many species in the sample and smaller and very varying chlorophyll concentrations. Since fluorescence methods are quick and rather easy to be mechanised they allow to obtain the spacial and temporal distribution of phytoplancton. The influence of poisons on a test culture can be checked by fluorescence too. Different colour classes (e.g. bluegreen, or green algae) can be identified to some extend by their excitation or fluorescence spectra.

INTRODUCTION

At present we have a good background knowledge on chlorophyll fluorescence methods and can observe their introduction into hydrobiological practise (Samuelsson et al. 1978). These methods are quick and easy to be mechanised as compared to classical limnological procedures, however, there is still a debate about interpretation.

At first I want to review very briefly a few facts of hydrobiology and state which parameters may be achieved by fluorescence measurements, then I shall refer the most important applications.

If we are concerned with adverse antropogenic effects in fresh water we may differentiate mainly between 1) events around eutrophication and O_2 depletion and 2) poisoning (agricultural runoff, chemical accidents). In this paper I want to concentrate mainly on fresh water applications.

In oceanography a further important question is the identification of water bodies which may move due to wind, tidal or seasonal influences or upwelling. Comparatative plancton studies from different stations, drift experiments and synoptical methods with airborn or satellite equipment are the appropriate techniques. Fluorescence methods may help here because they can easily cover larger areas (ESA 1986). Their chances to identify species are poor. Some hints are given in the section "colour class".

H. K. Lichtenthaler (ed.) Applications of Chlorophyll Fluorescence, 261–268.

LAKES

In lakes and reservoirs the main limnological viewpoints are O_2 balance, sources of plant nutrients and organic load, retention time and stratification. Classical parameters to be determined are: biological and chemical oxygen demand (BOD, COD) particulate and dissolved P, N and organic C (POC, DOC) - P being important because it usually is the first production-limiting element - and further: primary production, biomass, chlorophyll content. Fluorescence methods are giving access mainly to two parameters: chlorophyll content and production and their distribution in space and time.

RIVERS

In Flowing Waters the concerns are similar, stratification playing a minor role. O_2-depletion is important in lowland rivers and poisoning may be of particular concern. Besides the parameters mentioned the classification according to the saprobic system is important. Fluorescence methods are particularly effective in automated stations giving a continous record on chlorophyll (Noack 1983). In Germany with increasing efficiency of sewage treatment primary organic load (organic material from sewage) is becoming less critical. However, secondary load (phytoplancton) becomes important. In surveillance stations O_2 consumption is easily monitored by O_2-electrodes, which, however, can't discriminate between the two loads. A fluorescence probe can directly monitor the algae.

THE DETERMINATION OF CHLOROPHYLL CONTENT

Several fluorescence methods have been applied for this purpose:
In submersible or towable instruments (Suarez, Früngel, Hamburg) fluorescence is excited in the ambient water by a blue flash without further treatment. According to Nusch (in press) the signal is by 80% a function of chlorophyll content the rest depending on the physiological state of the algae.
If the electron transport chain is blocked by the addition of DCMU, CMU or other urea-derivatives the signal is by far better proportional to the chlorophyll content. Fig. 1 and 2 show rather good proportionalities achieved with a portable pulse fluorometer (Kleinfeld, Hannover). Such a technique, however, can at present not be carried out in towable instruments. Even after this blocking the factor transferring fluorescence into chlorophyll concentration might vary. Noack introduced species dependent conversion factors (Noack 1983). In contrast to e.g. agricultural or forest populations phytoplancton usually is a mixture of different species. Also the amount of chlorophyll in the samples is lower. Usual values are some dozens of ug/ltr extreme eutrophic waters have some mg/ltr. At the lower edge may be our recent determination of 2 ng/ltr (!) (Pätzold, unpublished, with a modified pulse fluorometer) in the Red Sea.

In this context it must be stated that the discussion about the extractive photometric chlorophyll determination is far from beeing settled. Different solvents (e.g. acetone, alcohole, DMSO) hot or cold extraction,

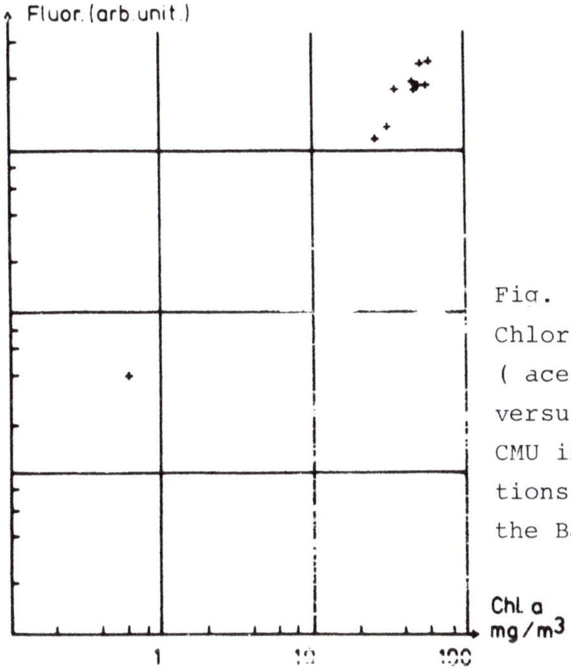

Fig. 1
Chlorophyllconcentration (acetone extraction) versus fluorescence under CMU in natural popula - tions from Schlei and in the Baltic

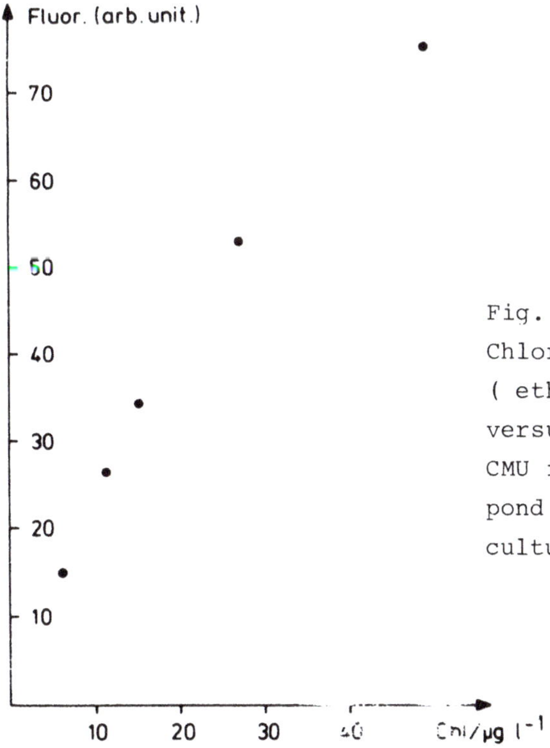

Fig. 2
Chlorophyllconcentration (ethanole extraction) versus fluorescence under CMU from an experimental pond and a Chlorella fusca culture

different wavelengths and chromatographic methods are competing. Remarkable species- dependent variations in extraction efficiency have been reported. (Rai 1982, Nusch 1980). We worked out particular quick procudure in which the algae are filtered on a membrane filter. Acetone is then poured on the filter which extracts the chlorophyll and dissolves the filter. So the pigment need not diffuse out of the cell and the second filtration step is saved. The extraction time may be as short as 5 to 10 minutes. The sample is then measured in the same portable pulse fluorometer (Schulze, unpublished, Güther 1987, Ernst, in press).

THE DETERMINATION OF PRIMARY PRODUCTIVITY

Once the chlorophyll content is known a fluorescence measurement in the living algae will give a figure depending on the quench phenomena which in turn depend somehow on primary production.Unfortunately the different quench effects show different time courses and dependencies concering production. So this approach depends very much on the instrument and method of evaluation (Renger, Schreiber 1986, Krause, Weis 1984). Usually the standardized difference between s- and p-level of the Kautsky-curve or s level and the level with blocked electron transport chain (called RFD, relat.fluoresc.depression) are used to estimate production. Phytoplancton shows an advantage as compared to the leaves of higher plants as DCMU diffuses into the organism almost momentaryly. So the level with blocked electron transport chain is readily accessible. In the pulse fluorometer the quench phenomena are rather different from the Kautsky-phenomenon (probably limited by the water-splitting system and a small contribution of e-quench). A parameter similar to RFD can be defined (RIFD, relat. impuls.-..). It apparently reflects potential production (production devided by light intensity) (Ernst 1986)(Fig. 3). RIFD shows characteristic changes during plancton blooms (Reckel 1984) and correlates somehow with the yellow-green-index (a senescence index) (Noack 1983, Margalef 1967).

A problem in all fluorescence measurements in water is the solube fluorescence and the proper background correction. The zero point has to be corrected with the same cuvettes as used in the measurement and with filtered sample water to correct for the soluble fluorescence. This fluorescence may be caused by dissolved chlorophyll decay products (mainly in marine samples) or by humic acids (limnic samples). It is usually small compared to the planctic signal but may become significant under high grazing pressure or in polluted waters. In a higher extend than usually expected it may be caused by very minute algae passing even a 0,45 um pore size filter. A check whether such "soluble" fluorescence is really dissolved or caused by algae may be carried out by blocking the electron transport chain. An increase in the signal reveals living algae.

Krause et al. (1987) found an apparent proportionality of delayed fluorescence to production.

FLUORESCENCE TESTS FOR POISONS

Fluorescence tests may also provide checks for a damage by even unknown substances. We must differentiate between acute and chronic tests. In an acute test the response is checked within minutes after addition of the substance. It reflects rather direct damage to the electron transport chain.In cronic tests algal cultures are followed over some generations.

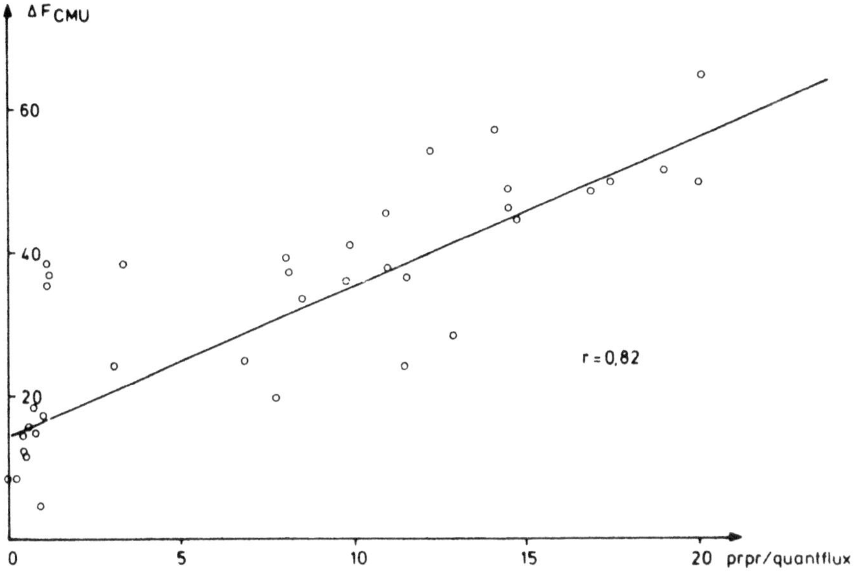

Fig. 3
correlation between fluorescence signal (ΔF_{CMU}) and potential primary production (prpr / quantflux). The fluorescence signal is the difference between the pulse fluorescence of viable pytoplancton (averaged over the 3rd to 7th pulse) and algae under CMU - block. Potential primary production is the ratio of pro - duction (^{14}C - method) over PAR - quantum flux in half Secchi - depth. Data are from a limno - corral experiment in lake Steinhuder Meer and comprise dif - ferent weather conditions and different artificially added nutrients.
r: correlation coeff. (from Ernst, 1986)

They reflect overall damage to phytoplanctic life. (Christoffers, Ernst 1983, Beneke et al. 1982).

COLOUR CLASSES

Different taxonomic groups (e.g. blue green, green, red algae etc.) display different excitation and emission spectra which were exhausted by Yentsch(1979) to determine the colour class of marine phytoplancton. Krause et al. (1987) published a similar trial using delayed fluorescence. However, it must be kept in mind that the pigment composition cannot be regarded to stay constant (see yellow green index) and fluorescence surely is a poor parameter concerning taxonomic questions.

CONCLUSION

Chlorophyll fluorescence methods are rather quick as compared to most of the classical limnological procedures. They, however, only give an indirect estimation of the two most important parameters: biomass (estimated by chlorophyll content) and primary productivity. The determination of chlorophyll content yields minor problems, one of the most important being the discrepancy between the different extractive photometric methods. The photosynthetic turnover and hence production are more difficult to achieve. Repetitive checks with the classical methods have to be carried out. The potential to overview spatial and temporal phytoplancton distributions and to correlate it with meteorological and other environmental factors will probably be the most valuable feature of fluorescence methods and open them numberless future applications.

REFERENCES

Author Group ESA (European Space Agency) - Study, 1986. The use of chlorophyll fluorescence measurements from space for separating constituents of seawater.GKSS, Geesthacht.

Beneke G, Falke W, Schmidt Chr, 1982. Use of algal fluorescence in an automated monitoring system. Bull Env Tox 28: 385

Christoffers D, Ernst D E W, 1983. The in-vivo fluorescence of Clorella fusca as a biological test for the inhibition of photosynthesis. Tox Env Chem 7: 61-71.

Ernst D, 1986. Comments on fluorometric chlorophyll determination in the field. Arch Hydrobiol 107: 521-527.

Ernst D. The Kinetics of Chlorophyll Extraction from Phytoplancton and a quick extraction method. Arch Hydrobiol Erg Limnol (in press).

Güther S, 1987. A quick method for extractive chlorophyll determinations in the fluorometer. Arch Hydrobiol Erg Limnol Beih 29: 93-98.

Krause G H, Weis E, 1984. Chlorophyl fluorescence as a tool in plant physiology. II Interpretation of fluorescence signals.Phot Syn Res 5: 139-157.

Krause H, Dirnhofer P, Gerhardt V, 1987: The application of delayed fluorescence. Arch Hydrobiol Erg Limnol Beih 29: 55-62.

Margalef R, 1967. Concepts relative to the organization of plancton. Oceanog Mar Biol Ann Rev 5: 257-289.

Noack U, 1983. Untersuchungen über Chlorophyll a-Gehalt, Chlorophyll-Fluoreszenz und weitere limnologische Parameter am Phytoplankton des Dümmers. Diplomarbeit, Inst.f.Biophysik,Uni Hannover.

Noack U, 1983. Bestimmung der Phytoplankton-Biomasse über Chlorophyll-a-Gehalt und Chlorophyllfluoreszenlz am Dümmer. Mitt Nieders Wasserunterungsuchsamt Hildesheim.

Nusch A, 1980. Comparison of different methods for chlorophyll and phaeopigment determination. Arch Hydrobiol Beih 14: 14-36.

Nusch A. In vivo fluorescence - a biomass - or a bioactivity indicator? Arch Hydrobiol (in press)

Rai H, Marker A, 1982: The measurement of photosynthetic pigments in freshwaters and standarization of methods. Arch Hydrobiol Erg Limnol Beih 16.

Reckel S, 1984. Vergleichende Untersuchungen von Primärproduktionsmessungen und Fluoreszenzverhalten von Algen in Versuchsteichen.Staatsexamensarbeit

Renger G, Schreiber U, 1986. Practical applications of fluorometric methods to algae and higher plant research. In: Light emission by plants and bacteria. Govindjee, Amesz J, Fork D C (eds) Acad Press, New York.

Samuelsson G, Öquist G, Halldal P R, 1978. The variable chlorophyll a fluorescence as a measure of photosynthetic capacity in algae. Mitt Int Verein Limnol 21: 207-215.

Yentsch Ch, Yentsch Cl, 1979. Fluorescence spectral signatures: The characterization of phytoplancton populations by the use of excitation and emission spectra. J Marine Res 37: 3.

Remote Sensing of Sunlight Induced Phytoplankton Fluorescence

Roland Doerffer
GKSS Forschungszentrum Geesthacht
D-2054 Geesthacht, F.R.G

Key words: Remote Sensing, phytoplankton, chlorophyll, fluorescence

Abstract

Remote sensing of ocean color for the measurement of phytoplankton chlorophyll in turbid coastal areas is problematic, since high concentrations of suspended matter and Gelbstoff also change the color from blue to green. The sunlight induced fluorescence of chlorophyll is a more specific signal compared to color ratio techniques. It can be used to map the phytoplankton distribution from aircrafts and satellites even in coastal areas. The paper demonstrates the principles of the method, summarizes the present experiences and discusses problems and the future potential of the method.

Introduction

Remote sensing is becoming a key technique for mapping phytoplankton distribution in the sea, since the need for information about the food web and the flow of matter within the earth-sea-atmosphere system has increased to a level where conventional techniques cannot cover the required scales in time and space (NASA, 1988). Dedicated sensors and evaluation procedures for phytoplankton remote sensing are under development and will be deployed on the next generation of satellites (NASA, 1986). This paper will discuss briefly the potential and the limits of the Fluorescence Line technique, which allows us to estimate the chlorophyll concentration remotely by its sunlight induced fluorescence.

We will first see, why we need this method for phytoplankton remote sensing of coastal areas, then give a short explanation of the basic algorithm to calculate the fluorescence from radiance spectra and finally discuss limitations and the resulting requirements for future work.

Why do we need the fluorescence technique?

Up to now radiometric and spectral characteristics of spaceborne imaging sensors have restricted chlorophyll remote sensing to sea areas where phytoplankton is the only factor determining the water color. The standard evaluation procedure is the color ratio algorithm: increasing phytoplankton concentration shifts the water color from blue to green caused by both chlorophyll absorption and scattering by the cells. Thus, the ratio of the radiances in a blue and a green spectral channel can be used to determine the concentration (Gordon & Morel, 1983). However, this ratio is also effected by suspended particles other than phytoplankton and by Gelbstoff (a fraction of the dissolved organic matter) due to their scattering and their absorption effects respectively (Doerffer, 1979; Fischer, 1985; Fischer, Doerffer & Grassl, 1986). The ambiguity of water color with respect to the influence of different substances is evident in images of turbid coastal water areas. An example is given in the Colour plate No. 13 (at the end of the book) which shows the German Bight (North Sea) as recorded by the Coastal Zone Color Scanner (CZCS),the only spaceborne ocean color scanner, which up to now has been deployed by NASA onboard the satellite NIMBUS 7 from 1978 to 1986. The image is passed through an atmospheric correction procedure: it subtracts the radiance which was specularly reflected by the water surface and scattered by the atmosphere, and thus reveals the water leaving radiance only (Sturm, 1980). The green/blue algorithm would indicate high chlorophyll concentrations in the center of the image; however, an analysis

H. K. Lichtenthaler (ed.) Applications of Chlorophyll Fluorescence, 269–274.

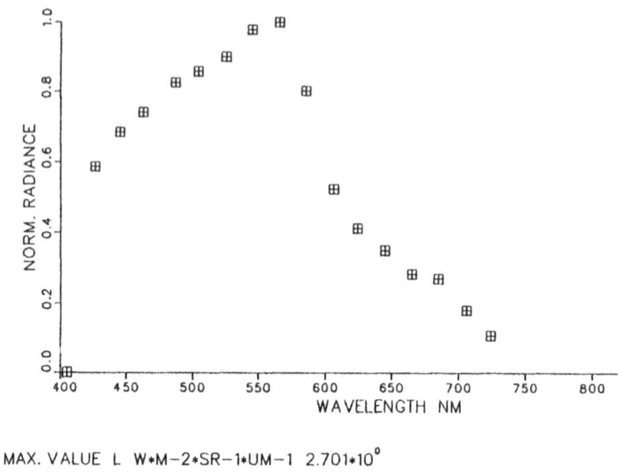

Figure 1: Spectrum of the water leaving radiance, measured with a filter-wheel radiometer during FLUREX 82, Kiel Bight, chlorophyll a concentration of $3 \mu g/l$

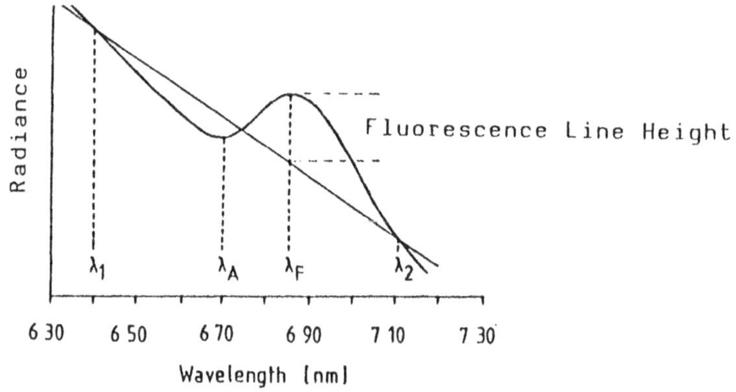

Figure 2: Scheme of the fluorescence line height: λ_F fluorescence maximum λ_A absorption of chlorophyll $\lambda_1 \lambda_2$ baseline wavelengths

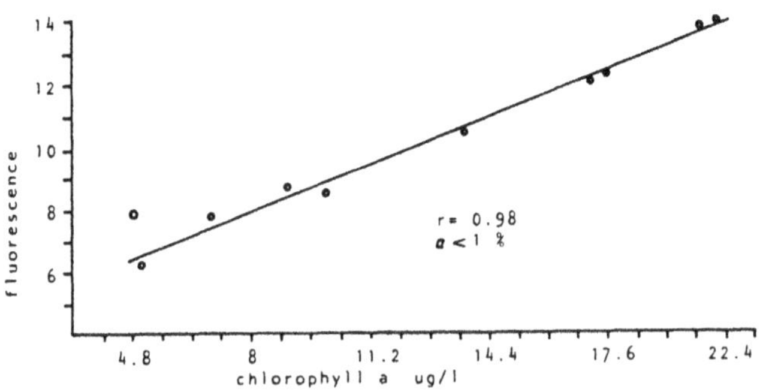

Figure 3: Relation between the fluorescence line height (rel. units), as measured from an altitude of 600 m, and the chlorophyll concentration in 2 m depth; data are from a 90 km profil in the Fladen Ground (North Sea), s. also colour plate 14

with a radiative transfer model (Fischer & Doerffer, 1987) demonstrates that the green patterns are areas with increased suspended matter concentration which also shifts the colour from blue to green by scattering.

One possibility to determine the chlorophyll concentration more specifically, is to measure the water leaving radiance around 685 nm, where the sunlight induced fluorescence of chlorophyll enhances the backscatterd radiation and where the influence of suspended matter and Gelbstoff is reduced (Gordon, 1979). This method was first applied by Neville & Gower(1977; Gower & Borstad, 1981) by using a diode array spectrometer. This instrument is equipped with a polarising filter in order to suppress the specularly reflected skylight when pointed at the sea surface under the brewster angle. Measurements carried out without a polariser by Doerffer during the FLEX Experiment in the North Sea 1976 demonstrate that the fluorescence signal can even be quantified in the presence of specularly reflected skylight (Doerffer, 1981; Amann & Doerffer, 1983). Colour plate 14 (end of book) shows the result of an aerial survey of the Fladen Ground area; the chlorophyll concentration was mapped by measuring the sunlight induced fluorescence with a radiance spectrometer from an altitude of 600 m.

The Fluorescence Line Algorithm

Part of the sunlight energy absorbed by an algal cell escapes in the form of fluorescence light, which is emitted mainly at 685 nm (Günther et al.,1986). This light enhances the radiance, which is scatterd backwards by the ocean, and causes a slight peak in the radiance spectrum within the 682-687 nm range (Fig. 1). The height of this peak over the baseline is determined as the fluorescence line height FLH (Fig. 2). Comparison with surface chlorophyll concentrations have shown that under constant irradiance FLH is a linear function of the chlorophyll concentration (Fig. 3), although modified by a number of factors which will be discussed below. The baseline can be calculated from the radiance of two neighbouring wavelengths using a linear interpolation. The choice of these baseline wavelengths is critical, since the spectrum in the range $\lambda\, 685nm$ is partly disturbed by absorption bands of the atmospheric gases oxygen and water vapor (Fischer, Kronfeld & Schlüssel, 1986). At the short side of the spectrum the chlorophyll absorption at 665-670 nm and the strong Fraunhofer line at 656 nm have to be avoided. Summarizing these facts one can state: The fluorescence peak can and should be measured within the wavelength intervall 680 - 687 nm, while the two baselines should be centered around 640 and 711 nm.

The variability of the fluorescence peak

External as well internal - physiological - factors modify the height and the shape of the fluorescence peak. The effect of some of these - particularly internal - factors are still under research and no final assessment is possible at present (Rabbani, 1984; Topliss, 1985; Topliss & Platt, 1986).

We first consider the external factors, which means a constant fluorescence efficiency (no physiological effects). In this case the specific fluorescence- per unit chlorophyll - is a factor of the exciting energy: it depends on the solar irradiance (sun elevation) and the transparency of the atmosphere and water mainly in the blue part of the spectrum around the maximum of the chlorophyll absorption peak at 440 nm. Particular high concentrations of suspended matter and Gelbstoff can limit the light penetration depth. However, since the fluoresence light, as measured above the water surface, stems only from the first 2 meters (due to the absorption of pure water in the red), this effect is only significant in the presence of very high concentrations as found in estuaries. In this case light backscatterd by suspended matter also amplifies the radiance at the baseline wavelength at 640 nm and thus causes a change in the baseline slope (Fischer & Kronfeld, 1986).

In the case of high concentrations of chlorophyll the reabsorption of fluorescence light can also modify the peak by causing a shift to the red part of the spectrum (Dirks & Spitzer, 1987). A contrary effect is the absorption of the red wing of the fluorescence peak by pure water due to the steep increase in the red part of the water absorption spectrum (Haardt & Maske, 1986). This latter effect can be important when the vertical chlorophyll concentration increases with depth. Both effects may shift the fluorescence maximum as observed in the radiance spectrum above the water by +/- 5 nm.

In coastal waters, high concentrations of suspended matter containing phaeopigments may also produce a fluorescence signal (Doerffer, 1986). This possible effect has been little investigated up to now, but statistical analysis of field data indicate that in some areas this effect may be a problem in calculating correct chlorophyll concentrations. The fluorescence efficiency, i.e. the ratio of the fluorescence light to the absorbed light or the photosynthetic active radiation (PAR) is a function of a large variety of factors and depends on the species composition of the phytoplankton population, on the irradiance, the light history and the nutrient conditions (Maske & Boje, 1986).

The influence of irradiance was investigated during the international Fluorescence Remote Sensing Experiment FLUREX 82: observations and model calculations by Doerffer & Fischer (1986) and Günther (1986) indicate that the fluorescence efficiency becomes constant with a light level (PAR) $> 300W/m^2$ corresponding to a sun elevation of 30^0.

Although the number of potential variables modifying the specific fluorescence efficiency is high, experiences from a large number of field observations from aircrafts and ship indicate that at least for a limited period and area these variables remain constant which thus allows us to apply the method successfully (Amann et al.,1986; Gower, 1986).

Future requirements and potentials

The open questions as discussed in the previous chapter require more basic research and improvements in data evaluation procedures and sensors. Future techniques will also increase the amount of information which can be retrieved from remote sensing data. Four aspects should be discussed here:

Imaging spectrometers based on a diode matrix technique will provide a much better spectral resolution ($\lambda < 5nm$), more and selectable channels (> 20) and an improved signal/noise ratio compared to optomechanical scanners (MBB, 1986; Gower et al.,1988). The programmable channels can be placed at wavelength of interest and can avoid critical spectral regions. However, the amount of data will increase considerably and will require efficient computing facilities.

The procedures to retrieve concentration parameters will be based on inverse modelling (Fischer & Doerffer, 1987). It will allow us to include all interesting wavelengths including the fluorescence channels and thus avoid ambiguities in the determination of chlorophyll, suspended matter and Gelbstoff concentrations. It will also offer the possibility of detecting exceptional plankton blooms such as red tides (Doerffer & Amann, 1986).

The data and experience gained from each experiment with respect to scattering and absorption coefficients of different plankton populations or of fluorescence efficiencies and typical vertical distributions can then be assimilated in information systems to help improve further the evaluation procedures.

The combination of remote sensing data with photosynthesis and phytoplankton distribution models will also allow us to estimate parameters such as biomass, productivity and the flow of matter as part of a global model, linking the carbon cycle with terrestrial and oceanic nutrients (Platt, 1986; Campbell, 1988).

References

Amann,V. and R. Doerffer, 1983: "Aerial Survey of the Temporal Distribution of Phytoplankton During FLEX'76". In: North Sea Dynamics, Sündermann and Lenz, Ed. (Springer-Verlag Berlin Heidelberg), pp. 517-529.

Amann, V., R. Doerffer, J.F.R. Gower, H.H. Kim, H. v.d. Piepen, 1986: "Aircraft Data Application Test". In: The Use of Chlorophyll Fluorescence Measurements from Space for Separating Constituents of Sea Water, ESA Contract No. RFQ 3-5059/84/ NL/ MD, Volume II.

Campbell, J.W. and J.E.O'Reilly, 1988: "Role of Satellites in Estimating Primary Productivity on the Northwest Atlantic Continental Shelf". Continental Shelf Research, 8(2), 179-204

Dirks, R.W. and D. Spitzer, 1987: " On the Radiative Transfer in the Sea, Including Fluorescence and Stratification Effects", Limnol. Oceanogr. 32(4), 942-953

Doerffer, R., 1979: "Untersuchungen über die Verteilung oberflächennaher Substanzen im Elbe-Ästuar mit Hilfe von Fernmessverfahren". Archiv f. Hydrobiologie 43, 119.

Doerffer, R., 1981: "Factor Analysis in Ocean Color Interpretation". In: Oceanography from Space, J.F.R. Gower, Ed.(Plenum, New York, 1981), pp. 339-345.

Doerffer, R., 1986; "The Influence of Phaeopigments on Remote Sensing of Phytoplankton Chlorophyll". In: The Use of Chlorophyll Fluorescence Measurements from Space for Separating Constituents of Sea Water, ESA Contract No. RFQ 3-5059/84/ NL/ MD, Volume II.

Doerffer, R. and V. Amann, 1986: "The Influence of Exceptional Phytoplankton Blooms on Remote Sensing of Chlorophyll". In: The Use of Chlorophyll Fluorescence Measurements from Space for Separating Constituents of Sea Water, ESA Contract No. RFQ 3-5059/84/ NL/ MD, Volume II.

Doerffer, R. and Fischer, J., 1987: "Measurements and Model Simulations of Sun Stimulated Chlorophyll Fluorescence within a Daily Cycle". Adv. Space Res., Vol. 7, No. 2, pp. (2)117-(2)120.

Fischer, J., 1985: "On the Information Content of Multispectral Radiance Measurements over an Ocean". Int. J. Remote Sensing 6, 773.

Fischer, J., Doerffer, R. & Grassl, H., 1986: "Factor Analysis of Multispectral Radiances over Coastal and Open Ocean Water Based on Radiative Transfer Calculations". Appl. Opt. 25, 448.

Fischer, J. and U. Kronfeld, 1986: "Chlorophyll Fluorescence within the Visible Spectrum: Measurements and Calculations". In: The Use of Chlorophyll Fluorescence Measurements from Space for Separating Constituents of Sea Water, ESA Contract No. RFQ 3-5059/84/ NL/ MD, Volume II.

Fischer J., U. Kronfeld and P. Schlüssel, 1986: "Masking of Chlorophyll Fluorescence by the Atmosphere". In: The Use of Chlorophyll Fluorescence Measurements from Space for Separating Constituents of Sea Water, ESA Contract No. RFQ 3-5059/84/ NL/ MD, Volume II.

Fischer J. and R. Doerffer, 1987: "An Inverse Technique for Remote Detection of Suspended Matter, Phytoplankton and Yellow Substance from CZCS Measurements". Adv. Space Res., Vol. 7, No. 2, pp. (2)21-(2)26.

Gordon, H.R., 1979: "Diffuse Reflectance of the Ocean: The Theory ot its Augmentation by Chlorophyll a Fluorescence at 685nm". Appl. Opt. 18, 1161.

Gordon, H.R. and Morel, A., 1983: "Remote Assessment of Ocean Color for Interpretation of Satellite Visible Imaginary". A Review, Springer Verlag, New York, Berlin, Heidelberg, Tokyo.

Gower, J.F.R. and Borstad, G.A., 1981: "Use of the *In Vivo* Fluorescence Line at 685 nm for Remote Sensing Surveys of Surface Chlorophyll a". In: Oceanography from Space, J.F.R. Gower, Ed.(Plenum, New York, 1981), pp. 329-338.

Gower, J.F.R., 1986: "Comparison with the Results of Other Areas". In: The Use of Chlorophyll Fluorescence Measurements from Space for Separating Constituents of Sea Water, ESA Contract No. RFQ 3-5059/84/ NL/ MD, Volume II.

Gower, J.F.R., G.A. Borstad, L.H. Gray and H.R. Edel, 1988: "The Fluorescence Line Imager: High-Resolution Imaging Spectroscopy over Water and Land". Proc. of the 4th International Colloquium Spectral Signatures of Objects in Remote Sensing held at Aussois, France, 18-22

January 1988 (ESA SP-287)

Günther, K., 1986: "Analysis of the Daily Cycle of In Vivo Chlorophyll a Stimulated by Artificial Light Sources". In: The Use of Chlorophyll Fluorescence Measurements from Space for Separating Constituents of Sea Water, ESA Contract No. RFQ 3-5059/84/ NL/ MD, Volume II.

Günther, K., D. Ernst, H. Maske, 1986: "Biophysical Processes of Chlorophyll a Fluorescence". In: The Use of Chlorophyll Fluorescence Measurements from Space for Separating Constituents of Sea Water, ESA Contract No. RFQ 3-5059/84/ NL/ MD, Volume II.

Haardt, H. and H. Maske, 1986: "Variability of Excitation-, Emission- and Absorption-Spectra Normalized to the Chlorophyll Concentration". In: The Use of Chlorophyll Fluorescence Measurements from Space for Separating Constituents of Sea Water, ESA Contract No. RFQ 3-5059/84/ NL/ MD, Volume II.

Maske, H. and R. Boje, 1986: "Effect of Environmental Factors and Species Composition on Chlorophyll Fluorescence of Phytoplankton". In: The Use of Chlorophyll Fluorescence Measurements from Space for Separating Constituents of Sea Water, ESA Contract No. RFQ 3-5059/84/ NL/ MD, Volume II.

MBB Messerschmidt-Bölkow-Blohm GmbH, 1986: "Chlorophyll Kartierung des Ozeans durch Detektor-Arrays aus dem Weltraum". Studie im Auftrag des Bundesministeriums f. Forschung und Technologie, F.R.G., Nr. R 4298-3500 R

NASA, 1988: "Earth System Science, A Closer View". Report of the Earth System Science Committee, NASA Advisory Council, NASA Washington D,C.

NASA, 1986: "MODIS, Moderate-Resolution Imaging Spectrometer ". Instrument Panel Report Earth Observing System Vol. IIb, NASA, Washington D.C.

Neville, R.A. and J.F.R. Gower, 1977: "Passive remote sensing of phytoplankton via chlorophyll a fluorescence". J. Geophys. Res. 82, 3487-3493.

Platt, T. 1986:" Primary Production of the Ocean Water Column as a Function of Surface Light Intensity: Algorithms for Remote Sensing". Deep-Sea Research, 33(2), 149-163

Prieur, L. and S. Sathyendranath, 1981: "An Optical Classification of Coastal and Oceanic Waters Based on the Specific Absorption Curves of Phytoplankton Pigments, Dissolved Organic Matter, and Other Particulate Materials". Limnol. Oceanogr. 26, 671

Rabbani, M.M., 1984: "Untersuchungen zum Fluoreszenzwirkungsgrad des Phytoplanktons und der damit assoziierten morphologischen und physiologischen Adaptationsmechanismen". Ph.D.Thesis, Institut für Meereskunde an der Christian-Albrechts-Universität zu Kiel

Sturm, B., 1980: "The Atmospheric Corrections of Remotely Sensed Data and the Qualitative Determination of Suspended Matter in Marine Water Surface Layers". In: Remote Sensing in Meteorology, Oceanography and Hydrology, ed. A.P. Cracknell, Chichester: Ellis Horwood Ltd. 163-197.

Topliss, B.J. 1985: "Optical Measurements in the Sargasso Sea: Solar Stimulated Chlorophyll Fluorescence". Oceanol. Acta, 8(3),pp.263-270

Topliss, B.J. and T. Platt, 1986: "Passive Fluorescence and Photosynthesis in the Ocean: Implications for Remote Sensing". Deep-Sea Research, 33: 849-864

QUANTUM YIELD OF IN SITU FLUORESCENCE OF PYHTOPLANKTON IN KIEL BAY UNDER DAYLIGHT, COMPARISON WITH PRIMARY PRODUCTION

H. Maske, Marine Planktologie, Institut für Meereskunde, Düsternbrooker Weg 20, D-2300 Kiel, F.R.Germany
H. Haardt, Alter Kieler Weg, D-2315 Klein Barkau, F.R.Germany

Key words: Natural chlorophyll fluorescence, quantum yield, primary production, marine phytoplankton, upwelling radiance.

ABSTRACT

We investigated the upwelling radiance signal between 650 and 750 nm in turbid coastal waters. The upwelling radiance is composed of elastically backscattered light and chlorophyll fluorescence. We calculated fluorescence upwelling radiance from the vertical distribution of chlorophyll, a standard specific absorption spectrum, the vertical spectral downwelling irradiance distribution and the diffuse downwelling attenuation coefficient. The subtraction of the calculated fluorescence radiance from the measured radiance yielded the elastic backscatter spectrum. In turbid coastal seawater at the surface the elastic backscatter spectrum is expected to be depressed near the chlorophyll absorption band at 675nm. The position of this depression at 675 nm was used as a sensitive criterion for the proper calculation of the fluorescence radiance. The quantum yield of fluorescence calculated this way correlated well with quantum yield of primary production measured by ^{14}C uptake in situ. Critical evaluation of the calculated quantum yields suggests the following average values: 0.02-0.03 ($Quanta_{em}/Quanta_{abs}$) for fluorescence and 0.03 (Mole C / $Einstein_{abs}$) for photosynthesis.

INTRODUCTION

Daylight excited chlorophyll fluorescence (also named natural or passive fluorescence) has been related to phytoplankton biomass (Gordon, 1979; Kattawar and Vastano, 1982; Kishino, Sugihara and Okami, 1984 a,b) and primary production (Kiefer and Chamberlin, 1988; Chamberlin, 1988). Natural fluorescence may prove to be a useful alternative approach to remote sensing of phytoplankton in coastal areas where high content of organic substance hinders the application of the blue/green method. The interpretation of the natural fluorescence depends to a large extent on a constant, or at least predictably variable, fluorescence efficiency (emission per chlorophyll content and scalar irradiance) or quantum yield (η, quanta emitted per quanta absorbed).

Great variability of fluorescence efficiency is documented in the literature. An increase in fluorescence efficiency or quantum yield is generally associated with reduced photosynthetic ability of the autotroph, and short term (minutes) fluorescence induction experiments can demonstrate an in-

H. K. Lichtenthaler (ed.) Applications of Chlorophyll Fluorescence, 275–284.

verse relationship of O_2 evolution and fluorescence efficiency. This relationship can be explained by a model of the dynamics of the photosystem II (Butler, 1966). Measurements in fluorometers seem to support the notion that limited growth increases in vivo fluorescence efficiency (Kiefer, 1973 a,b; Loftus and Seliger, 1975). Maske and Haardt (1987b) investigated the dependence of fluorescence efficiency (measured by means of a fluorometer) on the in vivo absorption efficiency (package effect) of the excited pigment and found no correlation of both properties. The question remains to what extent the laboratory data, obtained under generally reduced quantum flux and often preconditioned in the dark, are applicable to the interpretation and modelling of natural fluorescence. In the sea, in the top few meters the quantum flux available to photosynthetic organisms is considerably higher than in the typical laboratory fluorometer. In temperate latitudes during a considerable portion of the year in vivo fluorescence of marine phytoplankton near the surface is actually photoinhibited during noon hours, as can be judged from in situ pulsed excitation fluorometers.

There are few laboratory data available of quantum yield of fluorescence showing a range of 1.5 to 2.8 percent (Latimer et al., 1956). These values are corroborated by measurements of fluorescence lifetime under high light intensities (Porter et al., 1977). Most values of quantum yield of fluorescence in the literature were derived indirectly from models and in situ light data and they resulting in values between 0.15 and 10 percent (cf. Ernst, Günther and Maske, 1986). The major problem of many in situ models are the input data: specific absorption of chlorophyll a has mostly been overestimated due to the methodology applied in the older published investigations (Maske and Haardt, 1987a). The backscatter properties of the natural particles with respect to variable refractive index, size and shape was unknown for the specific station simulated. Therefore we address in this publication the calculation of quantum yield of fluorescence of natural phytoplankton under daylight in a very different way. Our interpretation is based on the wavelength shift of the chlorophyll fluorescence peak observed in upwelling radiance in turbid coastal waters.

MATERIAL AND METHODS

a) Downwelling spectral irradiance ($E_d(g, z)$, W m^{-2} nm^{-1}) was measured with a 24 channel, 360-730nm, half-band width 15 nm) time-integrating spectral radiometer (cosine response) (Haardt and Maske, in prep.).
b) Upwelling spectral radiance ($L_u(g, z)$, W m^{-2} sr^{-1} nm^{-1}) was measured with an in situ time-integrating diode array radiometer of 4 degree angle aperture. (Haardt and Maske, in prep.).
c) Downwelling irradiance in air (E_{ad}, W m^{-2}) was measured with a spectrally integrating irradiance meter (350 - 750 nm).
d) Chlorophyll was determined by means of the trichromatic method according to Jeffrey and Humphrey (1975). For the calculation of fluorescence radiance the chlorophyll concentration was recalculated with high resolution in depth by interpolating the values of in vitro trichromatic chlorophyll measurements at distinct depths with the help of in situ fluorometric depth profiles. High Performance Liquid Chromatography (HPLC) separation and quantification was performed according to the method of Mantoura and Llewellyn (1983).
e) ^{14}C samples were incubated in situ in 100 ml laboratory glass bottles for about 2 hours. Incubations were during or close to the time of in situ

radiance and irradiance measurement. Total inorganic carbon content of seawater was obtained by pH measurement before and after addition of a known amount of acid (Grasshoff, 1976). To calculate quantum efficiency of photosynthesis the average solar irradiance was calculated from integrated values of global irradiation (Kipp & Zonen, CC 1) obtained with a solarimeter (Kipp & Zonen, CM 5).

f) Fluorescence emission spectra were determined with a spectral-fluorometer (SHIMADZU RF 540) (Maske and Haardt, 1987b).

g) Model calculations

A simple one-dimensional model was used for the calculation of fluorescence radiance. This model calculated the amount of photosynthetic useable quanta absorbed, PQA (z) (Quanta m^{-3} s^{-1}) within successive depth intervals, dz from spectral downwelling quantum irradiance, $E_{Qd}(\lambda,z)$ (photons m^{-2} s^{-1} nm^{-1}) and the optical cross section, a_c defined by the product of the specific absorption coefficient, $a^*_c(\lambda)$ (m^2 mg Chl $\underline{a}^{-1}$) (Fig. 1) and the chlorophyll concentration, C(z) (mg Chl $\underline{a}$ m^{-3}). The specific absorption coefficient, a^*_c was taken from a representative fluorescence excitation spectrum and scaled to 0.011 m^2 mg^{-1} Chl $\underline{a}$ at 675nm at 10nm resolution (cf. Maske and Haardt, 1987a). PQA was multiplied by 1.2 to compensate for the use of downwelling irradiance instead of scalar irradiance in the calculations.

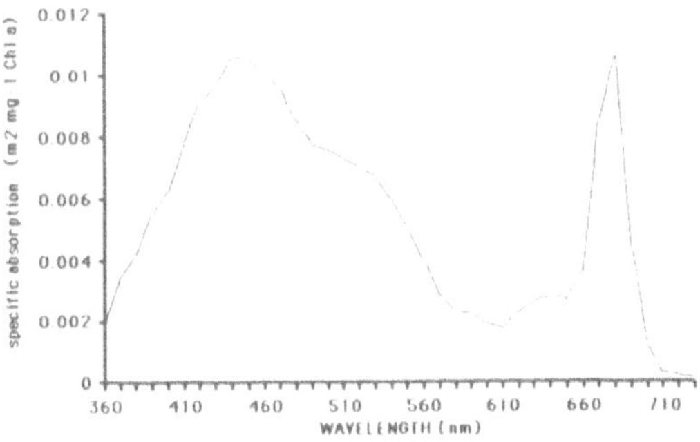

Figure 1. Spectrum of specific absorption coefficients (m^2 mg^{-1} Chl $\underline{a}$) used in our calculations.

A certain percentage (quantum yield, η) of PQA (z) (quanta m^{-3} s^{-1}) is re-emitted isotropically as fluorescence. The emitted quanta have an emission spectrum defined by a standard fluorescence emission spectrum, Em(λ) (Maske and Haardt. 1987b). The quanta are emitted isotropically in slabs of dz thickness em(λ,z) (photons m^{-3} s^{-1}). These can be viewed from the surface as horizontal lambertian planes of upwelling irradiance. This fluorescence irradiance is attenuated on the way from its source depth, z to the measuring depth according to the depth intervall and the diffuse attenuation coefficient, $K_d(\lambda,z)$ (m^{-1}) calculated from downwelling irradiance. Upward transport of fluorescence radiance is described by K_d because diffusivity of fluorescence radiance is comparable to downwelling light. The simulation was carried out with dz = 0.2 meter. The calculated upwelling fluorescence radiance was calculated with 5nm resolution. For comparison with measured radiance (Figs. 2,3,4) the fluorescence radiance was recalculated according to the spectral sensitivity of the radiometer channels.

RESULTS

In situ data from Kiel Bay were analysed from a typical spring bloom, a fall situation and an eutrophic fjord. From the available data sets we choose those that were taken on clear days with reasonable homogeneous vertical distribution of Chl *a* in the water column. This was verified by means of profiles on *in situ* fluorescence measured by flash excitation. None of these profiles showed significant inhibition of fluorescence near the surface.

Data from September 21st, 1983 (Kiel Bight) represent a typical fall population of dinoflagellates. The phytoplankton population at 3 m depth was dominated by *Prorocentrum micans* (6.6 10^4 cells l^{-1}), *Ceratium fusus* (2.9 10^4 cells l^{-1}) and by *Ceratium tripos* (6.9 10^3 cells l^{-1}). Nitrate, nitrite and ammonium ions were less than 0.01 µM in the top 5 meters. Other inorganic nutrients were nonlimiting. The Chl *a* concentration decreased in 9 m depth from 3 to about 2.5 mg Chl *a* m^{-3}. The upwelling radiance at 0.3 m (Fig. 2) showed a maximum in the channel with a peak wavelength of 683 nm. The fluorescence radiance was calculated with 2 and 3 percent quantum yield (Fig. 2). Difference spectra were calculated by subtracting the calculated fluorescence spectra from the measured radiance. For 2 percent quantum yield the maximum depression of backscatter, relativ to a straight line between 623 and 713nm, is at the 668nm-channel. With 3 percent quantum yield the maximum depression is found at the 683nm-channel. Therefore the most likely quantum yield is expected to be between 2 and 3 percent.

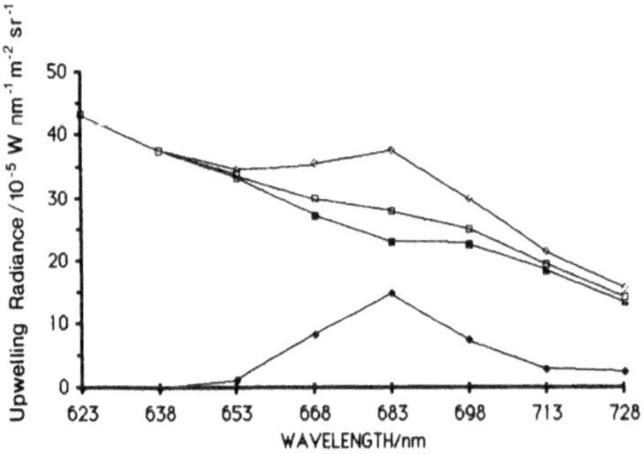

Figure 2. Upwelling radiance spectra from September 21, 1983, 0.3m depth. Rhombii: measured radiance; Filled rhombii: calculated values of fluorescence emission at η = 3%; Filled squares: difference spectra at η = 3%; Squares: difference spectra at η = 2%.

The data set of March 30, 1983, in Kiel Bight was obtained during a typical spring bloom of diatoms. Phytoplankton was dominated by *Detonula* spec. (0 - 3.9 10^6 cells l^{-1}), *Chaetoceros* spec. (1.6 - 3.8 10^6 cells l^{-1}) and *Thalassiosira* spec. (2 - 4.8 10^6 cells l^{-1}). Chlorophyll distribution varied between 7 and 8 mg Chl *a* m^{-3} at 0 to 12 m depth, chlorophyll concentration increased to 8.5 mg m^{-3} below 12 m. Between 0 and 11 m depth NO_3 concentrations were above 2.8 µM, but PO_4 ranged from 0.22 - 0.13 µM and SiO_4 from 1.38 - 0.07 µM. In 0.1 m depth the L_u spectrum showed a pronounced peak between 668 and 698 nm (Fig. 3). The expected difference spectrum lies between those calculated with η 1 and 1.5 percent. Therefore a quantum yield of 1.2 percent can be considered to be the most likely value.

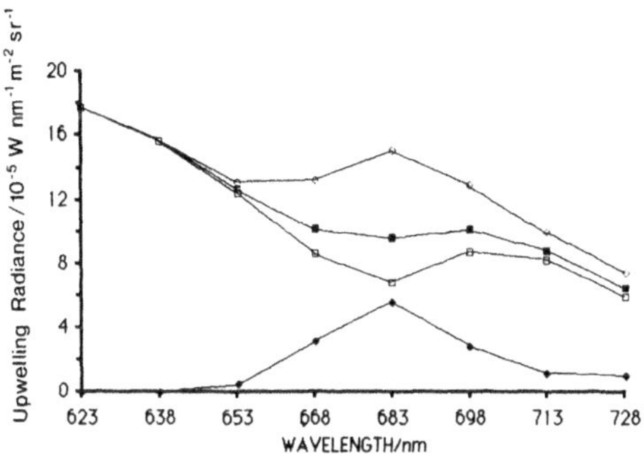

Figure 3. Upwelling radiance spectra from March 30, 1984, 0.1m depth. Rhombii: measured radiance; Filled rhombii: calculated values of fluorescence emission at η = 1%; Filled squares: difference spectra at η = 1.0%; Squares: difference spectra at η = 1.5%.

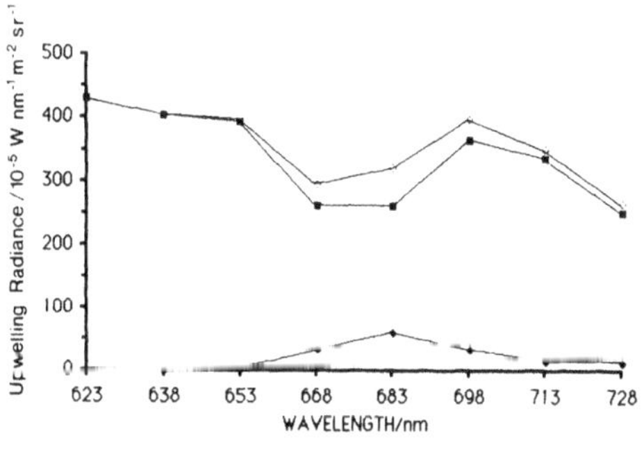

Figure 4. Upwelling radiance spectra, Schlei Fjord, September 22, 1983, 0.3 m depth. Rhombii: measured radiance; Filled rhombii: calculated fluorescence radiance at η = 4%; Filled squares: difference spectra at η = 4%.

On September 22nd, 1983, data were taken in the Schlei Fjord. Phytoplankton at 1.5 m depth was dominated by dinoflagellates: Prorocentrum minimum (8.78 10^6 cells l^{-1}) and cell-aggregate forming cyanobacteria: Merismopedia tenuissima (2.6 10^8 cells l^{-1}). Chlorophyll concentration was 25 mg m^{-3}. Major nutrients were nonlimiting with concentrations above 1 μM. 0.2 m below the surface the measured L_u spectrum presents a peak wavelength of about 698 nm (Fig. 4), and there is relatively high backscatter above 700nm. Quantum efficiency was 4 percent and resulted in a depression in calculated backscatter near 675nm.

DISCUSSION

Upwelling radiance spectra from beneath the sea surface show consistently a peak between 660 and 700 nm. The peak wavelength has been shown to increase with higher chlorophyll concentrations in coastal waters (Spitzer and Dirks, 1986). Haardt and Maske (in prep.) could show in a tank experiment that this peak in upwelling radiance is the result of superposition of two sources: a) the continuous elastic backscattered radiance spectrum is modified by absorption of chlorophyll bearing particles leading to a local decrease in upwelling radiance near 675nm and by water absorption at greater than 700 nm, producing a residual peak between the two absorption bands (Spitzer and Dirks, 1986); b) the chlorophyll fluorescence produces a fairly constant emission spectrum with a peak-wavelength of 681 nm (Maske and Haardt, 1987b).

Topliss (1985) and Topliss and Platt (1986) observed in the Sargasso Sea and the Canadian Arctic a peak in upwelling irradiance at 681 nm. The reason for the difference in spectral shape of upwelling radiance offshore and in coastal waters is probably found in the higher seston load of coastal waters and the generally lower chlorophyll concentration in offshore waters. We base our approach on the interpretation of the wavelength shift of the upwelling radiance peak between 680 and 700 nm, therefore our approach is restricted to coastal waters. Our concept is based largely on the fact that the elastic backscatter spectrum of particle assembly in suspension from Kiel Bay are nearly flat between 650 and 720 nm (Haardt, unpubl. data). This assumes explicity that anomaleous dispersion of backscatter due to the absorption band of chlorophyll at 675nm within particles (cf. Latimer and Rabinowitch, 1959) is not defining backscatter behaviour of the particle assembly _in situ_ because of the irregular shape and the wide range of the refractive index of its particles and their wide size spectrum. In addition, at a spectral resolution of 15nm the solar irradiance spectrum is considered to be linear with wavelength.

In coastal waters irradiance tends to narrow spectrally near 550nm with depth, therefore Raman scattering was considered as a possible source producing backscattered radiance at about 680 nm. Potentially this could lead to confusions with the fluorescence emission. However a simple comparison of signal strength by means of a spectral fluorometer on a seawater sample of 16 mg Chl _a_ m^{-3} showed that with monochromatic light of 562 nm the Raman signal was only about 3 percent of the fluorescence emission signal. With broader spectral excitation the fluorescence signal would increase in strength relative to the Raman signal.

We think that our interpretative approach for judging the contribution of fluorescence to the measured radiance, by parameterization of fluorescence and taking the backscatter depression wavelength as an indicator, is a sensitive method for turbid coastal waters. A change in quantum yield results in a distinct wavelength shift as can be observed in the Figures 2 and 3.

We found it helpful to investigate carbon uptake as a control parameter for the calculation of light absorption by phytoplankton in our model. At the same time as light measurements were performed in the water, _in situ_ ^{14}C incubations were performed. The assimilation data were adjusted such that average surface irradiance during incubation was equal to the reference surface irradiance used in the calculation of fluorescence radiance. Quantum efficiencies were then calculated using the same rate of absorption of quanta as in the calculation of fluorescence radiance. The resulting quantum yields (Table 1) of photosynthesis are close to expected values supporting our interpretation of the upwelling radiance spectra in coastal waters. The proportionality of the quantum yield of fluorescence and photosynthesis was not to be expected (Butler, 1966) and therefore we reanalysed the data sets.

Table 1. Quantum yield of photosynthesis (Mole C_{uptake} / $Einstein_{absorbed}$) and quantum yield of fluorescence ($Q_{emitted}$ / $Q_{absorbed}$). For both quantum yields the same procedure was used to calculate the absorbed quanta.

Depth	March 30, 1984	Sept. 21, 1983	Sept. 22, 1983
	Photosynthesis:		
0 meter	0.006	0.015	
0.5			0.036
1.5			0.060
3	0.015	0.032	0.040
5	0.013	0.041	
11	0.015	0.017	
average	0.014*	0.030*	0.045
	Fluorescence:		
	0.012	0.025	0.04

*Surface value was excluded from the average because of photoinhibition due to stationary incubation.

The quantum yield on March 30, was below average (Table 1). This can be explained by an overestimation of light absorption as a result of exaggerated pigment estimation by the standard trichromatic chlorophyll method. Below the pigment concentrations measured by the trichromatic method and the HPLC method are compared. On the surface: trichromatic method 8.0 mg Chl *a* m^{-3}, HPLC 4.1 mg Chl *a* m^{-3} and 4.9 mg m^{-3} Chl *a* degradation products. At 11 m depth: trichromatic method 8.0 mg Chl a m^{-3}, HPLC 4.4 mg Chl a m^{-3} and 5.1 mg Chl *a* m^{-3} degradation products. No HPLC data are available for the September data from Kiel Bight or the Schlei Fjord, but it can be assumed that normally the percentage of chlorophyll degradation products is low. Although chlorophyll degradation products themselves fluoresce in water they will nevertheless lead to overestimated absorption and fluorescence in our simulation results because the computation is based on a specific absorption spectrum, and this assumes a certain percentage of additional light harvesting by non-chlorophyllous pigments. These accessory pigments are presumably uncoupled from chlorophyll degradation products. If the data from March 30th are recalculated with the value of undegraded Chl *a* then quantum efficiency would be raised to 2.5 percent. It can be concluded that for remote sensing of the fluorescence signal care must be taken that the reference values of Chl *a* will not include degradation products. This could be important in coastal waters were chlorophyll degradation products can be as much or even several times that of the chlorophyll itself (Mantoura and Llewellyn, 1983).

The September 22 values of quantum efficiencies were probably somewhat overestimated because the scalar irradiance in the calculation of PQA may have been underestimated. Scalar irradiance was calculated to be 1.2 times the downwelling irradiance. This factor is probably too low in the more turbid water of high chlorophyll concentration. In addition the cyanophytes present in the water probably had higher light absorption in the 490 and 625nm range than assumed by the standard absorption spectrum (Fig. 1). Both processes would lead to increased calculated light absorption and therefore reduced quantum yield.

The values from the data sets March 30 and September 22 converge after correction to the value from September 21 and a likely quantum yield of 2-3 percent for fluorescence and 3 for photosynthesis. These values for fluorescence are similar to average quantum efficiencies of fluorescence measured in the laboratory on single cell algae (Latimer et al., 1956) and those calculated by other authors from in situ spectral irradiance values: Kishino et al. (1984b) reported for coastal surface values 1.8-2.4 percent and for offshore 4 percent, Topliss (1985) calculated 0.96-2.36 percent and Kiefer and Chamberlin (1988) and Chamberlin (1988) arrived at average values of 3.3 percent in tropical pacific waters. Kishino et al. (1984b) and Topliss (1985) measured specific absorption coefficients of particles from the water column after concentrating their samples on filters. This method does yield qualitative but not quantitative spectra according to our experience (Maske and Haardt, 1987a). Kiefer and Chamberlin (1988) assumed for their calculations an average specific absorption coefficient. The specific absorption spectrum entered into our calculation was taken from a fluorescence excitation spectrum and scaled to an average value at 675nm (Haardt and Maske, 1987). It has been shown, that after correction absorption and fluorescence excitation spectra are qualitatively similar (Maske and Haardt, 1987a). Neori et al. (1988) demonstrated that in marine microalgae the action spectra of photosynthesis and fluorescence excitation spectra are also similar, therefore our approach should provide a reasonable estimate. The calculation of light absorption with an absorption spectrum rather than a spectrally averaged value has the advantage that the change in spectral quality of downwelling irradiance is taken into account. In general it has to be pointed out, that as long as the light absorption properties of the natural populations are not measured the calculation of in situ quantum yield provides only an estimate.

Our calculated quantum yield of photosynthesis compares well with values reported in the literatur of 3 to 7% (cf. Bannister and Weidemann, 1984). Topliss and Platt (1986) interpreted their oceanic data to show an inverse relationship of quantum efficiencies of fluorescence and photosynthesis. They argued that this was to be expected from physiological models. Critical interpretation of our results (Table 1) actually suggests that quantum yields varied little for the three stations, despite the different environments. The variation between samples might have been small due to the limited geographic range investigated by us. It might also be that phytoplankton under natural conditions does not follow physiological models developed for short term changes in lighting conditions, but rather that the population adjusts such that similar values of quantum yield are maintained.

Acknowledgements

This research was funded by the Bundesministerium für Forschung und Technologie #MFG 00359, Fed.Rep.Germany. Dr. R. Boje provided the ^{14}C-uptake data and was responsible for the logistics of the field experiments.

REFERENCES

Bannister TT, Weidemann AD, 1984. The maximum quantum yield of phytoplankton photosynthesis in situ. Journal of Plankton Research, 6,2, 275-294.

Butler WL, 1966. Fluorescence yield in photosynthetic systems and its relations to electron transport. In: Sanadi DR, Current topics in bioenergetics. Academic Press, New York.

Chamberlin WS, 1988. A field test of the relationship between natural fluorescence and primary production. Oral presentation: Ocean Science Meeting, New Orleans, January 18-22.

Ernst D, Günther KP and Maske H, 1986. Biophysical processes of chlorophyll a fluorescence. In: The use of chlorophyll fluorescence measurements from space for separating constituents of sea water, Vol.2. ESA-Study Contract No. RFQ 3-5059/84/NL/MD.

Gordon H, 1979. Diffuse reflectance of the ocean: the theory of its augmentation by chlorophyll a fluorescence at 685nm. Appl. Opt. 18: 1161-1166.

Grasshoff K, 1976. Methods of seawater analysis. Verlag Chemie, Weinheim.

Haardt H and Maske H, 1987. Specific in vivo absorption coefficient of chlorophyll a at 675 nm. Limnol. Oceanogr. 32(3): 608-619.

Haardt H and Maske H, in prep. The contribution of chlorophyll fluorescence to the upwelling radiance under artificial illumination in seawater tanks.

Jeffrey SW and Humphrey GF, 1975. New spectrophotometric equations for determining chlorophylls a, b, c_1 and c_2 in higher plants, algae and natural phytoplankton. Biochem. Physiol. Pflanzen 167: 374-384.

Kattawar GW and Vastano JC, 1982. Exact 1-D solution to the problem of chlorophyll fluorescence from the ocean. Applied Optics 21: 2489-2492.

Kiefer DA, 1973a. Fluorescence properties of natural phytoplankton populations. Marine Biology 22: 263-269.

Kiefer DA, 1973b. Chlorophyll a fluoescence in marine centric diatoms: Responses of chloroplasts to light and nutrient stress. Marine Biology 23: 39-46.

Kiefer DA and Chamberlin WS, 1988. Natural fluorescence of chlorophyll a: Relationship to photosynthesis and chlorophyll a concentration in the Western South Pacific Gyre. Oral presentation: Ocean Science Meeting, New Orleans, January 18-22.

Kishino M, Sugihara S and Okami N, 1984a. Influence of fluorescence of chlorophyll a on underwater upward irradiance spectrum. La mer 22: 224-232.

Kishino M, Sugihara S and Okami N, 1984b. Estimation of quantum yield of chlorophyll a fluorescence from the upward irradiance spectrum in the sea. La mer 22: 233-240.

Latimer P, Bannister TT and Rabinowitch E, 1956. Quantum yields of fluorescence of plant pigments. Science 124: 585-586.

Latimer P and Rabinowitch E, 1959. Selective scattering of light by pigments in vivo. Arch. Biochem. Biophys. 84: 428-441.

Loftus ME and Seliger HH, 1975. Some limitations of the in vivo fluorescence technique. Chesapeake Science 16(2): 79-92.

Mantoura RFC and Llewellyn CA, 1983. The rapid determination of algal chlorophyll and carotinoid pigments and their breakdown products in natural waters by reverse-phase high-performance liquid chromatography. Analytica Chimica Acta 151: 297-314.

Maske H and Haardt H, 1987a. Quantitative in vivo absorption spectra of phytoplankton: Detrital absorption and comparison with fluorescence excitation spectra. Limnol. Oceanogr. 32(3): 620-633.

Maske H and Haardt H, 1987b. In vivo fluorescence yield (excitation 675nm, emission 730nm) and specific in vivo absorption coefficient of chlorophyll a at 675 nm: variations in quantum yield. Arch.Hydrobiol. Beih. 29: 123-129.

Neori A, Vernet M, Holm-Hansen O and Haxo FT, 1988. Comparison of chlorophyll far-red and red fluorescence excitation spectra with photosynthetic oxygen action spectra for photosystem II in algae. Mar. Ecol. Prog. Ser. 44: 297-302.

Porter G, Synowiec JA and Tredwell CJ, 1977. Intensity effects on the fluorescence of in vivo chlorophyll. Biochem. Biophys. Acta 459: 329-336.

Spitzer D and Dirks RWJ, 1986. Chlorophyll fluorescence effects in the red part of reflectance spectra of natural waters. Cont. Shelf Research 6(3): 385-395.

Topliss BJ, 1985. Optical measurements in the Sargasso Sea: solar stimulated chlorophyll fluorescence. Oceanologica Acta 8(3): 262-270.

Topliss BJ and Platt T, 1986. Passive fluorescence and photosynthesis in the ocean: implications for remote sensing. Deep-Sea Research 33(7): 849-864.

CHAPTER 4

Application in remote sensing of terrestrial vegetation

REMOTE SENSING OF CHLOROPHYLL FLUORESCENCE IN OCEANOGRAPHY AND IN TERRESTRIAL VEGETATION: AN INTRODUCTION

Hartmut K. LICHTENTHALER

Botanisches Institut II (Plant Physiology and Plant Biochemistry),
University of Karlsruhe, Kaiserstr. 12, D-7500 Karlsruhe 1, F.R.G.

KEY WORDS: fluorescence ratio F690/F735, laser-induced chlorophyll fluorescence, photosynthetic function, phytoplankton, reflection signatures, remote sensing, stress detection, vitality of plants, yellow organic matter (Gelbstoff).

ABSTRACT

The principles of detection of algal cultures (phytoplankton) in oceanography by remote sensing of the passive (sun light) and active (laser-induced) chlorophyll fluorescence are briefly summarized. The new and future possibilities of remote sensing of the physiological state of terrestrial vegetation via the laser-induced chlorophyll fluorescence, which requires other techniques than in oceanography, are pointed out. The limitations of remote sensing of terrestrial vegetation via reflectance signatures and the advantage of the in vivo chlorophyll fluorescence, which comes only from the green plant parts, are contrasted. Finally the successful application of chlorophyll fluorescence as a superior method of ground truth control related to remote sensing is presented.

INTRODUCTION

In the last years active and passive airborne fluorescence methods have been developed which permit the remote sensing and detection of algae populations in sea water via the in vivo chlorophyll fluorescence (fluorescence peak between 680 to 690 nm). The methods also allow determination of the yellow substance (Gelbstoff) which accumulates along the sea coasts, in the mouth and delta area of rivers and to monitor oil pollution (Bristow et al., 1981; Amann et al., 1986; Doerffer, 1988). Besides its application in oceanography attempts are made to apply chlorophyll fluorescence in the remote sensing of the physiological state of terrestrial vegetation (Lichtenthaler and Rinderle, 1988b; Zimmermann and Günther, 1986). In the ground truth control related to the remote sensing of the state of health of forest trees via the reflectance signatures, chlorophyll fluorescence has been applied with great success, since it permits a fast screening of the physiological state (Lichtenthaler and Buschmann, 1984; Rock et al, 1986; Schmuck and Lichtenthaler, 1986; Schmuck et al., 1987).

The chlorophyll fluorescence parameters to be measured and the techniques applied are quite different in oceanography and terrestrial remote sensing as well as in ground truth control. Unfortunately the literature on the application of the chlorophyll fluorescence in oceanography, in terrestrial remote sensing and in the ground truth control, related to remote sensing, is rather new and often only presented in working reports

H. K. Lichtenthaler (ed.) Applications of Chlorophyll Fluorescence, 287–297.

and not easily available. Thus the aim of this introduction is to point out some basic principles and give access to some original literature references where more details can be found.

A. REMOTE SENSING OF CHLOROPHYLL FLUORESCENCE IN OCEANOGRAPHY

In oceanography, limnology and hydrobiology chlorophyll fluorescence can be utilized to detect larger and widespread algae and phytoplankton populations ("algal bloom") from airborne systems. There are principally two systems which can be applied a) the passive system which utilizes sun light for excitation of the fluorescence and senses the natural chlorophyll fluorescence and b) the active system where chlorophyll fluorescence is excited by lasers (in general a blue-light laser). With the passive systems (see Doerffer, 1988; Gower et al., 1987) one can screen large areas and follow the development, spreading and drift of phytoplankton bloom along the coast. A certain disadvantage of this method is the fact that screening is only possible in full sunlight, which requires costy waiting time for the flying systems, and that the fluorescence intensity can change during the day due to the position of the sun. In the active systems in turn, only the chlorophyll fluorescence induced by a pulsed blue light eximer-laser is sensed (Günther et al., 1986; Amann et al., 1986). The advantage of the active system is its independence of the sun light (the system can be flown on cloudy days throughout the day). The disadvantage, however, is that the laser-induced chlorophyll fluorescence can be excited and sensed from airborne systems only in a rather narrow bandwidth (ca. 20-50 m) along the flying line. This requires several flights to remotely sense the size and extension of larger algal populations.

The chlorophyll fluorescence sensed in the active and passive systems is often masked by the much higher fluorescence intensity of the yellow organic matter (Gelbstoff) which represents anthropogenic water pollutants. This is particularly the case along the coast of warmer regions and near the river mouths and delta areas. The chlorophyll fluorescence is then often only a very small peak in the 680 nm region on top of the dissolved yellow matter fluorescence which has its maximum in the 550 nm region (see Doerffer, 1988, this volume and Spitzer and Dirks, 1985). To resolve the chlorophyll fluorescence this requires the simultaneous registration of several bands besides and in the center of the chlorophyll fluorescence peak.

In oceanography and hydrobiology the intensity of the measured chlorophyll fluorescence is directly correlated to the algae density. As a general rule one can state: "The higher the chlorophyll fluorescence intensity, the higher the density of the algal population". This statement, however, only applies to the rather diluted algae suspensions in lakes and the sea with up to 0.5 μg chlorophyll a+b per ml water. Only here can the fluorescence intensity be taken as a measure of the chlorophyll content. The situation is quite different for terrestrial vegetation with a much higher chlorophyll density, where the chlorophyll fluorescence decreases with increasing chlorophyll content of the leaves (see below chapter B).

Though the intensity of chlorophyll fluorescence is an indicator of the phytoplankton density, one cannot judge the algae density in absolute amounts from the remotely sensed chlorophyll fluorescence alone. This is due to the possibility of rather large variations in the very small chlorophyll fluorescence signal which can be caused e.g. by differential amounts of yellow matter, by smooth or rough sea as well as by the fact that the algae may drift near to the surface or somewhat deeper in the water, which can change during the course of a day. For these reasons the remotely sensed fluorescence data needs to be correlated with control measurements of water samples taken by boats at different parts of the remotely sensed water area (see Ernst, 1988; Maske and Haardt, 1988).

With the active and passive systems one can remotely sense the algal populations (phytoplankton bloom, algae carpets) along the coast and follow their development under longer sunshine periods in spring and summertime, their spreading and drift due to natural currents as well as wind and weather conditions. The importance of regularly screening and observing the developing algae populations is underlined by the fact that such algal carpets or girdles can accumulate heavy metals as well as toxic organic compounds and pollutants from the water which may be poisonous and toxic to fish as well as birds and other sea animals or cause fish to die by sticking and gluing to the fish as is assumed for the algal Chrysochromulina polylepis. The large scale accident of the dying of fishes, which started end of May 1988 from a rather dense and large phytoplankton girdle in the Kattegat before the Swedish coast and moved up to the Norwegian coast (Skagerrak) and the North Sea, demonstrates the necessity of screening regularly by airborne systems the rise and move of the algal populations in oceanography as well as in larger inland seas. This incident also underlines the necessity to drastically decrease the nitrogen and phosphorus content of sewage and waste waters, which are the main cause of the large-scale phytoplankton blooms.

Further details on the remote sensing of phytoplankton via chlorophyll fluorescence using active and passive systems are found in the references: Amann and Doerffer, 1983; Amann et al., 1986; Bristow et al., 1981; Doerffer, 1981, 1988; Gower, 1987; Günther, 1986; Hoge et al., 1986; Yadhav, 1987; Kondratyev and Poznyakov, 1987; Shu and Chen, 1987).

B. REMOTE SENSING OF TERRESTRIAL VEGETATION BY REFLECTANCE MEASUREMENTS AND CHLOROPHYLL FLUORESCENCE

1. Possibilities and limitation of the reflectance method

In the remote sensing of terrestrial vegetation and its state of health the airborne measurement of reflectance signatures of agricultural plants and forests has been used in the last ten years with great success. Examples of airborne reflectance measurements and their application in stress detection and the forest-decline research and forest-damage classification can be found in the following references: Huss, 1984; Lichtenthaler et al., 1987; Lichtenthaler and Rinderle, 1987; Rock et al. 1986a,b; Schmuck et al. 1987. Such airborne-reflectance measurements are based on the observation that healthy plants possess particular reflectance spectra in the visible region (400 - 700 nm) and the near infrared (700 - 900 nm) as is shown in Fig. 1. Damage effects in forests have been characterized in the last years by a specific increase of the reflectance in the visible range between 500 and 650 nm (which indicates a lower pigment content) and a considerable decrease of the reflectance in the near-infrared around 800 nm, which indicates that the arrangements and size of cells and aerial interspaces in the leaf of the damaged trees is differently organized (Lichtenthaler et al., 1987; Rock et al., 1986 a,b; Schmuck et al., 1987).

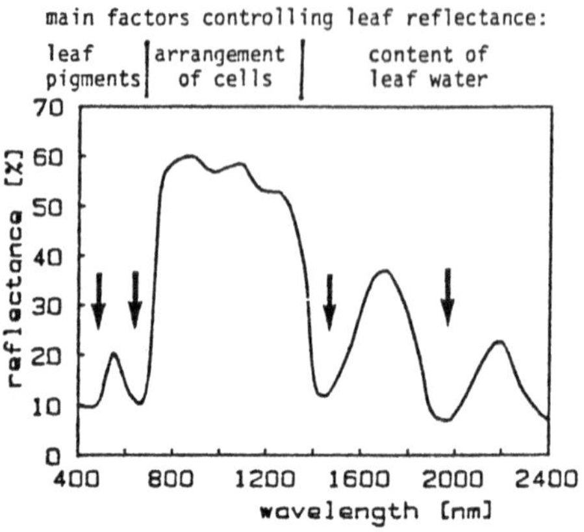

Figure 1. Reflectance spectrum of green leaves. The points of minimum reflection, which corresponds to absorption bands of photosynthetic pigments (chlorophylls and carotenoids) and water, are indicated by arrows.

Table 1. Results of the computer-aided classification of the test areas Althof (low damage site) and Mauzenberg (high damage site) from the Northern Black Forest by using reflection data obtained with a Bendix Scanner at the flight altitudes of 300 and 1000 m (from Schmuck et al., 1987).

flight altitude	300 m % of total		1000 m % of total
A. Althof (low damage site)			
damage class 0	20.2		
damage class 0-1	12.7	32.9	46.5
damage class 1	27.2		28.3
damage class 2	39.9		25.2
B. Mauzenberg (high-damage site)			
damage class 2	55.5		59
damage class 3-4	44.5		42

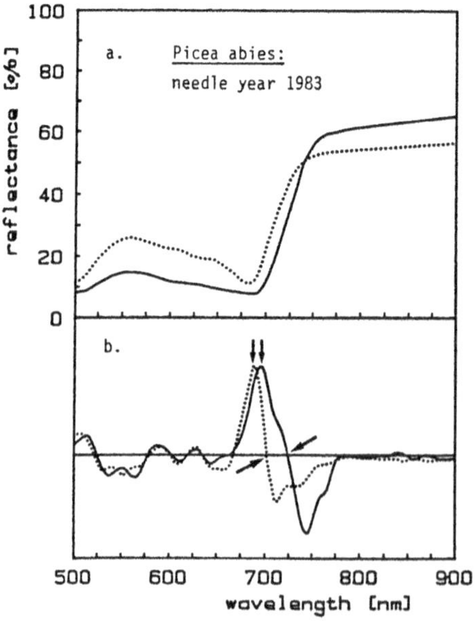

Figure 2. Reflectance (VIRIS) spectra (**a**) and 2nd derivative spectra (**b**) of 3rd year needles from spruce trees damaged to different degrees from Northern Black Forest. ——— green needles from trees of damage class 0-1 (Althof site),······ needles from trees of damage class 3-4 (Mauzenberg site; 60-90% needle loss; needle colour: light green to yellowish green). Mean of 7 measurements from 3 trees taken in August 1985. The differences in the reflectance in the visible and the NIR-plateau between healthy and damaged trees are significant (P = 0.01).

These changes in the reflectance spectra of needles of damaged spruces are associated with a "blue shift" of the long-wavelength inflection point of the red edge of the reflectance spectrum (Rock et al., 1987a; Schmuck et al., 1987; Lichtenthaler and Buschmann, 1987). This wavelength shift is seen in the 2nd derivative spectrum (arrows in Fig. 2). Examples of tree damage classification on the basis of remote by sensed reflectance data with a multispectral Bendix-11-channel scanner are shown in Table 1 and in the colour plates No. 3 at the end of this book.

In the last two years the reflectance decrease in the long-wavelength region (800 nm range) did not show up any longer in damaged needles, although the physiological activity of these needles and the amount of needles of the damaged trees was still low (Lichtenthaler et al., 1987). This brings difficulties in the future differentiation and classification of forest trees and agricultural crops. Reflectance data alone seem no longer to permit a classification, when the main difference (the reduced reflectance in the near infrared) is missing. This fact requires that the remote sensing of reflectance is complemented by an additional method which gives information on the physiology of the leaf material. This could be managed by remotely sensing the laser-induced chlorophyll fluorescence.

2. Possibilities for the remote sensing of the chlorophyll fluorescence

The remote sensing of chlorophyll fluorescence parameters of terrestrial vegetation could give a very accurate picture of the health state of the plants, since the chlorophyll fluorescence only comes from the green photosynthetically active plant parts. Though several parameters of the chlorophyll fluorescence can be used for stress detection in plants (Lichtenthaler 1988; Lichtenthaler and Rinderle, 1988), remote sensing must concentrate on those fluorescence parameters which can be sensed in a very fast way from airborne systems. This excludes the e.g. the possibility to measure the laser-induced chlorophyll fluorescence induction kinetics (either rise or decrease of the chlorophyll fluorescence) because the measuring time is too long. In addition measurements of the induction kinetics would require a predarkening of the vegetation, which is not possible. Flights in the night, e.g. with the slower airships (zeppelins), are not practical for several reasons and the navigation possibilities of an airship with respect to a straight flying line which can be reflown from time to time, are rather limited. There remain two possibilities for application in remote sensing, which have to be tested and developed:

- **a.** The measurement of the chlorophyll fluorescence ratio F690/F735 and
- **b.** the registration of the fluorescence life-time after a short laser pulse.

This requires in both cases an active laser-equipped airborne system which must have very sensitive detector systems.

a. The ratio F690/F735

It has been shown in detail that the ratio F690/F735 i.e. the ratio of the fluorescence in the two fluorescence maxima is increased under stress conditions due to a decrease or lower accumulation rate of chlorophylls as well as by a decline in the photosynthetic function (Lichtenthaler 1986 and 1987, Lichtenthaler and Rinderle 1988a,b, Rinderle and Lichtenthaler, 1988). Though the values of the ratio F690/F735 are quite different when the fluorescence is excited with blue or red light, the increase of the ratio due to stress is seen in both cases (Lichtenthaler and Rinderle 1988b, Rinderle and

Lichtenthaler, 1988). From this it is clear that a blue light or a red-light laser can be applied in future airborne system. Using red excitation light (e.g. a He Ne-laser 632.8 mm) one has to take certain precautions to exclude overlapping of excitation and fluorescence light, which is not the case using blue light. This would favour a blue-light laser for remote sensing. The remote sensing of the ratio F690/F735 of trees from a distant platform has already been achieved (Rosema et al., 1988). In addition, in a preliminary airborne experiment the laser-induced chlorophyll fluorescence in the 690 and 735 nm region could be sensed separately and differences were seen in the ratio F690/F735 between trees of different physiological state (Zimmermann and Günther, 1986).

b. Life-time measurements

With respect to the life-time measurements it has been shown that the life-time (decay) of the chlorophyll fluorescence can increase under certain stress conditions (Schneckenburger 1986). Whether the change in the fluorescence life-time is a general indicator of stress to plants has yet to be investigated. One has also to show whether it is possible to remotely sense the fluorescence life-time signal. Newer results on this topic are described by Bertolini et al., 1988. In any case the airborne registration of damage to terrestrial vegetation via measurements of the laser-induced chlorophyll fluorescence can be regarded as a valuable and promising tool, which opens new possibilities for stress detection in plants. The signals to look at are the fluorescence intensity, the ratio F690/F735 and the life-time.

One general feature of remote sensing of chlorophyll fluorescence of terrestrial plants must be kept in mind. In contrast to oceanography, where the fluorescence intensity increases with increasing chlorophyll (algae) content of the water, the chlorophyll density in leaves is several orders of magnitude higher (ca. 40 to 70 µg chlorophyll a+b cm^{-2} leaf area). As a consequence the chlorophyll fluorescence intensity of leaves decreases with increasing chlorophyll content (s. Fig. 3). This is due to reabsorption of the emitted fluorescence by the in vivo chlorophylls. Since the 690 nm fluorescence is reabsorbed to a higher degree than the 730 nm fluorescence, the ratio F690/F735 exhibits higher values in leaves with a lower chlorophyll content. This is also the basis of stress detection in plants, since stress in general leads to a lower chlorophyll content.

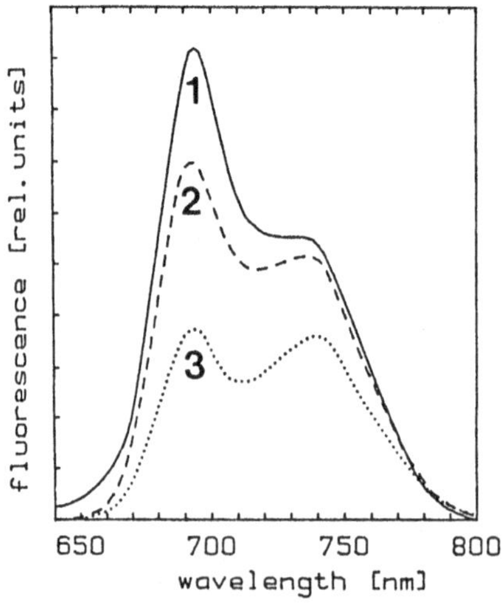

Figure 3. Chlorophyll fluorescence emission spectra (excited by blue light (470+30 nm) of leaves with different chlorophyll content. 1. light green, 2. green and 3. dark-green leaf. The ratio F690/F735 decrease with increasing chlorophyll content. The values for F690/F735 are: 1.7 (leaf 1) via 1.4 to 0.9 (leaf 3).

C. CHLOROPHYLL FLUORESCENCE AS THE METHOD OF PHYSIOLOGICAL GROUND TRUTH CONTROL IN REMOTE SENSING.

Without a correlation of remotely sensed fluorescence reflectance data with direct physiological measurements on the ground of either terrestrial vegetation or algae on the sea surface there is no clear interpretation of data acquired from airborne systems. In fact, remote sensing always requires a correct and efficient ground truth control. This statement is an essential feature of any airborne remote sensing and really needs to be emphasized. It applies to remote sensing of algae accumulation in sea water as well as in the stress detection of vegetation from airborne systems.

The absolute intensity of the chlorophyll fluorescence sensed via active or passive **airborne systems from the sea water or lakes** is matter of considerable variations due to sun or laser intensity, distance of the algae from the water surface wind and water movements content of yellow organic matter and other compounds interfering with the chlorophyll fluorescence. In oceanography, in hydrobiology and limnology the white light or laser-induced intensity of the chlorophyll fluorescence measured under standard conditions in water samples collected at different parts of the remotely screened sea area is then much better and a more correct measure of the density of the algae population. The latter can be further quantified and directly correlated to chlorophyll fluorescence by extraction and determination of the chlorophylls. In addition to measuring the fluorescence intensity one can also determine the physiological state of the algae population by registration of the chlorophyll fluorescence induction kinetics of predarkened samples (Kautsky effect). The shape of the fluorescence induction kinetics and the height of the variable fluorescence (ratio fmax/fo) and the fluorescence decrease ratio (Rfd-values as vitality index see Lichtenthaler 1986, 1988) indicate whether the algae possess a fully functional photosynthesis or are damaged. Thus the chlorophyll fluorescence method is an efficient instrumentation for the measurement of the density and vitality of algae populations required to correctly interpret remotely sensed chlorophyll fluorescence signals in terms of algae populations (Ernst, 1988).

In the ground truth control measurements of **terrestrial vegetation** (crop plants and forest trees) the physiological state of plants has to be determined and must be correlated to the reflectance or fluorescence signatures obtained from airborne sensing systems. The registration of the laser-induced chlorophyll fluorescence induction kinetics in our portable field fluorometer (Lichtenthaler and Rinderle, 1988a) eventually combined with a computer such as the LICAFF system (Kocsanyi et al. 1988) with determination of Rfd values provides valuable information on the health state and vitality of leaves and needles and on the intactness of the photosynthetic apparatus. The ground truth screening requires the measuring of many samples e.g. ca. 10 selected single trees from each test site with either healthy trees or trees with typical damage characteristics. Fast outdoor ground-truth measurements (e.g. a large number of samples per time unit) are possible via laser-induced chlorophyll fluorescence. The chlorophyll fluorescence method in screening of the physiology of plants is therefore superior to other much more time consuming ecophysiological methods such as the CO_2/H_2O porometer method. In one morning one can easily determine the Rfd-values of 40 leaf samples whereas in the same time the net CO_2 assimilation rates of only 4 to 6 samples can exactly be measured. Another advantage of the chlorophyll fluorescence method is the fact that it permits screening of single needles of conifer trees and in the case of broadleaf trees different parts of the same leaf in a non-destructive way. Furthermore the Rfd-values also function as an indicator of the potential photosynthetic activity of a leaf even with closed stomata, which is not the case with the CO_2/H_2O porometer.

In leaves with closed stomata and predarkened for 20 min prior to the registration of the laser-induced chlorophyll fluorescence induction kinetics there is enough CO_2 in the mesophyll from dark respiration to guarantee a normal fluorescence induction kinetic in the next 4 to 5 minutes. In our forest-decline research in the Black Forest (FRG) and in Vermont (USA) (Rock et al. 1986a, Schmuck et al. 1987, Nagel et al. 1987, Rinderle and Lichtenthaler 1987 and 1988) we were therefore able to screen the vitality of trees via Rfd-values during the whole day even on warm dry days when the stomata tended to be partially or almost fully closed. The differences in the Rfd-values between healthy and damaged spruce trees and trees from low and high stress sites were always clearly detectable. Once differences in the vitality of trees or crop plants are detected via field screening of the chlorophyll fluorescence induction kinetics (Rfd-values), one can apply further ecophysiological methods such as the CO_2/H_2O porometer system (net CO_2 assimilation, transpiration rate, opening state of stomata), the Scholander pressure device (determination of the leaf's water potential), determination of the chlorophyll (a+b) and carotenoid (x+c) content (including the ratio a+b/x+c) or the PAM fluorometer (qE quenching coefficient Q_A reoxidation capacity after Schreiber as outlined by Lichtenthaler, 1988; Lichtenthaler and Rinderle, 1988a).

The knowledge of the vitality of selected trees of smaller test areas is a basic requirement for the damage classification of the remotely sensed larger field or forest areas. Hitherto mainly the reflectance signatures has remotely been sensed and used for damage classification and stress detection in terrestrial vegetation. Remote sensing of damage via chlorophyll fluorescence (ratio F690/F735) will follow and possibly also airborne techniques based on radar cross section measurements(Rinderle et al. 1988). In all these cases as well as in oceanography is the chlorophyll fluorescence the superior method for a ground truth screening of the vitality of plants and algae.

CONCLUSION: The chlorophyll fluorescence method is a very suitable method for the detection of phytoplankton in oceanography, for ground truth control measurements as well as for the future determination of the state of health of terrestrial vegetation by remote sensing.

REFERENCES

AMANN V and DOERFFER R, 1983. Aerial Survey of the temporal and spatial distribution of phytoplankton during FLEX'76. In: North Sea Dynamics, Sündermann and Lenz ed., pp. 517-529, Springer Verlag, Berlin.

AMANN V, GUENTHER KP and SCHLITTENHARDT P, 1986. Remote sensing experiment Adria 1984: A comparison of passive and laser-induced chlorophyll-fluorescence. In: Proceed. Internat. Geoscience and Remote Sensing Symposium IGARSS Zürich, Vol. III, pp. 1615-1620, ESA Publications Division, Noordwijk.

BERTOLINI G, KOECHLER C, MOYA I and SCHMUCK G, 1988. Consideration on the possibility of in field remote sensing of health state of plants via registration of the chlorophyll fluorescence. In: Applications of Chlorophyll Fluorescence, Lichtenthaler HK ed., pp. 278-285, Kluwer Academic Publishers, Dordrecht.

ERNST DEW, 1988. Applications of chlorophyll fluorescence in hydrobiology. In: Applications of Chlorophyll Fluorescence, Lichtenthaler HK ed., pp.243 - 250 Kluwer Academic Publisher, Dordrecht.

BRISTOW M, NIELSEN D, BUNDY D, and FUSTEK R, 1981. Use of water Raman emission to correct airborne laser fluorosensor data for effects of water optical attennation. Appl. Opt. 20: 2889-2906.

DOERFFER R, 1981. Factor analysis in ocean colour interpretation. In: Oceanography from Space, Gower, J.F.R. ed., pp. 339-345, Plenum Press Corp.

DOERFFER R, 1988. Remote sensing of sunlight-induced phytoplankton fluorescence. In: Applications of Chlorophyll Fluorescence, Lichtenthaler HK ed., pp.251 - 256, Kluwer Academic Publishers, Dordrecht.

GOWER JFR, Borstad GA and Edel HR, 1987. The fluorescence line imager: first results from passive imaging of chlorophyll fluorescence. In: Proceed. Internat. Geoscience and Remote Sensing Symposium IGARSS Michigan, Vol. II, pp. 1605-1606, University of Michigan, Ann Arbor.

GÜNTHER KP, 1986. Photoinhibition of chlorophyll a fluorescence and its influence to remote sensing techniques. In: Proceed Internat. Geoscience and Remote Sensing Symposium, IGARSS Zürich, Vol. III, pp. 1603-1607,ESA Publications Division, Noordwijk.

HOGE FE, Swift RN and Jungel JK, 1986. Active-passive airborne ocean colour measurement 2: Applications. Appl. Opt. 25: 48-57.

HUSS J, 1984. Luftbildmessung und Fernerkundung in der Forstwirtschaft. H. Widmann Verlag, Karlsruhe.

JADHAV DB, 1987. On the method of detecting phytoplankton pigments by the property of their fluorescence. In: Internat. Geoscience and Remote Sensing Symposium, IGARSS Michigan, Vol. II, pp. 1601-1603, The University of Michigan, Ann Arbor.

KONDRATJEV KY and POZDNYAKOV DV, 1987. Passive and active techniques for remote sensing of phytoplankton of inland water bodies. In: Proceed. Internat. Geoscience and Remote Sensing Symposium, IGARSS Michigan, Vol. II, pp. 1607-1608, University of Michigan, Ann Arbor.

KOCSANYI L, HAITZ M, LICHTENTHALER HK, 1988. Measurement of the laser-induced chlorophyll fluorescence kinetics using a fast acoustooptic device. In: Applications of chlorophyll fluorescence, Lichtenthaler HK ed., pp. 91- 99, Kluwer Academic Publishers, Dordrecht.

LICHTENTHALER HK, 1986. Laser-induced chlorophyll fluorescence of living plants. In: Proceed. Internat. Geoscience and Remote Sensing Symposium, IGARSS Zürich, Vol. III, pp. 1571-1579 ESA Publications Division, Noordwijk.

LICHTENTHALER HK, 1988. In vivo chlorophyll fluorescence as a tool for stress detection in plants. In: Applications of Chlorophyll Fluorescence , Lichtenthaler HK ed., pp. 119 - 132, Kluwer Academic Publishers, Dordrecht.

LICHTENTHALER HK and BUSCHMANN C, 1984. Photooxidative changes in pigment composition and photosynthetic activity of air-polluted spruce needles (Picea abies L.). In Advances in Photosynthesis Research Vol. IV, Sybesma C ed., pp. 245-250, Martinus Nijhoff Publisher, The Hague.

LICHTENTHALER HK and BUSCHMANN C, 1987. Reflectance and chlorophyll fluorescence signatures of leaves. In: Proceed. Internat. Geoscience and Remote Sensing Symposium, IGARSS Michigan, Vol. II, pp. 1207-1213, University of Michigan, Ann Arbor.

LICHTENTHALER HK and BUSCHMANN C, 1988. Changes in the chlorophyll fluorescence spectra during the Kautsky induction kinetics. In: Proceed. 4th Internat. Colloq. Spectral Signatures of Objects in Remote Sensing, Aussois, Vol. I , pp. 245-250, ESA Publications Division, Noordwijk.

LICHTENTHALER HK and RINDERLE U, 1988a. The role of chlorophyll fluorescence in the detection of stress conditions in plants. CRC Critical Reviews in Analytical Chemistry 19, Suppl. I, S 29 - S 85.

LICHTENTHALER HK and RINDERLE U, 1988b. Chlorophyll fluorescence spectra of leaves as induced by blue light and red laser light. In: Proceed. 4th Internat. Colloq. Spectral Signatures of Objects in Remote Sensing, Aussois, Vol. I, pp. 251-254, ESA Publications Division, Noordwijk.

LICHTENTHALER HK, SCHMUCK G, DÖLL M & BUSCHMANN C, 1985. Untersuchungen über die Funktionsfähigkeit des Photosyntheseapparates bei Nadeln gesunder und geschädigter Koniferen. In: PEF-Bericht Kfk-PEF 2, PEF-Projektleitung ed, pp. 81-105, Kernforschungszentrum Karlsruhe.

LICHTENTHALER HK, SCHMUCK G, NAGEL E & BUSCHMANN C, 1986. Direct and indirect damages of photosynthesis as one of the main causes of the large scale forest die-back. In: Premier colloque scientifique des universités du rhin superieur: "Recherche sur l'environnement dans la region", Rentz et al., eds. pp. 261-280, Université Louis Pasteur et Conseil de l'Europe, Strasbourg.

LICHTENTHALER HK, RINDERLE U, KRITIKOS G and ROCK B, 1987. Classification of damaged spruce stands in the Northern Black Forest by airborne reflectance and terrestrial chlorophyll fluorescence measurements. In: 2nd DFVLR Seminar Untersuchung und Kartierung von Waldschäden mit Methoden der Fernerkundung, pp. 238-252, DFVLR Oberpfaffenhofen/München.

MASKE H and HAARDT H, 1988. Quantum yield of in situ fluorescence of phytoplankton in Kiel Bay under daylight, comparison with primary production. In: Applications of Chlorophyll Fluorescence, Lichtenthaler HK ed., pp. 257-266, Kluwer Academic Publishers, Dordrecht.

NAGEL EM, BUSCHMANN C and LICHTENTHALER HK, 1987. Photoacoustic spectra of needles as an indicator for the activity of the photosynthetic apparatus of healthy and damaged conifers. Physiol. Plantarum 70: 427-437.

NEVILLE RA and GOWER JFR, 1977. Passive remote sensing of phytoplankton via chlorophyll a fluorescence. J. Geophys. Res. 82: 3487-3493.

RINDERLE U and LICHTENTHALER HK, 1988. The chlorophyll fluorescence ratio F690/F735 as a possible stress indicator. In: Applications of Chlorophyll Fluorescence, Lichtenthaler HK ed. pp. 176-183, Kluwer Academic Publishers, Dordrecht.

RINDERLE U, HAITZ M, LICHTENTHALER HK, KÄHNY DH, SHIZ and WIESBECK W. Correlation of Radar reflectivity and chlorophyll fluorescence of forest trees. In: Proceed. Geoscience and Remote Sensing Symposium, IGARSS Edinburgh, ESA Publications Division, Noordwijk, (in press).

ROCK BN, HOSHIZAKI T, LICHTENTHALER HK & SCHMUCK G, 1986a. Codmparison of in situ spectral measurements of forest decline symptoms in Vermont (USA) and the Schwarzwald (FRG). In: Proceed. Internat. Geoscience and Remote Sensing Symposium. IGARSS Zürich, Vol. III, pp. 1667-1672, ESA Publications Division, Noordwijk.

ROCK BN, VOGELMANN JE, WILLIAMS DL, VOGELMANN AF and HOSHIZAKI T, 1986b. Remote detection of forest damage. Bioscience 36: 439-445.

ROSEMA A, CECCHI G, PANTANI L, RADICATTI B, ROMULI M, MAZZINGHI P, van KOOTEN O and KLIFFEN C, 1988. Results of the "LIFT" project: Air pollution effects on the fluorescence of Douglas fir and Poplar. In: Applications of Chlorophyll Fluorescence, Lichtenthaler HK ed., pp. 286-296, Kluwer Academic Publishers, Dordrecht.

SCHMUCK G & LICHTENTHALER HK, 1986. Application of laser-induced chlorophyll fluorescence in the forest decline research. In: Proceed. Internat. Geoscience and Remote Sensing Symposium, IGARSS Zürich, Vol. III, pp. 1587-1590, ESA Publications Division, Noordwijk.

SCHMUCK G, LICHTENTHALER HK, KRITIKOS G, AMANN V and ROCK B, 1987. Comparison of terrestrial and airborne reflection measurements of forest trees. In: Proceed. Internat. Geoscience and Remote Sensing Symposium IGARSS Michigan, Vol. I, pp. 1201-1206, University of Michigan, Ann Arbor.

SHU S and CHEN J, 1987. A new method to measure chlorophyll concentration in different depth water using spectroradiometer. In: Proceed. Internat. Geoscience and Remote Sensing Symposium IGARSS Michigan, Vol. II, pp. 1589-1594, The University of Michigan, Ann Arbor.

SCHNECKENBURGER H and FRENZ M, 1986. Time-resolved fluorescence of conifers exposed to environmental pollution. Radiat. Environm. Biophysics 25: 289-295.

SPITZER D and DIRKS RWJ, 1985. Contamination of the reflectance of natural waters by solar-induced fluorescence of dissolved organic matter. Applied Optics 24: 444-445.

ZIMMERMANN R and GÜNTHER KP, 1986. Laser-induced chlorophyll a fluorescence of terrestrial plants. In: Proceed. Internat. Geoscience Remote Sensing Symposium, IGARSS Zürich, Vol. III, pp. 1609-1613, ESA Publications Division, Noordwijk.

CONSIDERATION ON THE POSSIBILITY OF IN FIELD REMOTE SENSING OF THE HEALTHY STATE OF PLANTS VIA REGISTRATION OF THE CHLOROPHYLL FLUORESCENCE

G. Bertolini, C. Koechler*, J. Moya** and G. Schmuck**

* Commission of the European Communities - Joint Research Centre. I-21020 Ispra (VA) Italy

**L.U.R.E. (Laboratoire pour l'utilisation du rayonnement électromagnétique), Université Paris XI, Bât. 209 C, 91405 Orsay, France.

ABSTRACT

The time decay of the various components of 685 nm chlorophyll fluorescence emission excited by picosecond lasers is analyzed as a possible approach to determine the state of stress of the green matter.
Two possible experimental arrangements for in field remote measurement are described and discussed in terms of spectral and temporal resolution. The total number of fluorescence counts per single laser shot and the solar light background are evaluated.

INTRODUCTION

The present applications of the different existing remote sensing techniques, oriented towards agriculture and forestry, include land use, crop type classification, early warning of plant stress, assessment of crop photosynthesis and prediction of yield. All these techniques have in common, that the obtained results are not sensitive only to vegetation responses, because interferences exist with the optical properties of the surrounding environment.

A new technique for remote detection of plant stress, due to biochemical and physiological changes, seems to be the detection of the light induced chlorophyll fluorescence. Since the chlorophyll fluorescence is only due to emission of green vegetation, it can be considered a complementation of the existing techniques.

The solar energy incident on plants (leaves, canopies) is reflected, transmitted and absorbed by the photosynthetic pigments. A small part of the absorbed light (less than 5%) is reemitted as chlorophyll fluorescence (650-800 nm). Because of the inverse relationship between photosynthetic activity and the in-vivo chlorophyll fluorescence, the study of the spectral and temporal characteristics of this fluorescence emission appears to be a promising tool for assessing the physiological status of plants.

As a passive remote sensing system, a Frauenhofer Line Discriminator (FLD), operating at the 656. 3nm Frauenhoferline, was used to investigate water stress in lemon trees. With this instrument the sun-induced chlorophyll fluorescence was measured from a platform 12 m above the ground. It was shown that an increase in fluorescence was correlated with water stress of the investigated lemon trees (Mc Farlane et al., 1980). A possible application of this passive device for airborne and orbital remote sensing of in-vivo chlorophyll fluorescence has been analyzed in a previous study (Plascyk, 1975).

Beside this passive technique, a possible application of the registration of laser-induced chlorophyll fluorescence from remote is under discussion. The possibility of airborne detection of laser-induced fluorescence from trees, bushes and grasses has been demonstrated (Hoge et al., 1983).

H. K. Lichtenthaler (ed.) Applications of Chlorophyll Fluorescence, 299–306.

Zimmermann and Guenther reported about a first test flight over different damaged forested areas with a modified oceanographic lidar system. The laser-induced chlorophyll fluorescence was detected at 685 nm and 735 nm wavelength. The results show that the fluorescence emission ratio (685/735) is a possible method for classification of damaged and undamaged trees (Zimmermann and Guenther, 1986).

Additional information about the physiological state of vegetation has been observed in laboratory by investigation of the "blue" fluorescence excited by UV light. The emission band, around 450 nm, has been discussed in terms of stress conditions (Chapelle and Williams, 1986).

In recent years special effort was devoted in the characterization of time decay components of red fluorescence emission as an index of the energy transfer process in photosynthetic systems.

A program has been set-up to verify if the time decay components and their relative yield are correlated to the state of stress of vegetation. The advantages of fluorescence technique are indeed due to its high sensitivity and non-destructive nature.

STATE OF THE ART

In a photosynthetic system the light energy, absorbed by a large array of antenna pigments is trapped on a time scale of tens to hundreds of picoseconds by the reaction centers. The energy transfer processes and therefore the functional organization of the photosynthetic apparatus can be investigated by time-resolved fluorescence methods in a time scale of picoseconds to some nanoseconds.

Chloroplasts and green algae

Fluorescence lifetime measurements performed on green algae and on chloroplasts from higher plants provide clear evidence of the heterogeneity of the room temperature fluorescence emission.The overall fluorescence time decay is multiexponential and can usually be statistically defined by three kinetic components (Gulotty et al. 1982; Haehnel et al. 1983) with lifetimes in the order of 50 ps, 200-850 ps and 450-2000 ps according to the state of the primary stable election accepter of Photosystem II Q_A (Moya et al. 1986). Fluorescence lifetimes of chlorella pyrenoidosa and of isolated spinach chloroplast at different states of Q_A are shown below. At F_0-state all the reaction centers of PS II are open (Q_A is oxidized). Closing the reaction centers of PS II (Q_A is reduced, F_{max}-state) produced two variable lifetime components and two constant components between initial and maximal fluorescence levels.

Lifetimes (ps)					
Chlorella pyrenoidosa	F_0: τ, A (%)	59 (12)	223 (44)	447 (44)	-
	F_{max}: τ, A (%)	57 (2)	224 (4)	1444 (74)	2589 (20)
Spinach chloroplasts	F_0: τ, A (%)	43 (13)	193 (59)	581 (28)	-
	F_{max}: τ, A (%)	44 (1)	250 (5)	1432 (50)	2778 (44)

(Hodges and Moya, 1986)

More recent studies suggest that this is an oversimplistic model (Gulotty et al. 1985; Hodges and Moya 1986) and that the overall decay could contain, at least, five individual decay components (Hodges et al. 1986).

This is due to the complexity of the light absorbing system of photosynthetic organisms, containing two types of reaction centre-core antenna complex (PS I and PS II) each associated with their own peripheral antenna systems (LHC I and LHCP2) (Barber, 1986). The situation is further complicated by the reported structural heterogeneity of PS II into α and β centres (Glazer and Melis, 1987) and two pools of LHC2 differing in polypeptide composition (Larsson et al. 1986).

For a description of the fluorescence time decay in terms of energy transfer and trapping processes, it is necessary to assign each decay component to a functional constituent of the light harvesting system. By analysing the decay of well-defined mutants, lacking one or more of the pigment-protein complexes it is suggested that the two long-lived components (2 and 1 ns) arise from LHC2, while PS I produces two rapid decays (< 200 ps) and the PS II core-complex a fast variable decay (40-250 ps) (Hodges and Moya, 1987).

Higher Plants

Only little information is available about the chlorophyll fluorescence decay of intact leaves.

Pellegrino, 1981 described the influence of excitation pulse intensity on the lifetimes of leaves of spinach and Norway maple. At low intensity, the fluorescence decays as a single exponential (150-250 ps), at higher excitation intensities (> 10^{13} photons/cm^2 s) the decay was characterized by two components 50 and ~ 200 ps. He attributed this effect to singlet-singlet annihilation, which provided a radiationless deactivation pathway for excited state.

In another approach, the decay of variable chlorophyll fluorescence in leaves of maize and spruce was determined to be a single exponential with a lifetime of 2.0 ns (Senonev, 1986).

Schneckenburger and Frenz, 1986 reported about their studies on fluorescence lifetimes of spruce and pine needles after picosecond laser pulse excitation. Green needles with intact photosynthesis were characterized by 2 components of τ_1 (0.1 ns) and τ_2 (0.5 ns). An increase of τ_2 and a further time constant τ_3 (2.5 ns) were found, when the photosynthetic system was damaged (discolouring) and after exposition of green healthy needles to high ozone doses (100 μgr/m^3 of 0_3, g h per day, during two months) during longer time periods (Schneckenburger and Frenz, 1986). Similar measurements with 40 μgr/m^3 of 0_3 and 150 μgr/m^{-3} of SO_2 give no time decay variations.

DESIGN OF AN "IN-FIELD" INSTRUMENTATION

In spite of the necessity to perform more laboratory measurements on the various spectral and temporal characteristics of the excited chlorophyll fluorescence of vegetation, it is useful to analyze the parameters which have to be taken into account in the design of a remote in-field instrumentation. In the case in which the time decay of the various time components of the red emission of the chl-a is to be determined, a certain basic considerations have to be done.

The development of mode-locked lasers enabled time resolved fluorescence measurements of chl. in vivo in the time-domain of picoseconds.

Various experimental set-up have been utilized: single photon counting, phase-shift technique and synchroscan streak camera are used with high repetition rate (~ MHz) picosecond lasers, single shot streak camera and transient analyzers are used with single shot (~ 10 Hz) picosecond lasers.

Through considerations based on the mobility of the leaves and the deepness of a vegetation target we will analyze an experimental system based on single shot picosecond laser.

The following experimental parameters habe to be taken into account:

- The choice of the excitation wavelength
- The FWHM of the time width of the light pulse
- The maximum intensity of the light pulse
- The optical and detection system
- The signal on the noise ratio.

The well known absorption spectra of chlorophylls show that λ_{exc} in the blue and in the red regions have the maximum fluorescence conversion factor. The choice between these two wavelength bands is given by the available mode locked lasers. The Nd-Yag laser makes available λ emission equal to 532 nm (II HG) and 355 nm (III HG). At 355 nm the fluorescence yield is maximum and it will permit the analysis of the spectrum also up to the blue band.

In a mode-locked configuration pulse width of about 20 ps and 10 mJ of energy are obtainable.

The energy of this light pulse is even to much if the diameter of the spot on the target cannot be higher than about 10 cm for timing reasons. In this case the photon flux will be of the order of 10^{14} hv.cm^{-2}. Several authors showed that at this photon flux a shortening of the lifetime and a decrease in the fluorescence yield (Campillo et al. 1976) will happen. Fluorescence lifetimes from the alga chlorella pyrenoidosa as a function of the excitation intensity is given below and show clearly that light energy must be limited to 3.10^{13} photons.cm^{-2} as a maximum to avoid an artificial shortening of the lifetimes. The measurements were performed at F_0-state (PS II reaction centers are open, Q_A is oxidized) and after closing the reaction centers of PS II (Q_A is reduced, F_{max}-state).

Intensity (photons.cm^{-2})	τ (ps) F_0	τ (ps) F_{max}
3×10^{13}	450	1800
5×10^{14}	280	660
8×10^{15}	50	220

(Porter et al. 1977)

The design of the optical and detection system depends from the fluorescence parameters to be measured.

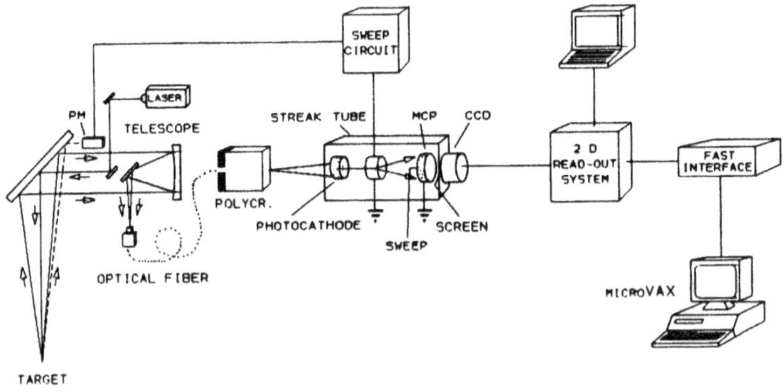

Fig. 1 Conceptual layout of an in-field time resolved Lidar Fluorosensor

Fig. 1 shows the conceptual design of a time-resolved airborne fluorosensor which has been conceived to characterize the oils at sea. We are actually assembling a similar instrumentation which will have also the possibility to perform measurements on targets at 100 m distance. In the new configuration the laser will be a mode-locked Nd-Yag, 20 ps pulse width and about 10 mJ at 355 and 30 mJ at 532 nm, at 10 Hz of repetition rate. The main features of this facility are:

- The detector is an intensified streak camera with 2 dimensional CCD array which gives simultaneous spectral and temporal resolution with at least 30 and 90 channels respectively.
  The spectral range will be between 350 to 760 nm and the temporal range from 0.2 to ~ 50 ns. The highest temporal resolution is much higher than the effective time resolution which is possible to achieve taking into account the photocatode diameter, the optical limitations and the laser pulse width ($\Delta T > 50$ ps).
- The starting time of the streak camera sweep is given by the reflection of the excitation light pulse from the target. In this way the t = 0 of the time analysis is independent from source-target distance.

The CCD output is directly interfaced to a Microvax computer via a fast parallel DMA interface with buffer memory, allowing the transfer of ten images per second.

The camera dynamic range in single sweep, defined as the peak count on the rms noise is better than 1000.

The configuration shown in Fig. 1 is one of the most complete set-up to characterize the spectral and temporal behaviour of the laser-induced chlorophyll fluorescence in single shot.

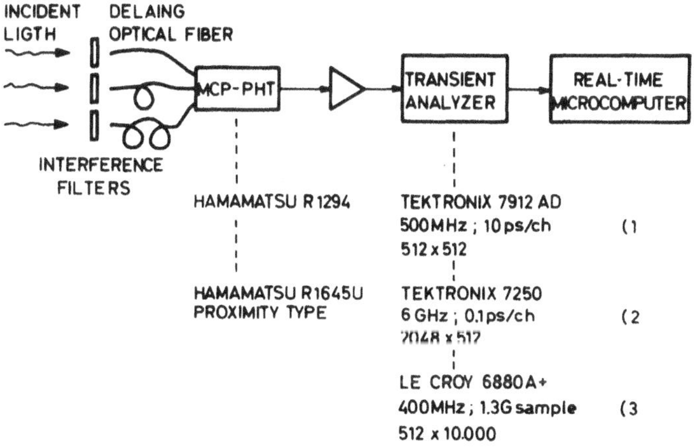

Fig. 2 Layout of the analogical time measurement system

Fig. 2 shows a measurement configuration actually under test in our laboratory using a PM Hamamatsu R1294 U and a transient analyzer model (1). The main characteristics of the two PM shown in the figure are:

	Rise Time (ns)	FWHM (ns)	Spectr.Range	G
MCP - R 1294 V	0.23	0.68	250 ÷ 750	5.10^5
MCP - R 1564 V (proximity type)	0.25	0.75	140 ÷ 850	5.10^5

Working in analog way the response functions of the two PM are not different and therefore it is not possible to evaluate the shortest time decay components in the chl. fluorescence spectrum.

On the contrary the appearance of a longer component in the cases of energy transfer inhibition is measurable as shown in Fig. 3 where the fluorescence time decay of two organic liquids, excited by a 0.1 ns laser pulse at 337 nm, are compared.

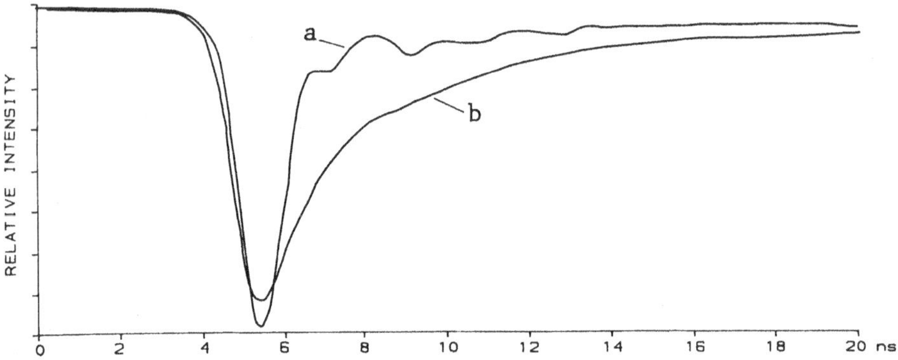

Fig. 3 Time decay of the fluorescence of erithrosine (τ = 0.017 ns) and oxazine (τ = 0.82 ns) excited with a 337 nm, 0.1 ns, light pulse. The curves have been obtained using an Hamamatsu PM mod. R1294V and a transient analyzer (1) of the Fig. 2

CONSIDERATIONS ON PRACTICAL LIMITATION OF REMOTE SENSING

Two main considerations must be done on the possibility to use the time decay approach for in-field measurements at a certain distance. The first problem derives from the difficulty to identify well defined targets, the second problem is due to the signal to noise ratio along a single decay curve.

a) Problems related to the target
Higher plants (trees) seem not suitable for a remote measurement, in particular from an airplane. The leaves have a too small dimension and, also in the case of a low divergence laser beam (~ 0.2 mrd), the light spot at an average distance of 200 meters will have a diameter not less than ~ 10 cm. In this case a high probability exists that the light pulse will excite the fluorescence of more leaves at different relative distances. Even by assuming that the time decay of the 685 nm

emission from more leaves of the same tree will have the same trend, the depth equivalent of this target will be such to perturb, or destroy, the timing information of the global fluorescence signal. Certain crops seem to be more suitable from this point of view but, in any case, a careful analysis of this problem have to be performed.

b) Problems related to the fluorescence over solar photons ratio
The main limitation is due to the maximum specific intensity of light pulses which is equal to some 10^{13} photons.cm^{-2}, to avoid the shortening of decay times due to annihilation effects.

Assuming a spot diameter of about 10 cm the total number of photons is equal to some 10^{15}.

Therefore the timing data must be extracted from a limited number of photons. This means that an average of many laser shots must be done and this excludes a flying platform.

A qualitative determination of this ratio gives the follow:

1. Laser pulse energy: 1 mJ at 355 nm.
   Number of photons = 10^{-3} J x 355 x $10^{16}/1.987 = 1.8 \times 10^{15}$ photons
   $\phi_{spot} = 100$ cm^2 gives $1.8 \cdot 10^{13}$ photons.cm^{-2}
2. Fluorescence quantum efficiency = 0.02
   Number of fluorescence photons = 3.6×10^{13}
3. Solar irradiance at 690 nm at nadir assuming an atmospheric transmittance ~ 0.8 gives:
   $S_I = 112$ mJ . cm^{-2} . µm^{-1} . s^{-1}
4. Assuming a leave reflectance coefficient equal to 0.08 and an isotropic diffused reflectance:
   $S_R = 112 \times 0.7 \times 0.08$ mJ . cm^{-2} . µm^{-1} . s^{-1}
5. Assuming a bandwidth of the fluorescence emission equal to 0.1 µm and $\phi_{spot} = 100$ cm^2:
   $S_R = 63$ mJ . s^{-1}
6. Assuming that the fluorescence emission last for $20 \cdot 10^{-9}$ s the ratio in this time interval of the number of fluorescence photons to the reflected solar photons is equal to:

$$3.6 \cdot 10^{13} / 4.2 \cdot 10^{9} \sim 10^{4}$$

Given this ratio the noise due to the solar reflection on vegetation is a minor factor which must not influence the time measurements along the fluorescence decay.

Assuming a target distance of 100 m and a collector with a diameter of 30 cm (installation already existing) the total number of photoelectrons released by the photocathode of the streak camera will be of the order of 10^5.

CONCLUSIONS

Many laboratory studies will be necessary to prove that the time decay analysis of the laser induced fluorescence under natural illumination condition, is a suitable approach to monitor vegetation stress conditions.

The utilization of this technique to perform remote measurements will be mainly conditioned by the geometry of the plant canopy.

ACKNOWLEDGMENTS

The authors acknowledge Messrs. G. Maracci and A. Pedrini for stimulating and useful discussions.

One of the authors, G. Schmuck, would like to thank the Commission of the European Community for financial support.

LITERATURE

Plascyk, J.A. (1975), Opt. Eng. 14, 339

Mcfarlane, J.C., Watson, R.D., Theisen, A.F., Jackson, R.D., Ehrler, W.L., Pinter, P.J., Jaso, Jr. S.B. and Reginato, R.J. (1980), Applied Optics 19, 3287

Hoge, F.E., Swift, R.N. and Yungel, J.K. (1983), Applied Optics 22, 2991

Zimmermann, R., Guenther, K.P. (1986), Proc. Int. Geoscience and Remote Sensing Symposium, Vol. IV, 1609

Gulotty, R.J., Fleming, G.R. and Alberte, R.S. (1982), BBA 682, 322-331

Haehnel, W., Holzwarth, A.R. and Wendler, J. (1983), Photochem. Photobiol. 37, 435-443

Moya, I., Hodges, M. and Barbet, J.C. (1986), FEBS Lett. 198, 256-262

Gulotty, R.T., Metz, L., Alberte, R.S. and Fleming, G.R. (1985), Photochem. Photobiol. 41, 487-496

Hodges, M. and Moya, I. (1986), BBA 849, 193-202

Hodges, M., Moya, I., Briantais, J.M. and Remy, R., (1986), in: Progress in Photosynthesis Research (Biggins, J., ed.), Vol. I, 115-118, Martinus Nijhoff, Dordrecht

Barber, J. (1986), in: Topics in Photosynthesis (Barber, J. and Baker, N.K., eds), Vol. 5, 91-134, Elsevier Science Press

Glazer, A.N. and Melis, A. (1987), Annu. Rev. Plant Physiol. 38, 11-46

Larsson, U.K., Sunby, C. and Andersson, B. (1986), in: Progress in Photosynthesis Research (Biggins, J., ed.) Vol. 2, 677-680, Martinus Nijhoff, Dordrecht

Hodges, M. and Moya, I. (1987), Photosynthesis Res. 13, 125-141

Pellegrino (1981), Phd-Thesis

Senoner, M. (1986), BBA 849, 374-380

Schneckenburger, H. and Frenz, M. (1986), Radiat. Environ. Biophys. 25, 289-295

Porter, G., Synoviec, J.A. and Tredwell, C.J. (1977), BBA 459, 329-336

Campillo, A.J., Sharpiro, S.L., Kollman, V.H. and Winn, K.R. (1976), Biophys. J. 16, 93-97

RESULTS OF THE 'LIFT' PROJECT: AIR POLLUTION EFFECTS ON THE FLUORESCENCE OF DOUGLAS FIR AND POPLAR

A. ROSEMA
Environmental Analysis and Remote Sensing (EARS) Ltd, PO Box 449, 2628 EB Delft, The Netherlands
G. CECCHI, L. PANTANI, B. RADICATTI, M. ROMULI, P. MAZZINGHI
Institute of Research on Electromagnetic Waves (IROE), National Research Centre (CNR), Firenze, Italy
O. VAN KOOTEN
Agricultural University of Wageningen, Department of Plant Physiological Research, Wageningen, The Netherlands
C. KLIFFEN
Institute of Plant Disease Research, Wageningen, The Netherlands

ABSTRACT

A brief review is given of literature on remote sensing of chlorophyll fluorescence. The objective of the project 'Laser Induced Fluorescence of Trees' (LIFT) is to investigate the possible use of this technique to observe and monitor vegetation stress by air pollution. The instrumentation and set up is briefly described and measuring results are presented. Relative highs in air pollution appear to be associated with anomalies in fluorescence, implying a stress situation to the plant. Particularly the effect of rather low concentrations ozone (20-30 ppb), possibly in combination with SO_2, is very strong.

Keywords: laser induced fluorescence, air pollution, vegetation stress, remote sensing, ozone.

INTRODUCTION

The LIFT project (Laser Induced Fluorescence of Trees) is a research project of the Working Group on Remote Sensing of Plant Fluorescence, in the Netherlands, in cooperation with the Institute of Research on Electromagnetic Waves, Florence, Italy. The objective is to investigate the potential of chlorophyll fluorescence for remote sensing and monitoring of photosynthetic activity in relation to environmental stress, air pollution in particular. The focus is on the problem area of 'acid rain' and its detrimental effects on the vitality of European woods.

REMOTE FLUORO-SENSING

Instruments applied for remote sensing of chlorophyll fluorescence may be discerned in passive and active systems, using the sun or an artificial exitation source, respectively. Possibly the first instrument of this kind was the Fraunhofer Line Discriminator, a passive method based on a radiometric measurement in and next to a Fraunhofer line (656.3 nm). Measurements on Pinus Ponderosa, growing in soil with high Cu and Zn

H. K. Lichtenthaler (ed.) Applications of Chlorophyll Fluorescence, 307–317.

content, showed enhanced fluorescence, but also a considerable dependance on light and weather conditions (Watson et.al. 1974,1976). The same instrument was later used for measurements on drought stressed Citrus trees (Mac Farlane et.al. 1980). It was concluded that, in absence of any visual signs, plant fluorescence was very sensitive in evaluating water stress.

It has also been tried to infer chlorophyll fluorescence, as superimposed on the plants reflection spectrum. Nevil and Gower (1977) measured the spectral radiance of the sea and demonstrated a small peak at 685 nm, that was attributed to chlorophyll fluorescence. Peak height correlated well to the algae concentration. On the same basis Kim et.al. (1985) mapped chlorophyll patterns in the Elbe River. Rosema and Werner (1983) investigated the possible use of this approach on land vegetation, by means of a Kubelka-Munk leaf transmission model, that was modified to include fluorescence. They found that increased fluorescence yield could notably effect the 'apparent reflection' and proposed the Fluorescence Index (FI = (R680-R660)/R660) as an approximate measure of plant fluorescence. Indeed the Kautsky induction effect can be observed in plant reflection, although to a much lesser extent (Rosema, Kliffen, unpublished data). It is believed, that the so called blue shift of the red edge in the plant spectrum, observed with a high resolution spectrometer over area's of Pb,Cu mineralization (Collins et.al. 1983, Chang and Collins 1983) could be explained by increased fluorescence as well.

Laser induced fluorescence (LIF) is considered the most powerfull technique, but requires very advanced instrumentation. Early work was mainly related to sea applications. Hickman et.al. (1972) investigated the relation between laser induced fluorescence and algae concentration in a laboratory tank. Hoge and Swift (1980,1981,1983), Gehlhaar et.al. (1981) and Castagnoli et.al (1986) report the development of airborne oceanographic lidar systems. However, the use of such systems for land vegetation was demonstrated as well (Hoge, Swift and Yungel 1983, Cecchi et.al. 1984). Chapelle et.al. (1984a,1984b,1985) studied LIF in relation to water stress, chlorosis, nutrient deficiences and plant types. In 1986 and 1987 te LIFT project has been carried out, using the IROE-CNR fluorescence lidar system (FLIDAR) and investigating the effects of air polluting gasses on plants and trees.

EXPERIMENTAL SET-UP

The FLIDAR system (Castagnoli et.al. 1986) consists of a laser source, a telescope collector, a spectrometer and a data acquisition system. The primary laser source is a Xenon-Chloride excimer laser, producing 80 mJ pulses of 15 ns pulse width at a wavelength of 308 nm. This laser is pumping a dye laser in front of it. The dye laser produces a pulsed beam with a wavelength depending on the dye, 480 nm in the present case, which is reflected along the longitudinal axis of the telescope. The radiation collected by the telescope is spectrally analysed by means of an OMA-3 system, which uses a grating spectrometer and a high performance, gateable intensified diode array. Data acquisition and processing is performed by the the main control unit of this system. The fluorescence spectrum is obtained by subtraction of a measurement with and without laser action.

Forest rim at the Garderen field showing LIF target trees.

IROE's fluorescence lidar (FLIDAR) at the Garderen field

The FLIDAR system was used for measurements both in the laboratory and in the field. The laboratory measurements were of two types. At the Institute of Plant Disease Research, two years old Douglas firs and dense garden bean plants were subjected to SO_2 gas treatment during 6 weeks in 8 fully climatized rooms at 15 °C and 75% relative humidity. Gas concentrations were 0, 25, 50, 75, 100, 150, 200 and 300 ug/m^2 SO_2 respectively. The plants had to be taken out of the climate rooms for the LIF measurements, which was done immediately before the actual measurement.

At the Department of Plant Physiological Research LIF measurements were carried out on Poplar cuttings, which had been exposed for 16 days to 200 ug/m^3 SO_2 and 100 ug/m^3 NH_3. For the experiment a leaf was enclosed in a gas exchange leaf cuvette, which allows controlled gas treatment and simultaneous measurement of CO_2 and O_2 exchange. The leaf was exitated and measured via a glas window at various light levels, decreasing from PAR = 60 W/m^2 to dark. Thereafter the leaf was exposed to full light and the SO_2 concentration was increased to 600 ug/m^3. After reaching the same CO_2 assimilation level the set of LIF measurements at decreasing light levels was repeated.

The field measurements were carried out at an 'acid rain' measuring field near Garderen on Douglas fir at a distance of about 30 m and a height of approximately 15 m. The fir stood at a forest rim with southern exposition, near a meteorological and a gas analyzing mast of the Department of Air Pollution of the Agricultural University of Wageningen. Here global radiation, windspeed, air temperature and air humidity are monitored, as well as the gasses NO_x, NO, SO_2 and O_3. LIF measurements were taken half hourly or hourly, from about 04.00 am to 22.00 pm. A good data set was collected during 1½ day.

RESULTS

From the experimental point of view the laboratory measurements on whole plants or small trees appeared problematic and did not give very satisfactory results. The main cause is believed to be the variable leaf orientation and shadowing. Small displacements of the plants could even give an almost 2-fold increase of fluorescence. Apperently the measuring area was not large enough to obtain a good statistical average. However, with the rank correlation test of Spearman, Kliffen could show a significant negative correlation between LIF and the SO_2 concentration applied, in the case of young Douglas fir. This confirms the observation that they seemed to be stimulated by SO_2 treatment. On the contrary there was a positive correlation between fluorescence and SO_2 concentration in the case of Garden Beans. This corresponds with the observation that these plants were negatively effected in their development. It was found that the ratio F685/F720 less variable and possibly compensates for the variations in canopy structure.

The experiment with the leaf cuvette provides a much better defined measuring environment. The position of the leaf and the laser are not changed during the experiment. The measuring results on the Poplar leaf are presented in fig. 1. On the vertical axis there is the fluorescence at 685 nm, on the horizontal axis the CO_2 assimilation. Measurements at various light intensities (and corresponding CO_2 assimilation) are indicated by the time they were taken, beginning at 9.56 hrs and ending at 15.40 hrs.

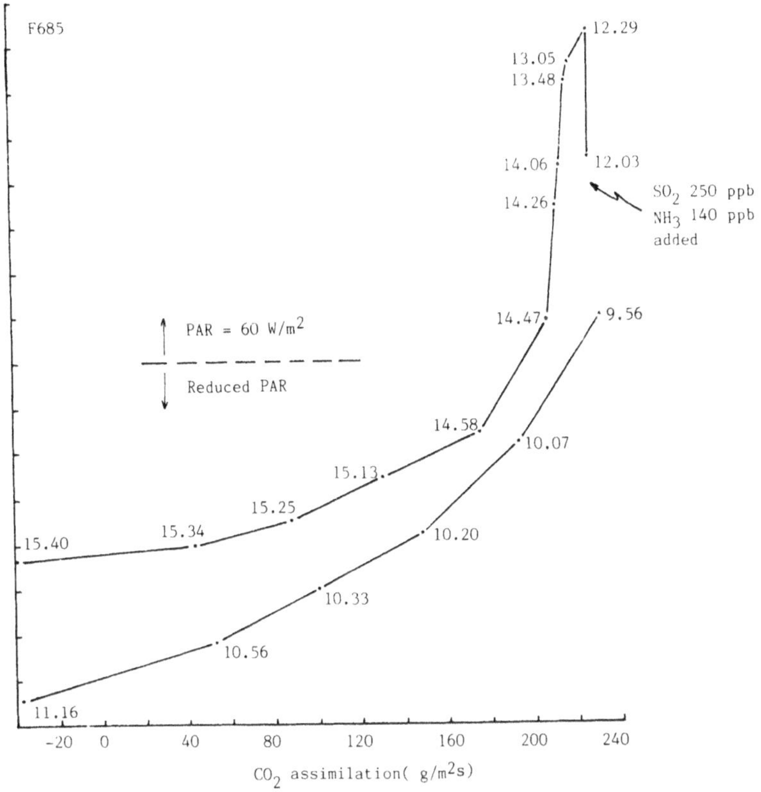

Fig 1: Laser induced fluorescence at 685 nm versus CO_2 assimilation. Measurements were taken on a leaf of a Poplar cutting, enclosed in a gas exchange cuvette, at various light intensities, before and after gas treatment.

The experiment started with a measuring series at decreasing light intensity until dark (60, 41, 25, 15, 8.4, 0 W/m^2 PAR), from 9.56 to 11.16 hrs. Then the light was turned on again and the higher SO_2 concentration was added to the air flow. At 12.03 measurements started again, when the CO_2 assimilation was at the same level as at the beginning of the experiment. Between 14.47 and 15.50 hrs, a new series of measurements with decreasing light intensity was taken.

Fig. 1 shows a very consistent picture. After the additional SO_2 treatment the F685-CO_2 assimilation relation is at a higher level.There is a decrease of 10 to 20% in CO_2 assimilation. However at high light conditions there is little difference in fluorescence, while at low light conditions there is a strong relative increase of F685. The dark adapted level (F_0) is 6 times higher than before gas treatment. This suggests decreased exiton transfer from the light harvesting complex to photosystem 2.

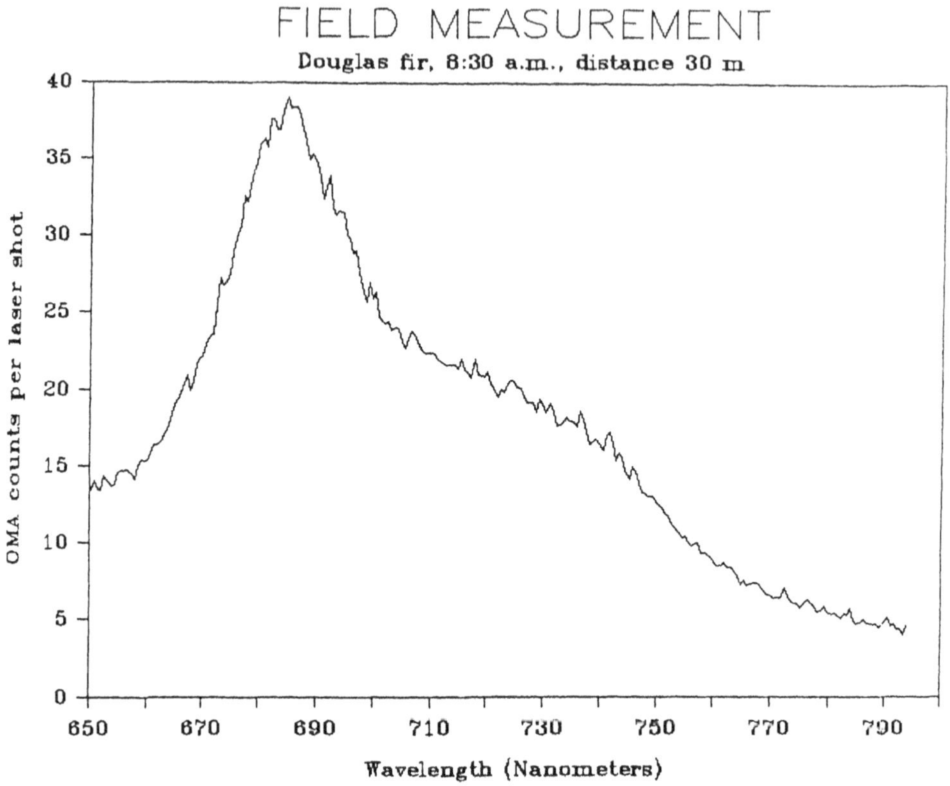

Fig. 2: Example of a laser induced fluorescence spectrum taken with the FLIDAR system at the Garderen field.

The observed differences in fluorescence before and after gas treatment are very significant and can be considered an indication of stress due to SO_2. The plant physiological explication, however, is not completely clear and open for discussion. In terms of practical remote sensing it seems that low light conditions, e.g. late afternoon, would be the best moment for observation.

Figs 2-4 present the results of the field measurements on Douglas fir at Garderen. Fig 2 is an example of a fluorescence spectrum taken with one laser shot at the measuring field in Garderen. Fig. 3 presents an overview of the daily course of global radiation, air temperature and windspeed on the measuring day, 28 june 1987, as well as the fluorescence measured at 685 nm. Fig. 4 shows the corresponding measurements of O_3, SO_2, NO and NO_x.

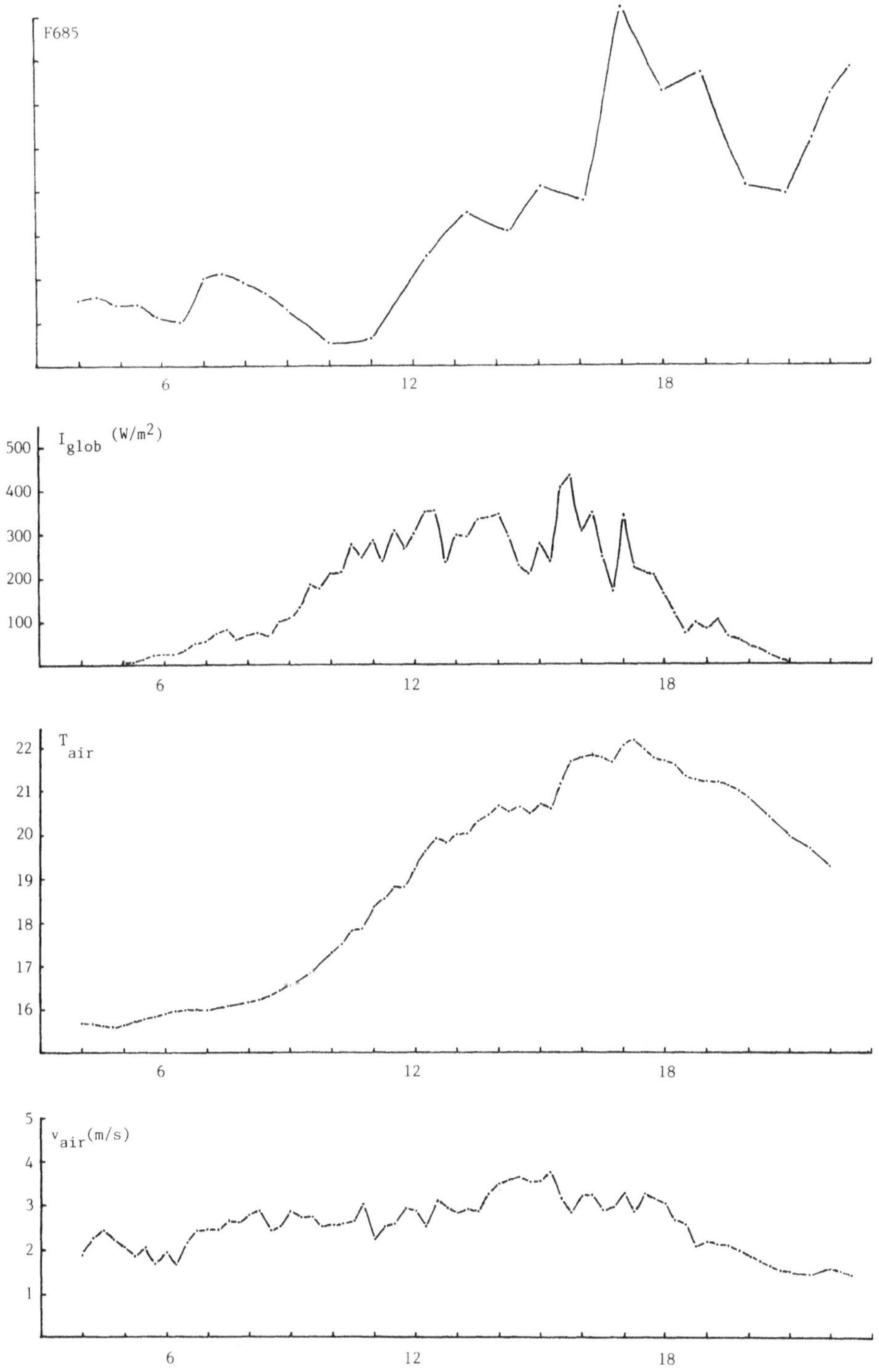

Fig. 3: Laser induced fluorescence at 685 nm, global radiation, air temperature and windspeed at the Garderen field on 28 june 1987.

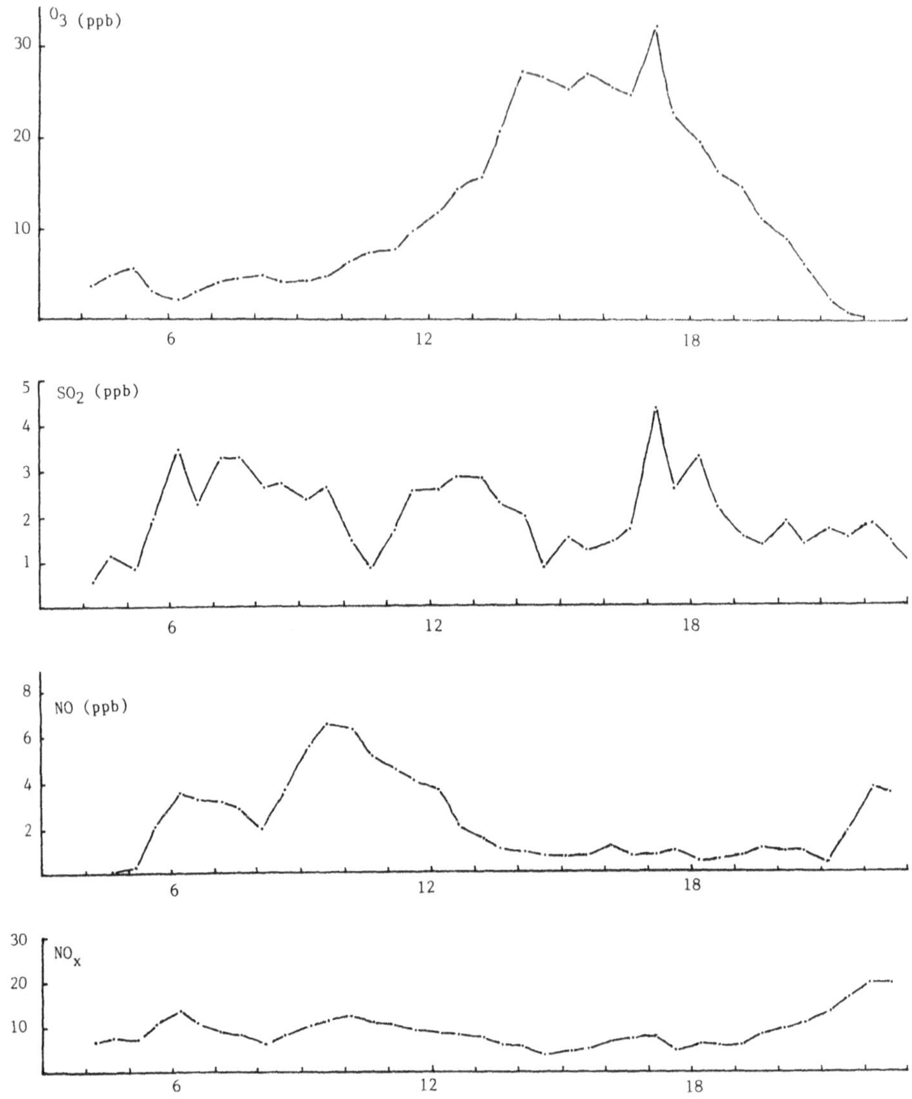

Fig. 4: O_3, SO_2, NO and NO_x, measured at the Garderen field on 28 june 1987.

At first sight there is little correspondance between the F685 level and global radiation. The average trend, however, between 5 and 17 hrs should, and seems to reflect radiation. Superimposed on this trend, there are some remarkable anomalies. First, there is a clear negative anomaly between 8 and 12 hrs, which seems well correlated to relatively high NO (3-7 ppb) and somewhat increased NO_x content (10-15 ppb). In the late afternoon between 16 and 20 hrs there is a very strong positive anomaly, which corresponds to relatively high O_3 (20-30 ppb) possibly in combination with somewhat increased SO_2 levels (2-4 ppb). Finally, at the end of the daylight period, after 21 hrs, there is again a strong positive anomaly, which seems to correspond to high NO_2 (15-20 ppb) and somewhat increased NO (2-4 ppb), while at the same time O_3 is neglegible.

DISCUSSION

Schreiber et.al. (1978) investigated the effect of ozone (250 ppb) on the fluorescence induction curves of bean, during variable periods of exposure, up to two hours. They found an increase in terminal (stationary) fluorescence, which was attributed to a decrease in energy-dependent quenching and inhibition of electron transport between the photosystems.

More then additive detrimental effects of O_3 in combination with SO_2, at low concentrations, have been reported in the literature. It has also been reported that the damaging effect of NO_2 in the dark is much stronger than during daytime, since the toxic nitrites that are formed, cannot be reduced to ammonia by the reducing force of photosynthesis. At daytime the reduction of NO_x to NH_3 seems to compete with that of CO_2 and cause growth inhibition (Den Boer and Bastiaans 1984).

Questions remain with respect to the plant physiological explanation of the fluorescence anomalies observed. It is evident, however, that the laser induced fluorescence of Douglas fir is very sensitive to variations in air pollution at low concentrations (O_3: 20-30 ppb + SO_2: 2-4 ppb, NO: 4-6 ppb, NO_x: 15-20 ppb). The high sensitivity of these trees may be a general sign of bad 'health', caused by their exposure to a more or less continuous background stress. The fluorescence anomalies would then represent specific and immediate stress conditions, which, in time, increasingly affect their vitality.

Particularly the effect of ozone seems serious. The observed values of 2-30 ppb are not notably higher than the background concentration, which is presently 20-25 ppb. At the end of the last century, however, this value was only 10-15 ppb (Slooff et.al. 1987). This suggests that ozone could be a main factor in the deterioration of European woods. Very recently, Manning (1988) claimed that ozone is the most important air pollutant that adversely effects trees in North America.

CONCLUSION

The possibility of using laser induced fluorescence for remote sensing of plant stress, caused by relatively low concentrations of air pollution, has been demonstrated. The number of data, however, is still small and there is a need for plant physiological explanation of the observations. Continuity of research is necessary to produce larger data sets and allow for a better establishment and modeling of the relations in question.

ACNOWLEDGEMENT

We gratefully acnowledge Drs. A.W.M. Vermetten, Department of Air Hygienics and Air Pollution of the Agricultural University of Wageningen for making available the meteo and air pollution measurements of the research field at Garderen.

This project could be realised thanks to the financial support of the Netherlands Remote Sensing Board and the Italian Special Programme on Improvement of Agricultural Productivity.

REFERENCES

Castagnoli F., G. Cecchi, L. Pantani, I. Pippi, B. Radicati, P. Mazzinghi (1986) "A fluorescence lidar for land and sea remote sensing", IROE-CNR, Florence.

Cecchi G., L. Pantani, I. Pippi, R. Magli, P. Mazzinghi (1984) "Vegetation Remote Sensing: a new field for lidar applications", ECOOSA 84, Amsterdam, october 1984.

Chang S.H. and W. Collins (1983) "Confirmation of the airborne biogeophysical mineral exploration technique using laboratory methods". Economic Geology, vol. 78, 1983, page 723-736.

Chapelle E.W., F.M. Wood, W.W. Newcomb and J.E. McMurtrey (1985) "Laser induced fluorescence of green plants. 3: LIF spectral signatures of five major plant types", Applied Optics, vol. 24, no. 1, page 74-80, 1 january 1985.

Chapelle E.W., F.M. Wood, J.E. McNurtrey and W. W. Newcomb (1984a) "Laser induced fluorescence of green plants. 1: A technique for the remote detection of plant stress and spies differentiation", Applied Optics, vol. 23, no. 1, page 134-138, 1 january 1984.

Chapelle E.W., J.E. McMurtrey, F.M. Wood and W.W. Newcomb (1984b) "Laser induced fluorescence of green plants. 2: LIF caused by nutrient deficiences in corn", Applied Optics, vol. 23, no. 1, page 139-142, 1 january 1984.

Collins W., S.H. Chang, G. Raines, F. Canney and R. Ashley (1983) "Airborne biophysical mapping of hidden mineral deposits". Economic Geology, vol. 78, 1983 page 737-749.

Den Boer and Bastiaens (1984) "Acidification by Atmospheric Deposition: Vegetation", Publikatiereeks Milieubeheer no. 84, Ministry of Housing, Physical Planning and Environment, The Netherlands, januari 1984 (in Dutch).

Gehlhaar U., K.P. Gunther en J. Luther (1981) "Compact and highly sensitive fluorescence lidar for oceanographic measurements", Applied Optics, vol. 20, page 3318-3320, October 1981.

Hickman G.D., J.E. Hogg, E.J. Friedman en A.H. Ghovanlou (1972) "Application of a pulsed laser for measurements of bathymetry and algal fluorescence", Proc. 8th Int. Symp. Remote Sensing Environment, vol. 1, October 1972, Ann Arbor, Michigan, page 617-637.

Hoge F.E. and R.N. Swift (1983) "Airborne dual laser excitation and mapping of phytoplanton photopigments in a Gulf Stream Warm Core Ring", Applied Optics, 1 august 1983, vol. 22, no. 15., page 2271-2281.

Hoge F.E. en R.N. Swift (1980) "Application of the NASA airborne oceanographic lidar to the mapping of chlorophyll and other organic pigments", in Chesapeake Bay Plume Study, Superflux 1980, Janet W. Campbell (ed.), NASA Conference Publication 2188.

Hoge F.E. en R.N. Swift (1981) "Airborne simultaneous spectroscopic detection of laser-inducted water Raman backscatter and fluorescence from chlorophyll a and other naturally occuring pigments", Applied Optics, vol. 20 no. 18, page 3197-3205, September 1981.

Hoge F.E., R.N. Swift and J.K. Yungel (1983) "Feasibility of airborne detection of laser induced fluorescence emissions from green terrestrial plants", Applied Optics, 1 october 1983, vol. 22, no. 19, page 2991-3000.

Kim H.H., H. van der Piepen, V. Amann and R. Doerffer (1985) "An evaluation of 685 nm fluorescence imagery of coastal waters", ESA Journal 1985, vol. 9, number 1, page 17-27.

Larssen, R.I., A.S. Heagle and W.W. Heck (1983) "An air quality data analysis system for interrelating effects, standards and needed source reductions. Part 7: An O_3-SO_2 leaf injury mathematical model"., J. Air Pollut. Control Assoc. 3:198-207.

MacFarlane G., R.D. Watson, A.F. Theisen, R.D. Jackson, W.L.Ehler,J. Pinter, Jr., S.B. Idso en R.J. Reginato (1980) "Plant stress detection by remote measurement of fluorescence", Applied Optics, Vol. 19 no. 19, page 3287-3289, October 1980.

Manning W.J. (1988) "Effects of ozone and acidic deposition on forest trees in North America", in Abstracts Third US-Dutch International Symposium on "Atmospheric ozone research and its policy implications", Nijmegen, The Netherlands, 9-13 may 1988, Ministry of Housing, Physical Planning and Environment, The Netherlands.

Neville R.A. en J.F.R. Gower (1976) "Passive remote sensing of phytoplankton via chlorophyll-a fluorescence", Journal of Geophysical Research, vol. 82 no. 24,page 3487-3493) August 1977.

Rosema A. and W. Werner (1983) "An investigation of leaf fluorescence in relation to environmental pollution", Report to the Ministry of Education and Science, 20 july 1983, EARS bv, Delft, Netherlands (in Dutch).

Schreiber, U, W. Vidaver, V.C. Runeckles en P. Rosen (1977) "Chlorophyll fluorescence assay for ozone injury in intact plants", Plant Physiol. (1978) 61, pag. 80-84.

Slooff W., R.M. van Aalst, E. Heijna-Merkus and R. Thomas (1987) "Design Basic Document Ozone", RIVM, Bilthoven, Report no. 758474002, september 1987 (in Dutch).

Watson R.D. and W.R. Hemphill (1976) "Use of an airborne Fraunhofer line discriminator for the detection of solar stimulated luminescence". U.S. Geological Survey, open file report 76202.

Watson R.D., W.R. Hemphill, T.D. Hessin en R.C. Bigelow (1974) "Prediction of the Fraunhofer Line Detectivity of luminescent materials", Int. Symposium on Remote Sensing of Environment, 9th, Ann Arbor, Michigan Proc., page 1959-1980.

COUPLED FLUORESCENCE AND REFLECTANCE MEASUREMENTS TO IMPROVE CROP PRODUCTIVITY EVALUATION

F. BARET (1), G. GUYOT (1) and D. MAJOR (2)

(1) INRA, Bioclimatologie, BP 91, 84140 MONTFAVET (France)
(2) Agriculture Canada, Lethbridge, Alberta TIJ 4B1 (Canada)

ABSTRACT

The productivity of a plant canopy depends on the amont of photosynthetically active radiation (PAR) absorbed and on the efficiency of the transformation of absorbed PAR into dry matter.

The combination of reflectance measurements in red and near infrared spectral bands enables estimation of the absorbed PAR but photosynthetic efficiency can only be determined from chlorophyll fluorescence measurements. Since fluorescence intensity depends on chlorophyll concentration it is necessary to determine this first. High spectral resolution measurements for determination of the wavelength of the inflection point of the red edge can be used for determining leaf chlorophyll content.

Using this approach, characterizing the potential productivity of a plant canopy is best done by coupling the measurements of reflectance with high spectral resolution and chlorophyll fluorescence.

KEY-WORDS : Reflectance, Fluorescence, Crop Productivity, Red edge.

INTRODUCTION

In recent years the development of relationships for evaluating the biomass or the productivity of plant canopies has been paramount in many remote sensing programs. The feasibility of using multispectral reflectance data for biomass estimation was initially based on direct regressions of instantaneous spectral reflectance on total dry aerial biomass (Aase and Siddoway, 1981). But in this approach, the physical and physiological processes was not explicitely taken into account. More recently a new approach based on integrated radiometric data related to the potential photosynthetic activity of plant canopies has been proposed (Asrar et al., 1985 ; Tucker, 1985).

In this paper we will discuss the potentials of reflectance and fluorescence measurement for estimating the absorbed PAR and the photosynthetic efficiency of plan canopies.

I. EVALUATION OF PLANT CANOPIES PRODUCTIVITY

MONTEITH (1977) proposed a formula describing primary plant canopy productivity :

$$DM = \int_t \varepsilon_b \cdot \varepsilon_i \cdot \varepsilon_c \cdot S \cdot dt$$

H. K. Lichtenthaler (ed.) Applications of Chlorophyll Fluorescence, 319–324.

in which

DM is total dry matter production during the time interval t,

S is incoming solar radiation,

ε_i is the fraction of incident photosynthetically active radiation (PAR : 400 – 700 nm) absorbed by the canopy,

ε_b is the photochemical efficiency of the canopy, and

ε_c is the fraction of the total incoming radiation in the photosynthetically active radiation (PAR) region (400 – 700 nm). This factor is relatively constant and its value ranges from 0.43 to 0.48, (Gosse et al., 1986 ; Olioso, 1987).

The factor ε_i can be derived from remote sensing data. From an analysis based on radiative transfer models, Sellers (1985) showed that the combination of reflectance measurements performed in red (ρ_r) and near – infrared (ρ_{ir}) wide spectral bands (Landsat MSS 5 and 7 or TM 3 and 4 or SPOT 2 and 3) such as the normalized difference $ND = (\rho_{ir} - \rho_r)/(\rho_{ir} + \rho_r)$, could be used to estimate the proportion of photosynthetically active radiation absorbed by a plant canopy.

Therefore, The relationship between ε_i and the vegetation index (ND) is complex because ND and ε_i do not correspond to the same time scale. ND is generally derived from one measurement taken near solar noon while ε_i is integrated over the entire day. Thus the relationship between ND and the daily value of ε_i will depend on the position of the sun at the moment of ND measurement, and the daily evolution of irradiance and incident PAR.

Models describing the light interaction within a plant canopy provide a useful tool for analysing the relationship between ε_i and ND and its variability. Baret and Major (1988) used the SAIL model (Verhoef, 1984) to obtain the relationship :

$$\varepsilon_i = \varepsilon_{i\infty} \frac{(ND - ND_s)}{(ND_\infty - ND_s)}$$

in which ND_s is the value of ND for bare soil and depends on the soil type and on its roughness and moisture. It ranges from 0.4 for dark soils to 0.15 for light soils.
The upper limit of ND, ND_∞, occurs when the soil is completely covered by the vegetation, and $\varepsilon_{i\infty}$ is the upper limit of ε_i both of which occur when the reflectance reaches its saturation level. The parameters $\varepsilon_{i\infty}$ and ND_∞ are not very sensitive to leaf inclination or sun position and Baret and Major (1988) proposed values of 0.94 and 0.90 for these two parameters respectively.

The expression proposed by Baret and Major (1988) is in good agreement with the Sellers' (1985) results which were based on simulations using a two stream model.

$$\varepsilon_i = 1.124\ (ND - 0.080)$$

It is also in good agreement with the experimental results of Hatfield et al. (1984) and ASRAR et al. (1985) :

$$\varepsilon_i = 1.200\ (\mathrm{ND} - 0.153)$$

The introduction of the values of ND_∞ and $\varepsilon_{i\infty}$ in the expression proposed by Baret and Major (1988) leads to a slope for the linear relationship between ε_i and ND of 1.14 for the Sellers results and 1.26 for the Hatfield results.

The main problem of assessing ε_i from ND is its sensitivity to ND_s which plays an important role for low leaf area index, low solar zenith angle, erect leaves and dark soils.

If ε_i is available the only factor needed for estimating plant canopies productivity is ε_b. Monteith (1977) and later Varlet-Grancher et al. (1982) and Gosse et al. (1986) among others found that the photochemical efficiency coefficient ε_b, when calculated over the entire growth cycle, was relatively constant for crops like winter wheat (with a value of about 2.0 $g.Mj^{-1}$). But Varlet-Grancher et al. (1982) and Green (1987) demonstrated that the photochemical efficiency is not constant over the season because it depends on the chemical characteristics of the products of photosynthesis, as well as on climatic factors such as temperature or stresses (water stress, diseases, ...). Thus it would be promising to use the chlorophyll fluorescence for estimating ε_b.

II. FLUORESCENCE OF PLANT CANOPIES

Leaf fluorescence depends on pigment concentration, ultracellular structure of the tissues and photosynthetic activity (Lichtenthaler et al., 1986 ; Lichtenthaler and Buschmann, 1987a, 1987b).

For a given level of photosynthetic activity, fluorescence intensity decreases when the leaf chlorophyll content increases. The chlorophyll fluorescence spectra of green leaves show two maxima near 690 nm and 735 nm (Fig. 1). The ratio of the fluorescence at the two maxima F690/F735 ranges between 0.8 and 1.2 for fully green leaves (Lichtenthaler and Buschmann, 1987a). With decreasing chlorophyll content the total fluorescence intensity increases (leaves 1, 2, 3 in Fig. 1) and reaches a maximum for a chlorophyll concentration per leaf area unit of about 2 $\mu m.cm^{-2}$. When the chlorophyll content is lower than 2 $\mu g.cm^{-2}$ the fluorescence intensity decreases since the chlorophyll amounts are too low (leaf 4, in Fig. 1).

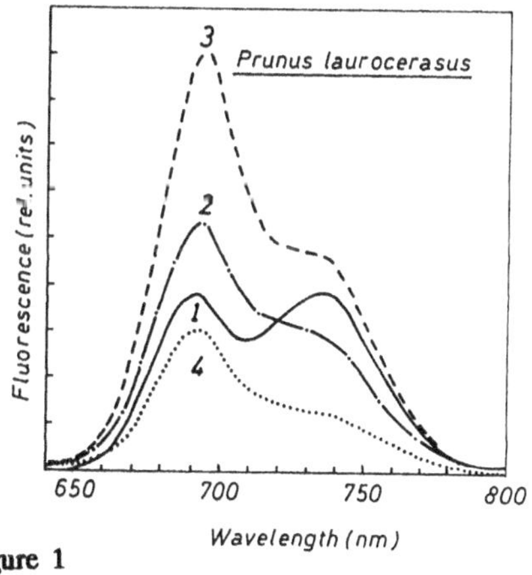

Figure 1
Chlorophyll fluorescence emission spectra of intact cherry-laurel leaves with decreasing chlorophyll content from leaf 1 to 4 (chlorophyll a + b, 1 : 52, 2 : 31, 3 : 2.9, 4 : 1.1 $\mu g.cm^{-2}$) (after Lichtenthaler and Buschmann, 1987)

The fluorescence is not as strongly re-absorbed by the leaf in the 735 nm region as it is in the 690 nm region so the shape of the fluorescence spectra changes when the leaf chlorophyll content decreases showing a maximum near 690 nm and a shoulder near 735 nm. A reduction in photosynthetic activity also increases the chlorophyll fluorescence (Lichtenthaler et al., 1986). This phenomenon has the same effect on fluorescence spectra as a decrease in chlorophyll content. Lichtenthaler and Buschmann (1987a) have shown that chlorophyll fluorescence is significantly correlated with leaf reflectance near the minimum in the red region (near 690 nm) and can simultaneously induce a shift of the inflection point of the red edge towards the shorter wavelengths. The experimental results of Horler et al. (1983) and the model simulations of Baret et al. (1988) on the effect of leaf chlorophyll content on the wavelength of the inflection point of the red edge show a similar effect (Fig. 2).

This brief review highlights the major problem which occurs when one wants to assess the photosynthetic efficiency ε_b from fluorescence ratio measurements. It is necessary to separate the effects of the photosynthetic activity from the effect of the leaf chlorophyll content on the optical properties of plant canopies. The problem is again complicated when we will extract ε_b at canopy level because of crop geometry effects.

III. ACTUAL POSSIBILITIES FOR DETERMINING CHLOROPHYLL CONCENTRATION AND PHOTOSYNTHETIC ACTIVITY

III.1. Determination of chlorophyll content

The use of high spectral resolution has potentials to infer leaf chlorophyll content of a plant canopy. Experimental results (Horler et al., 1983 ; Baret et al., 1987) and model simulations (Guyot et al., 1988) show that the inflexion point of the red edge shifts progressively towards the shorter wavelengths when the chlorophyll concentration decreases (Fig. 3). Fig. 3 also shows that the LAI has a large effect on the position λ_i of the inflexion point.

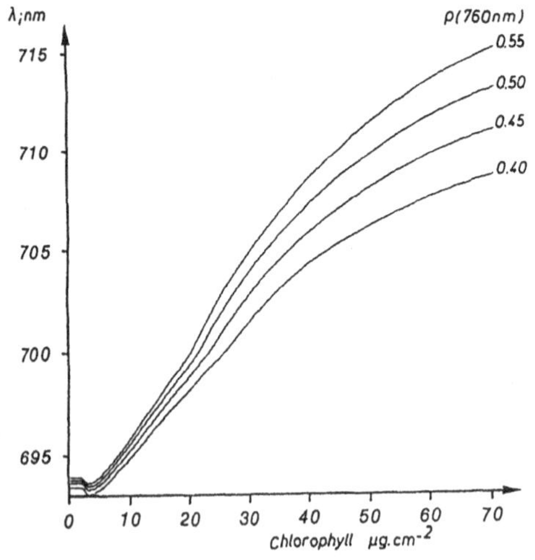

Figure 2
Variation of the position of the inflection point λ_i of the red edge as a function of chlorophyll content of three leaves with different near infrared reflectances (after GUYOT et al., 1988)

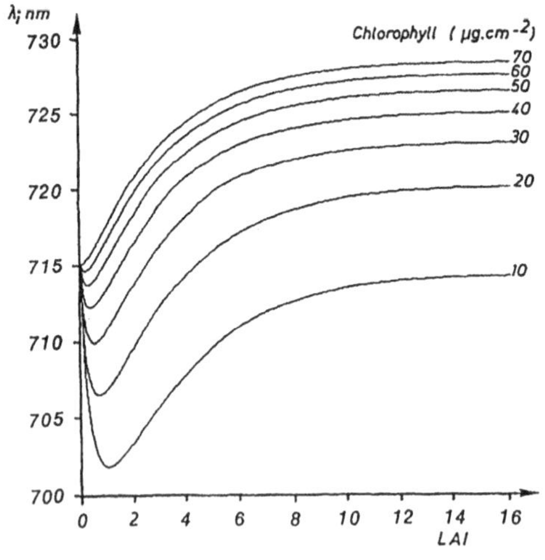

Figure 3
Effect of leaf chlorophyll content and LAI on the position of the inflection point λ_i of the red edge (after GUYOT et al., 1988)

For small LAI, there is a displacement of λ_i towards the short wavelengths. This displacement corresponds to the change in the curvature of the spectrum between the red and the near infrared. For a bare soil, the curvature of the spectrum is very small and λ_i is very close to 715 nm which corresponds to the average of the limits : 670 and 760 nm (715 nm).

III.2. Determination of chlorophyll fluorescence

We have seen that chlorophyll fluorescence can have the same spectral effect as a variation in chlorophyll concentration. For this reason we might expect to use the position of the inflexion point of the red edge to determine the variation of the photosynthetic activity within a homogeneous plant canopy such as a forest (Schmuch et al., 1987).

But this does not correspond to the general conditions of plant canopies where leaf chlorophyll content and photosynthetic activity vary independently. Thus it will be necessary to measure chlorophyll fluorescence and to associate this measurement with a reflectance measurement, in order to distinguish the effects of chlorophyll concentration fluorescence effects which reveals the state of the photosynthetic apparatus.

CONCLUSION

This short presentation showns that it is necessary to use simultaneously reflectance and fluorescence measurements for characterizing the productivity of plant canopies. Reflectance measurements are needed to estimate the fraction of the absorbed photosynthetic active radiation by a plant canopy. Combining these measurements with fluorescence techniques will allow direct estimation of photochemical efficiency and subsequently the potential production of a plant canopy.

Since fluorescence intensity depends not only on the photosynthetic activity but also on leaf chlorophyll content and canopy geometry, it is necessary to separate these differing effects. For this reason, a promising new technique, the measurement of the life time of chlorophyll fluorescence should be useful because this factor is not affected by the chlorophyll amount and is directly related to the quantum yield of photosynthesis.

REFERENCES

Asrar G, Kanemasu ET, Jackson RD, Pinter PJ, 1985. Estimation of total above ground phytomass production using remotely sensed data. Remote Sens. Environ. 17 : 211 – 220.

Baret F, Champion I, Guyot G, Podaire A, 1987. Monitoring wheat canopies with a high spectral resolution radiometer. Remote Sens. Environ. 22 : 367 – 378.

Baret F, Major D, 1988. Significance of the normalized difference. In Remote Sensing Workshop on "Vegetation indices and their interpretation". Nottingham (UK). 25 May 1988.

Baret F., Andrieu B., Guyot G., 1988. Significance of the normalized difference. In Remote Sensing Workshop on "Vegetation indices and their interpretation". Nottingham (UK). 25 May 1988.

Gosse G, Varlet – Grancher C., Bonhomme R, Chartier M., Allirand JM, Lemaire G, 1986. Production maximale de matière sèche et rayonnement solaire intercepté par un couvert. Agronomie 6(1) : 47 – 56.

Green CF, 1987. Nitrogen nutrition and wheat growth in relation to absorbed solar radiation. Agric. Forest. Meteor. 41 : 207 – 248.

Guyot G, Baret F, Major D, 1988. La haute résolution spectrale : détermination des déformations spectrales entre le rouge et le proche infrarouge. In Proc. 6ème Congrès de l'Association Québecoise de Télédétection. Sherbrooke (Canada). 3 – 6 May 1988.

Hatfield JL, Asrar G, Kanemasu ET, 1984. Intercepted photosynthetically active radiation in wheat canopies estimated by spectral reflectance. Remote Sens. Environ. 14 : 65 – 75.

Horler DNH, Dockray M, Barber J, 1983. The red edge of plant leaf reflectance. Int. J. Rem. Sens. 4 : 273 – 288.

Lichtenthaler HK, Buschmann C, Rinderle U, Schmuck G, 1986. Application of chlorophyll fluorescence in ecophysiology. Radiat. Environ. Biophy. 25 : 297 – 308.

Lichtenthaler HK, Buschmann C, 1987a. Reflectance an chlorophyll fluorescence signatures of leaves. In Proc. IGARSS'87 Symp. Ann. Arbor. MI (USA). 18 – 21 May 1987. 1201 – 1206.

Monteith JL, 1977. Climate and the efficiency of crop production in Britain. Phil. Trans. Roy. Soc. London. B 281 : 277 – 294.

Sellers PJ, 1985. Canopy reflectance, photosynthesis and transpiration. Int. J. Remote Sens. 6(8) : 1335 – 1372.

Schmuck G, Lichtenthaler HK, Kritikos G, Amann U, Rock B, 1987. Comparison of terrestrial and airborne reflection measurements of forest trees. In Proc. IGARSS'87 Symp. Ann. Arbor. MI (USA). 18 – 21 May 1987 : 1207 – 1212.

Varlet – Grancher C, Bonhomme R, Chartier M, Artis P, 1982. Efficience de la conversion de l'énergie solaire par un couvert végétal. Acta Oecologica Plant 3(17) : 3 – 26.

Verhoef W, 1984. Light scattering by leaf layers with application to canopy reflectance modeling : the SAIL model. Remote Sens. Environ. 16 : 125 – 141.

REFLECTANCE AND CHLOROPHYLL FLUORESCENCE SIGNATURES OF LEAVES

Claus BUSCHMANN and Hartmut K. LICHTENTHALER

Botanisches Institut II (Plant Physiology and Plant Biochemistry), University of Karlsruhe, Kaiserstr. 12, D-7500 Karlsruhe, F.R.G.

KEY WORDS: Chlorophyll fluorescence, forest decline, photosynthesis, reflection spectra, VIRAF-spectrometer.

ABSTRACT

The reflection signature of vegetation is the basis for the detection of tree damage by remote sensing. Damaged trees are characterized by (a) an increase of the signal betwenn 500 and 680 nm, (b) a decrease of the signal above 750 nm and (c) a "blue shift" of the inflection point of the rise towards 750 nm. The comparison of reflection and fluorescence spectra demonstrates that fluorescence emission might influence the shape of the reflection spectra. Especially in the region of the rise of the reflection signal towards 750 nm ("red edge") the detection of fluorescence cannot be excluded in reflection measurements. The measurement of fluorescence signatures parallel to reflection signatures is proposed for remote sensing. A spectrometer (VIRAF-spectrometer), which has recently been developed to detect absorption, fluorescence and reflection spectra as well as fluorescence induction kinetics with one leaf sample, is presented as a help for the interpretation of remote sensing signatures more based on physiological data.

INTRODUCTION

Remote sensing of plants today is mainly based on the measurement of sunlight reflected by plants and in particular by leaves (Huss 1984, Rock et al. 1986, Schmuck et al. 1987). The reflection spectrum of leaves is determined by different factors depending on the spectral range. In the visible part of the spectrum (400 - 700 nm) the pigments, e.g. chlorophylls and carotenoids, in the near infra-red (750 - 1300 nm) the cell structure and later (at about 1400 and 1900 nm) the water content of the tissue influence the reflection characteristics. Most detectors used for remote sensing analyze the spectrum of the reflected light by measuring it through filters in few different fixed spectral ranges (channels). Only recently have monochromators been used; these give a higher spectral resolution but need a much faster transfer and processing of data.

The reflection signature of trees affected by forest decline were characterized by a higher reflection in the green band at 550 nm and a lower reflection in the near infra-red (Fig. 1). The sharp increase of the reflection towards the near infra-red ("red edge") has a shift in its inflection point towards shorter wavelengths ("blue shift"), which becomes visible in

H. K. Lichtenthaler (ed.) Applications of Chlorophyll Fluorescence, 325–332.

the second derivative as a shift of the point of intersection with the zero line (Fig. 1). In the past two years the decrease of the reflection in the near infra-red and the "blue shift" has not been so pronounced in damaged leaves or needles, even though ground-truth measurements of these leaves still showed lower physiological activity (Lichtenthaler et al. 1988). Thus the need was felt to look for further characteristics of leaf damage and to put the interpretation of reflection signatures on a more physiological basis. We therefore developed a measuring system (Buschmann et al. 1988), which allows measurement of absorption and fluorescence spectra as well as fluorescence induction kinetics parallel to reflection spectra (VIRAF-spectrometer: **V**isible, **I**nfra-red, **R**eflection, **A**bsorption, **F**luorescence).

Reflection spectra are measured in conventional spectrometers by illuminating the sample with a low intensity monochromatic light. In the VIRAF-spectrometer, however, the leaf is illuminated with strong white light in analogy to the sunlight used in remote sensing. Under these conditions the emission of chlorophyll fluorescence in the spectral range between 640 and 800 nm must be taken into acount for the interpretation of reflection signatures. The newly developed VIRAF-spectrometer is able to measure the spectral parameters with the leaf left fully intact, attached to the plant and without removing it during the measuring procedure.

MATERIALS AND METHODS

Plants. Spruce needles (Picea abies Karst.) were taken from healthy trees (Althof, damage class 0/1: 0-15% needle loss) and damaged trees (Mauzenberg, damage class 3/4: 60-90% needle loss) growing in the Northern Black Forest at an altitude of 450 to 700 m above sea level. Leaves from cherry-laurel plants (Prunus laurocerasus L.) growing in the Karlsruhe area were chosen according to their different colour, ranging from fully green to almost fully yellow. An etiolated bean leaf (Phaseolus vulgaris L.) was used from a plant grown for 4 weeks in total darkness on peat. Leaves of tobacco (Nicotiana tabacum L.) were taken from fully green plants grown for about 12 weeks in the greenhouse of the Botanical Garden of the University of Karlsruhe.

Instrumentation. The reflectance spectra of spruce needles were measured using the VIRIS (Visible Infrar-red Intelligent Spectrometer) of the JET Propulsion Laboratory (JPL, Pasadena, USA). Reflectance spectra of cherry-laurel leaves were taken using a spectrometer (Shimadzu UV 200) with its integrating sphere (Ulbrichtkugel). Fluorescence emission spectra of cherry-laurel leaves were determined at room temperature using the fluorescence attachment of a spectrometer (Shimadzu MPS 5000). A broad-band blue light excited the leaves, which were preilluminated for 5 min in order to measure under steady state conditions. Multiple spectral data acqisition of bean and tobacco leaves was carried out with the VIRAF-spectrometer (VIRAF = Visible, Infra-red, Reflection, Absorption, Fluorescence). Further details of this newly developed equipment are found under results (see also Buschmann et al. 1988).

RESULTS AND DISCUSSION

Reflection characteristics of needles affected by the forest decline

Needles affected by forest decline are characterized by an increase of the reflection in the visible region between 500 and 680 nm (Fig. 1). This can be explained by the lower chlorophyll content found in the damaged needles compared to the healthy needles, and agrees with the increase of the reflection in this region observed in cherry-laurel leaves parallel to the decreasing chlorophyll content (Fig. 2). A further characteristic of damaged needles has been the decrease of the reflection in the near infra-red region above 750 nm, which is an indicator of cellular damage in the needle tissue.

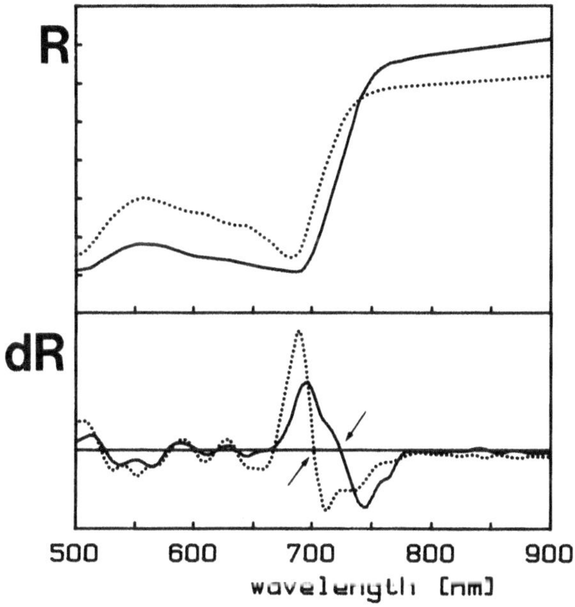

Figure 1. Reflection spectra (R) and their second derivative (dR) of three-year-old spruce needles (Picea abies Karst.) measured in August 1985 using the VIRIS-system. Green needles (———) of a healthy tree (damage class 0-1: 0-15% needle loss) are compared to light-green needles (.....) of a damaged tree (damage class 3-4: 60-90% needle loss). Modified after Schmuck et al. (1987).

The rise of the spectrum towards 750 nm proceeds in damaged needles at lower wavelengths than in healthy needles. This is seen in the second derivative of the reflection spectrum as a "blue shift" of the inflection point of this rise, which becomes clearly visible as a shift of the point of intersection with the zero-line towards shorter wavelengths (Fig. 1). This "blue shift" may be induced only by the lower chlorophyll content of the needles but can also be caused by a higher fluorescence intensity observed in leaves with a decreased chlorophyll content (see Fig. 3). Both the decrease of the reflection signal in the near infra-red region and the "blue shift" have not been so pronounced in the past two years, although the physiological activity of the the needles from damaged trees was still lower than that of needles from healthy trees (Lichtenthaler et al. 1988). Therefore these characteristics alone are no longer sufficient as indicators of forest damage.

Reflection and fluorescence spectra of leaves with different chlorophyll content

The reflection in the visible region between 500 and 680 nm increases with decreasing chlorophyll content of a leaf. This can be shown by comparing the reflection spectra of cherry-laurel leaves with different colour ranging from fully green to almost fully yellow (Fig. 2). The more chlorophyll a leaf contains, the more light is absorbed in this region and the less light can be reflected. The decrease of the chlorophyll content also causes a shift of the inflection point of the spectral rise towards 750 nm ("red edge") and a shift of the maximum of the second derivative. Both characteristics found in the second derivative are shifted towards shorter wavelengths ("blue shift") as in needles of damaged forest trees (confer Fig.1).

The fluorescence spectra overlap with the reflection spectra in their short wavelength region around 690 nm (Fig. 3). Fluorescence spectra of leaves with a decreased chlorophyll content show a higher signal than spectra of leaves which are fully green. In the fully green leaf the fluorescence, especially in the 690 nm region, is partially reabsorbed, whereas in the yellow-green leaf the 690 nm fluorescence becomes predominant (Fig. 3). Since during reflection measurements the determination of fluorescence cannot be excluded (see also below), fluorescence emission could influence the rise of the reflection spectrum towards 750 nm and could in part contribute to the "blue shift".

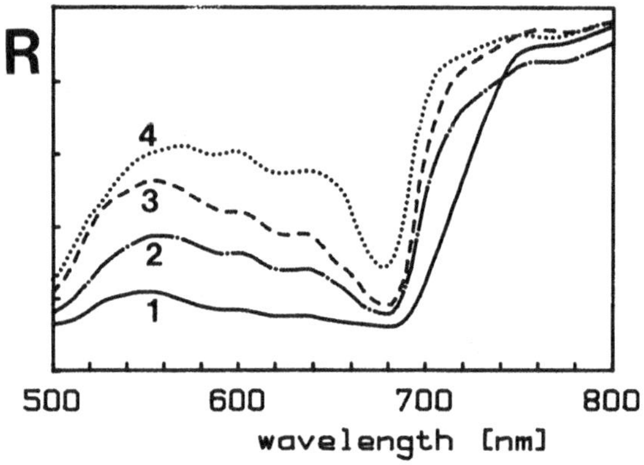

Figure 2. Reflection spectra of cherry-laurel leaves (Prunus laurocerasus L.) with decreasing chlorophyll content from leaf 1 to 4. The concentration of chlorophyll per leaf area is given in Table 1. Modified after Lichtenthaler and Buschmann (1987).

Table 1. Wavelength position of the inflection point of the "red edge" and of the red maximum in the second derivative of cherry-laurel leaves (Prunus lauracerasus L.) with decreasing chlorophyll content from leaf 1 to 4. The original reflection spectra are given in Figure 2. Modified after Lichtenthaler and Buschmann (1987).

sample	leaf colour	chlorophyll (μg a+b cm^{-2})	inflection point of the "red edge"	red maximum of the 2nd derivative
leaf 1	green	52	723 nm	692 nm
leaf 2	light-green	31	696 nm	688 nm
leaf 3	yellow-green	2.9	692 nm	685 nm
leaf 4	yellow	1.1	692 nm	684 nm

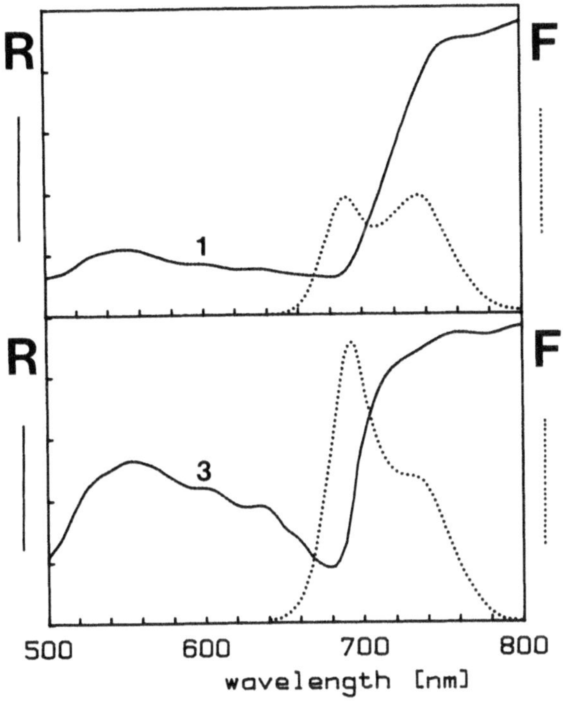

Figure 3. Comparison of the reflection (——— R) and the fluorescence spectra (..... F) of cherry-laurel leaves (Prunus lauracerasus L.) with a high (leaf 1) and a low chlorophyll content (leaf 3). The chlorophyll concentration is given in Table 1. Modified after Lichtenthaler and Buschmann (1987).

Multiple spectral data acquisition with the VIRAF-spectrometer

A scheme of the newly developed VIRAF-spectrometer is given in Fig. 4. This spectrometer allows the parallel detection of absorption, reflection and fluorescence spectra as well as fluorescence induction kinetics. The leaf sample remains intact, attached to the plant and is not removed during the measurement, which makes it possible to repeat measurements with the same sample after longer intervals or to carry out long-term measurements with one leaf only (e.g. Fig. 5). A personal computer controls the function of the monochromator (Jobin Yvon H 25, Instruments SA, D-8025 Unterhaching), stores the digitized data and carries out calculations, e.g. division with reference signals, difference spectra and second derivatives (software: Instruments SA).

The leaf sample is fixed at the outside of the measuring compartment. By means of a 50% transmittant mirror light from light source I illuminates the sample perpendicularly for reflection and fluorescence measurements. Reflected light and fluorescence passes the 50% transmittant mirror and is detected by the photomultiplier after having passed the monochromator. Light source II illuminates the sample from behind and is used for the determination of absorption (transmittance) spectra. For the excitation of fluorescence a broad-band blue filter is moved in front of light source I. Induction kinetics of fluorescence are carried out after exactly 5 min of dark adaptation after the photosynthesis of the leaf has reached its steady state during the measurement of absorption and reflection. An example of a multispectral analysis carried out with the VIRAF-spectrometer is given in Figs. 6 and 7 for a green tobacco leaf. The entire measuring procedure took about

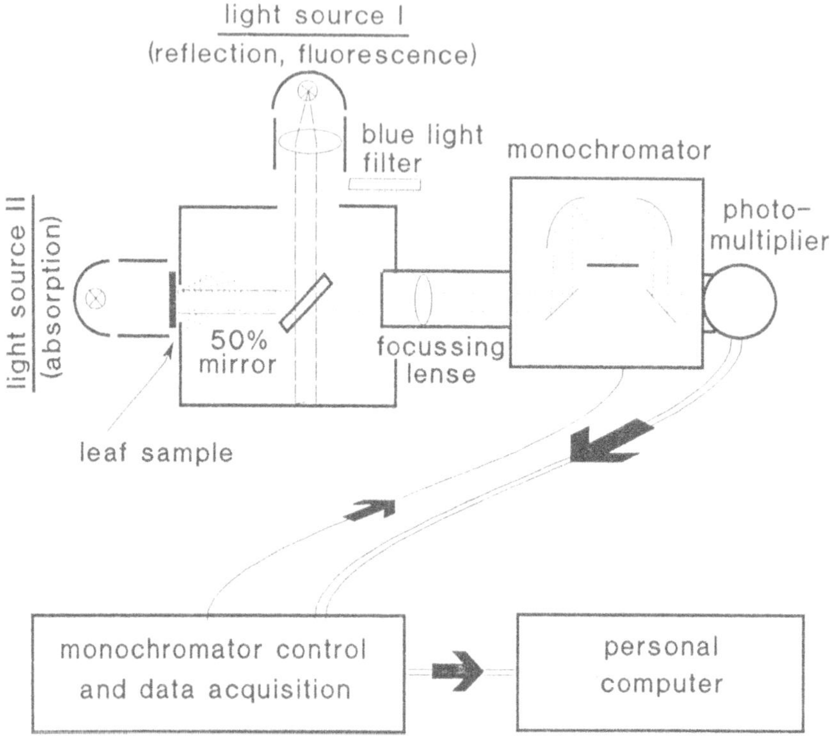

Figure 4. Scheme of the VIRAF-spectrometer explained in detail in the text under results (see also Buschmann et al. 1988).

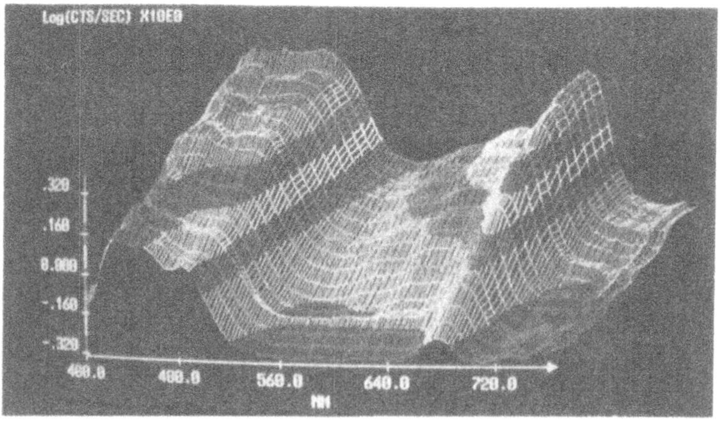

Figure 5. (see colour plate No. 11 at the end of this book) Absorption spectrum of an etiolated bean leaf (<u>Phaseolus vulgaris</u> L.) taken with the VIRAF-spectrometer during 17 h of greening in the light (1 spectrum every 55 min). The leaf remained always attached to the plant. Light source II (Fig. 4) served for inducing the greening process and for measuring the spectrum.

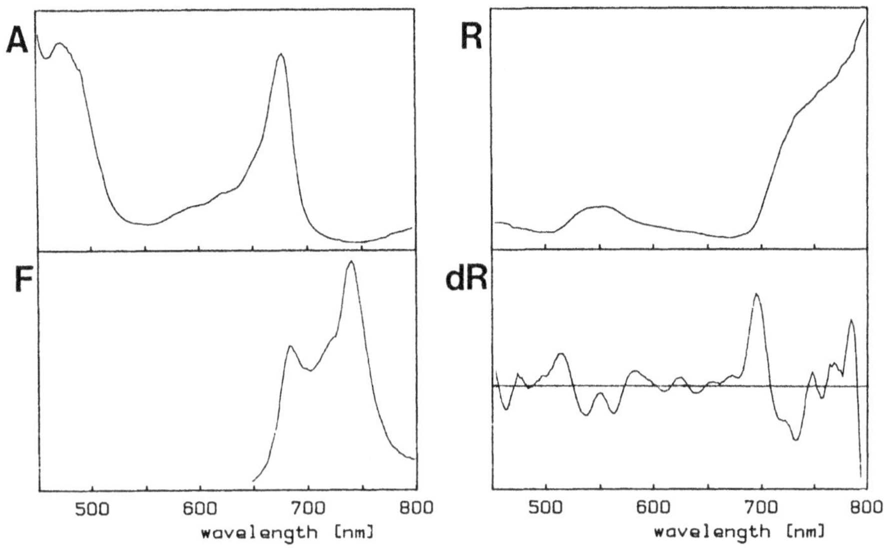

Figure 6. Spectra taken from one fully green tobacco leaf (Nicotiana tabacum L.) with the VIRAF-spectrometer. A: absorption spectrum, F: fluorescence emission spectrum, R: reflection spectrum, dR: second derivative of the reflection spectrum R.

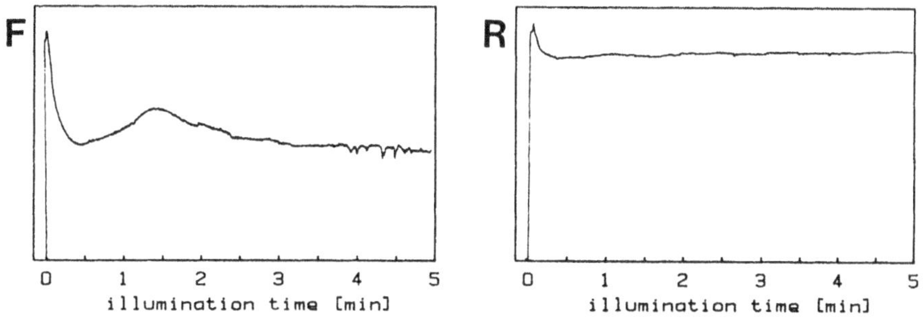

Figure 7. Induction kinetic of fluorescence (F) and of reflection (R) measured at 690 nm with a fully green tobacco leaf (Nicotiana tabacum L.) dark-adapted for 5 min using the VIRAF-spectrometer. For the reflection measurement the sample was illuminated with white light, whereas blue light was used for the excitation of the fluorescence.

45 min. The overlap of the long-wavelength part of the absorption spectrum with the short-wavelength part explains the low short-wavelength maximum of the fluorescence spectrum. The inflection point at the "red edge" of the reflection spectrum is situated at 710 nm as is seen from the second derivative of the reflection spectrum (Fig. 6). The fluorescence signal at 690 nm

shows a clear Kautsky induction kinetic (Fig. 7, left part). The reflection of a leaf (predarkened as for the measurement of the fluorescence induction kinetic) measured at 690 nm during illumination with white light also exhibits a small induction kinetic (Fig. 7, right part). This clearly shows that the fluorescence must contribute to the reflection signal.

CONCLUSION

Fluorescence emission might influence the shape of the reflection spectra especially in the rise towards 750 nm ("red edge"), since fluorescence can not be excluded in reflection measurements. Reflection signatures widely applied in remote sensing of forest damage should be complemented by parallel measurements of the fluorescence. The newly developed VIRAF-spectrometer makes it possible to detect absorption, fluorescence and reflection spectra as well as fluorescence induction kinetic with one leaf sample. It can help to establish a better interpretation of remote sensing signatures on a more physiological basis.

Acknowledgement: Financial support of the Projekt Europäisches Forschungszentrum für Maßnahmen zur Luftreinhaltung (PEF, KfK Karlsruhe) given for the construction of the VIRAF-spectrometer is gratefully acknowledged. We thank Mr. Stefan Rang and Mr. Fred Stober for technical assistance.

REFERENCES

BUSCHMANN C, RANG S, STOBER F and KOCSÁNYI L, 1988. Reflection spectra of leaves and needles as a basis for the physiological evaluation of tree damage. - I. Instrumentation. In: PEF-Report (Projekt Europäisches Forschungszentrum), KfK Karlsruhe, KfK-PEF 35, pp. 191-196.

HUSS J, 1984. Luftbildmessung und Fernerkundung in der Forstwirtschaft. H Wichmann, Karlsruhe.

LICHTENTHALER HK and BUSCHMANN C, 1987. Reflection and chlorophyll fluorescence signatures of leaves. In: Proceedings of the IGARSS '87, Vol II, pp 1201-1206, University of Michigan, Ann Arbor.

LICHTENTHALER HK, BUSCHMANN C, RINDERLE U and SCHMUCK G, 1986. Application of chlorophyll fluorescence in ecophysiology. Radiation and Environmental Biophysics 25: 297-308.

LICHTENTHALER HK, RINDERLE U, KRITIKOS G and ROCK BN, 1988. Classification of damaged spruce stands in the Northern Black Forest by airborne reflectance and terrestrial chlorophyll fluorescence measurements. In: Proceedings of the 2nd DFVLR Statusseminar, pp 238-252, DFVLR, Oberpfaffenhofen.

ROCK BN, VOGELMAN JE, WILLIAMS DL, VOGELMAN AF and HOSHIZAKI T, 1986. Remote detection of forest damage. Bioscience 36: 439-445.

SCHMUCK G, LICHTENTHALER HK, KRITIKOS G, AMANN V and ROCK BN, 1987. Comparison of terrestrial and airborne reflection measuremnets of forest trees. In: Proceedings of the IGARSS '87, vol II, pp 1207-1212, University of Michigan, Ann Arbor.

A CONCEPT FOR STRESS AND ITS APPLICATION IN REMOTE SENSING

Reto J. Strasser

Bioenergetics Laboratory, University of Geneva
Station de Botanique Expérimentale, CH 1254 Lullier, Geneva

INTRODUCTION:

There is an increasing interest to analyse big areas of biotopes in order to get information about the health of its vegetation. The goal is to detect early effects of stress. The optical techniques today are so highly adapted that remote sensing for these purposes will be possible in the near future. Reflectance and fluorescence signals can be detected in a passive manner, that means with the sun as only light source. Multichannel techniques have still to be developed so that by means of natural fluctuations of light intensity (due e.g. to clouds) information about the variable fluorescence can be obtained. With the amplitude of reflectance and fluorescence as a function of wavelength, incident light intensity and time there are enough data to classify the biological state and the biological behaviour of a big vegetation area relatively to a stress concept. Technically the data acquisition from satellites will soon be possible. A theory of behaviour of plants in a given environment is still awaited. An approach to such a theory has been given elsewhere (Strasser, Graf 1988).

THE JKB-TRILOGY:

As a first approach every very complex biological system is reduced to the three symbols J-K-B, the so called JKB-trilogy. J stands for all energetic inputs from the environment into the biological system called BIOS. K stands for all molecules and rate constants of the biological system and so represents the architecture of the organism. B stands for the dynamic "BEHAVIOUR" of the system. It is a measure of the overall flow level in the open system.

The JKB-trilogy therefore represents

	the environmental INPUT TERM	**J**
	the hardware EXTENSIVE TERM	**K**
	the behaviour INTENSIVE TERM	**B**

THE STATE-CHANGE CONCEPT:

The biological activity is a function of B and K which both depend on J. However J is a direct function of the environment I. Already with this very simple JKB-trilogy it is possible to formulate a state-change of a biological system, what represents an adaptation process upon a stress (Fig.1 and 2). The idea is that for any given conformation K it is possible to adjust the relative flow level B between zero and one by changing the input parameters J which is a function of the environmental parameters I. Therefore the efficiency of the system (η (i) = output flux i/input fluxes) is a function of B with one or more values of B where the efficiency is optimal. As long as K is constant, the system can only work according to its state function.

H. K. Lichtenthaler (ed.) Applications of Chlorophyll Fluorescence, 333–337.

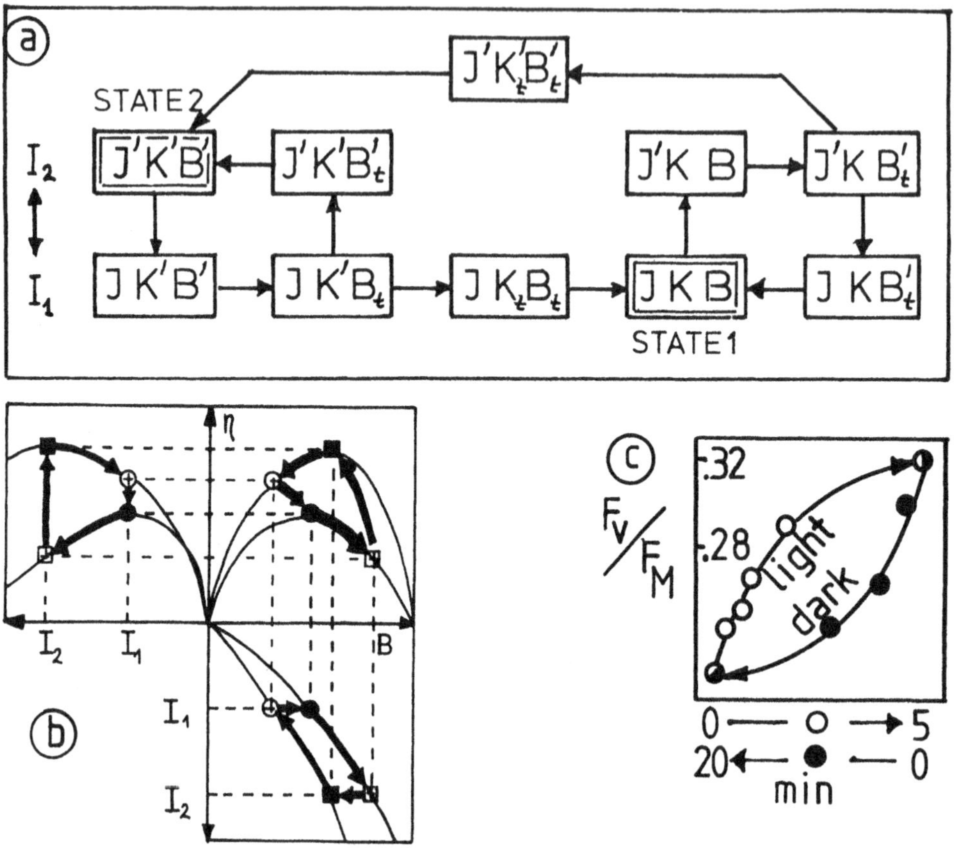

Fig.1: a) The JKB-trilogy describing a state-change from a dark adapted leaf (state 1) to red light adaptation (state 2). b) The theoretical function η vs B vs Inputparameter I. c) F(v)/F(M) at 690 nm and 77 K during state 1 to state 2 to state 1 transition.

SUBOPTIMALITY AS DRIVING FORCE FOR ADAPTATION:

The concept of the state-change theory is that every SUBOPTIMALITY creates a force which alters the statistical distribution of the microstates what appears as a change in the conformation of the system. Every change in K to K' is called a state-change. To the new state function K' there belongs a new η vs B with a new optimum.

- moving within one state function corresponds to intensity changes of the biological activity (B-change), leading to suboptimality.
- moving from one state function to another corresponds to conformational changes (K-change), leading to optimality.

Changing the environmental parameter from I(1) to I(2) and back to I(1) (e.g. clouds - full sunshine - clouds) forces the system into suboptimality and therefore into a state-change cycle (Fig.1b and 1c).
That means the way of adaptation to new conditions is different to the way of readaptation to the original conditions. The theoretical and experimental state-change function is shown in Fig.1b and 1c.

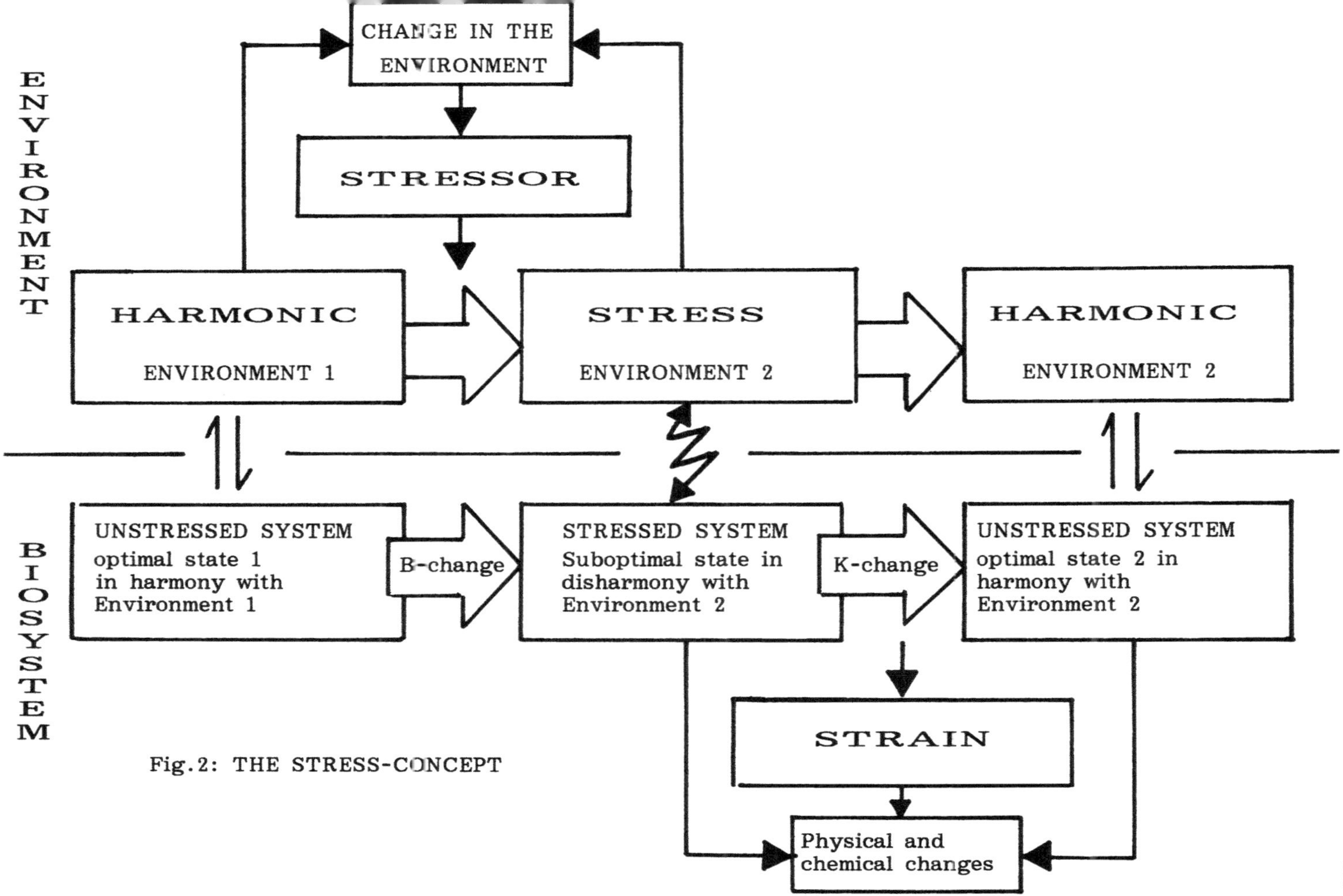

Fig.2: THE STRESS-CONCEPT

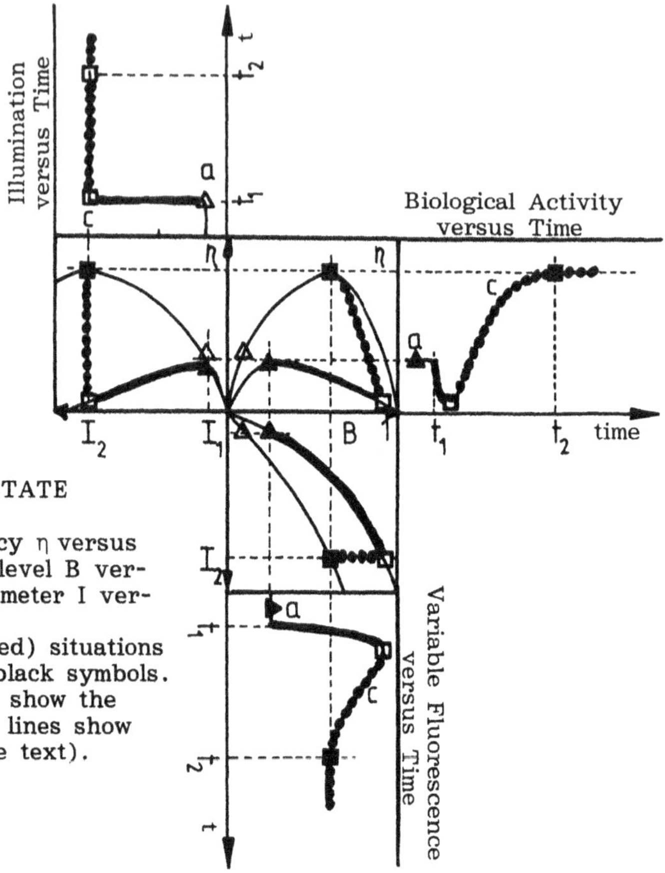

Fig.3: THE INTERNAL STATE FUNCTION:
Biological efficiency η versus the internal flow level B versus the inputparameter I versus time.
Optimal (unstressed) situations are marked with black symbols.
Heavy black lines show the B-change, dotted lines show the K-change (see text).

THE STRESS CONCEPT:

Every state-change is a consequence of stress. The following definitions are used for the stress concept:

STRESSOR is called every factor which provokes a STRESS.

STRESS is called every condition established which forces a system away from its thermodynamic OPTIMAL state.

OPTIMAL state is called when a biological system is in full HARMONY with its environment.

HARMONY of a biological system with its environment is achieved when the system does not tend to change any activity or conformation what so ever ---> ECOSYSTEM ---> STABILITY ---> DISSIPATIVE STRUCTURE ---> MINIMUM ENTROPY PRODUCTION.

STRAIN is any physical or chemical change produced by a STRESS.

The stress concept deals always with the interaction of environment and the biological system in this environment as shown in Fig.2 and 3.
The stress concept given here is based on nonequilibrium thermodynamics and dissipative structures. It offers the possibility of analytical description and quantification. Nevertheless my concept is in full agreement with the

definition of stress given by W.Larcher of the University of Innsbruck as follows:

"Every organism experiences stress, although the way in which it is expressed differs according its level of organisation. From the botanist's point of view, stress can be described as a state in which increasing demands made upon a plant lead to an initial destabilization of functions, followed by normalization and improved resistance. If the limits of tolerance are exceeded and the adaptive capacity is overtaxed, permanent damage or even death may result. Stress thus contains both destructive and constructive elements: it is a selection factor as well as a driving force for improved resistance and adaptive evolution."

Correlation between stress concept Larcher and Strasser:

Level of organization	--->	the organized hardware and conformation term K
Destabilization	--->	Stability means no change which corresponds to thermodynamic optimal state or harmony with the given environment. Destability means thermodynamic suboptimality manifested so that the actual value of B is not the optimal value of B. Destabilization = B-change from optimal to suboptimal state.
Normalization	--->	Conformational adaptation. Suboptimal B creates a driving force which changes the suboptimal K until B and K are again optimal (or normal). Normalization = K-change from suboptimal to optimal
Driving force	--->	Suboptimality. Difference of the actual state and the optimal state the system is searching for at the given environmental conditions.

The environmental conditions are always changing what forces the system to readapt. It is a perpetual search to attain the optimal state. Only at this point the system could dream with the words of Goethe: "Oh Augenblick, verweile doch, du bist so schön...".

The essence of the optimalization theory is that it is always the changed environment which determines the future optimal state of the biological system. If the new environment requires irrealistic values of B (e.g. $B > 1$) then the state-change driving force will push the system in direction of the attraction point of $B > 1$. The adaptive capicity however is overtaxed and permanent damage or even death may result (destructive element of stress according to Larcher). If the attraction point is within realistic limits adaptation and improved resistance will result (constructive element of stress according to Larcher).

REFERENCES

Larcher, W.: Naturwissenschaften 74, 158-167 (1987)

Strasser, R.: ESA-SP-254, 1581-1585 (1986)

Strasser, R.: Ber.Deutsch.Bot.Ges. 98, 52-72 (1985)

Strasser, R.: ESA-TT-988, 17-25 (1986)

Strasser, R.; Graf, J.: Int.Symp.Air Pollut.Plant Metab., Blauk et al. (Eds.), De Gruyter, Berlin (1988)

POSSIBILITIES FOR REMOTE SENSING OF P-700

J. HARBINSON AND C.L. HEDLEY

John Innes Institute and AFRC Institute of Plant Science Research
Colney Lane,
Norwich NR4 7UH.
United Kingdom.

P-700, PSI

ABSTRACT
The usefulness of in vivo absorbance changes caused by P-700 oxidation are discussed and the principles of the measurement of these changes described. The possible application of this technique to remote sensing of photosynthetic performance is described.

INTRODUCTION

Chlorophyll fluorescence is used almost exclusively as a probe of photosystem II (PSII) processes (Krause and Weis, 1984); except at very low temperatures its application to photosystem I (PSI) has been very limited (Bradbury and Baker, 1981; Kyle, Baker and Arntzen, 1983). Optical absorbance and electron spin resonance changes that occur as P-700 (the specialized chlorophyll molecule(s) that form the reaction centre of PSI) is oxidized are the usual techniques by which PSI has been probed in vitro and, to a much lesser extent, in vivo (e.g. Melis, 1982; Ruhle and Wild, 1979a; Tikhonov, Khomutov and Ruuge, 1984).

It is clear from the results obtained in vitro and in vivo that P-700 is a potentially valuable native probe of thylakoid and photosynthetic function. During irradiation of leaves or chloroplasts with photosynthetically active radiation (PAR) a proportion of the available pool of P-700 becomes oxidized and following the cessation of that excitation the 'holes' produced in PSI are filled by electrons derived from suitable electron donors. The native donor pool is principally plastoquinol (PQH_2) from which electrons reach $P\text{-}700^+$ via the cyt b_6/f complex and plastocyanin with a half time in the order of 10^{-2} s (Haehnel, 1984). Plastoquinol is in turn generated by the reduction of plastoquinone by electrons derived from water via PSII. $P\text{-}700^+$ is as good quencher of the excitation energy present within the PSI pigment bed as P-700 (Nuijs, Shuvalov, van Gorkom, Plijter and Duysens, 1986) (and no inter-connectivity is observed between PSI units (Melis, 1982)). Therefore the percentage oxidation of P-700 is a measure of the relative quantum efficiency of PSI (Weis, Ball and Berry, 1987; Harbinson and Hedley, in prep.) provided no limitation of P-700 oxidation occurs due to restricted acceptor pool activity. In the absence of an appropriate electron acceptor no P-700 oxidation can occur and yet the quantum efficiency for linear electron flow is zero (Harbinson and Hedley, in prep.). A reduction in PSI quantum efficiency by this means will not be directly apparent to a measuring system which relies on P-700 oxidation, nevertheless it is possible to infer its presence indirectly. However a restriction of PSI activity due to a limitation of electron acceptor activity only seems to occur in air during photosynthetic induction or during treatment with inhibitors that abolish or restrict $NADP^+$ regeneration (e.g. DLglyceraldehyde). In summary using P-700 as an in situ probe of photosynthesis offers a technique which, under most conditions, allows a measurement of PSI quantum efficiency to be made. It also allows the half time of electron donation from PQH_2 to $P\text{-}700^+$ to be measured either directly

H. K. Lichtenthaler (ed.) Applications of Chlorophyll Fluorescence, 339–344.

or indirectly: this half time is determined principally by the rate of reaction between PQH^- and the Reiske Fe S centre of the cyt b_6/f complex (Bendall, 1982) and is the rate limiting reaction of thylakoid electron transport. It is the balance between the photochemical creation of P-700$^+$, and its destruction via thylakoid electrochemistry that determines the steady state pool of P-700$^+$ and therefore the quantum efficiency of PSI.

In this article I shall attempt the following:-

i) to describe the general principles of measuring the oxidation state of P-700 using near infra red radiation.

ii) to describe a possible system that might allow the $t_{1/2}$ for P-700$^+$ reduction to be measured non-invasively and remotely.

iii) to describe some of the uses to which such measurements can be put.

The measurement of P-700$^+$ generation and destruction using measuring wavelengths around 820nm

The broad, weak absorbance charge displayed by P-700/P-700$^+$ around 820nm (Ke, 1972) is shared with other chlorophyll species following oxidation (Borg, Fayer, Felton and Dolphin, 1970), reduction (Nuijs et al., 1986), or excitation (Nuijs et al., 1986) and phaeophytins following reduction (Klimov, Klevanik, Shuvalov and Krasnovsky, 1977). It also overlaps with absorbance changes associated with plastocyanin reduction (Katoh, Shivalori and Takamiga, 1962) and the S1/S2 transition (Velthuys, 1988). In addition during the irradiation of leaves with PAR a large absorbance change of unspecified origin can occur. This absorbance change develops and decays slowly ($t_{1/2}$ ~10-10^2 s) and may be due to conformational changes of the leaf. In practice only the latter slow absorbance changes cause any significant interference in vivo. To isolate the absorbance change due to P-700 oxidation/reduction from this 'background' change it is necessary to remove the actinic light at intervals following which the P-700$^+$ pool is reduced with a $t_{1/2}$ of ~10^{-2}s. It is therefore easy to separate the fast phase of the absorbance decay due to P-700$^+$ reduction from the slow phase which is due to other processes (Harbinson and Woodward, 1987).

Chlorophyll fluorescence incident on the detector will swamp the small (<0.5%) signal changes due to P-700 oxidation/reduction and to eliminate this interference it is necessary to modulate the measuring beam.

Within a leaf the near infra-red measuring beam is extensively scattered so the effective path length of beam is extended by 5-10 times (Ruhle and Wild, 1979b). This results in an absorbance change proportionally larger than would be expected from Lambert-Beer law. This scattering also allows the absorbance change caused by P-700 oxidation/reduction to be measured using either transmitted or backscattered radiation . Using a modulated measuring beam it is therefore possible to measure the oxidation and subsequent dark reduction of P-700 in the presence of an actinic beam, and to do so using either transmitted or backscattered radiation. Obviously with this technique the time resolution is limited by the modulation frequency.

It is also possible to measure the absorbance change associated with P-700$^+$ reduction using a non-modulated measuring beam. By removing the actinic light quickly ($t_{1/2}$ < 0.1ms) in the presence of a non-modulated measuring beam the absorbance change associated with the reduction of P-700$^+$

can be resolved. (Fig. 1)

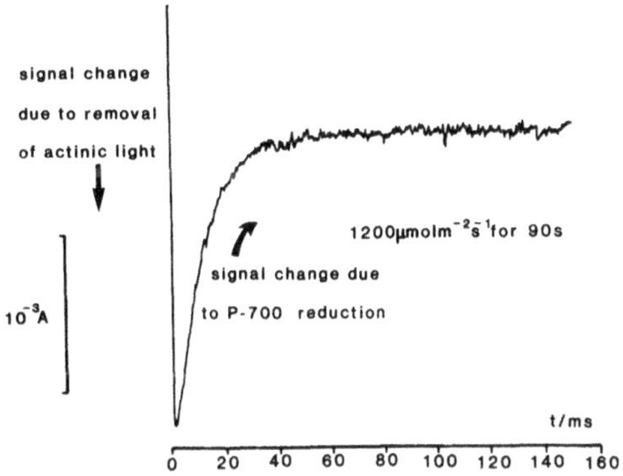

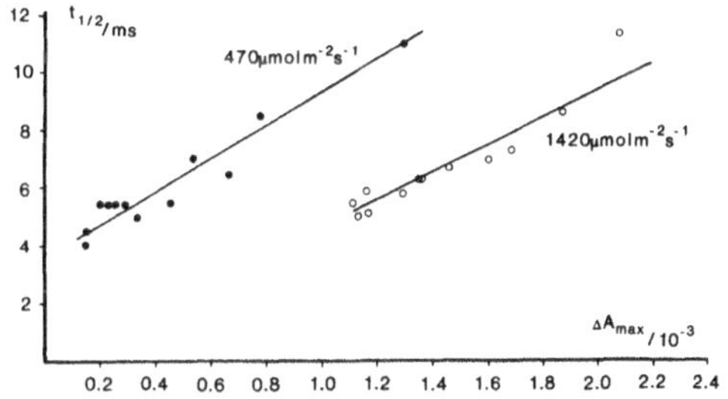

Figure 1 (L). The dark reduction ofthe photochmically oxidized pool of $P\text{-}700^+$ following a light dark transition.
Figure 2 (R). The Correspondence between the halftime for $P\text{-}700^+$ reduction and the absorbance at the point of the light-dark transition.

Delayed fluorescence can corrupt the first millisecond of the measurement but subsequently the decay of the absorbance and therefore the $P\text{-}700^+$ pool ocurs monotonically with a half time of $\tilde{}10^{-2}$s (Maxwell and Biggins, 1976; Harbinson and Hedley, as submitted). This kinetic phase of $P\text{-}700^+$ reduction is caused by donation from PQH_2 (Haehnel, 1976), faster phases of $P\text{-}700^+$ reduction from reduced plastocyanin and cyt.f can occur, but the cyt.f and plastocyanin pools are oxidized in the light and so can be largely discounted as significant primary sources of reductant following a light-dark transition. The absorbance at zero time (the time of the light-dark transition) corresponds to steady state absorbance and the $t_{1/2}$ for the decay is the 'resistance' for electron flow from PQH_2 to $P\text{-}700^+$. This technique has the advantage over the modulated technique in that it provides a measure of $t_{1/2}$ as well as absorbance, but there is a more limited capacity to monitor the detail of the absorbance change during induction etc. for which a modulated measuring beam is required. Of course by increasing the modulation frequency it is possible to monitor the decay of the $P\text{-}700^+$ pool following a light-dark transition using a modulated measuring beam.

A technique to measure P-700 oxidation/reduction non-invasively and remotely in the field would probably function best if it relied on measurements of kinetics rather than attempted to monitor the steady state population of $P\text{-}700^+$. With individual leaves in a laboratory it is possible to get an absorbance corresponding to complete non-oxidation of P-700 by simply removing the active light. Equally it is possible to get an estimate of the absorbance corresponding to complete oxidation of P-700 using far-red light whch preferentially excites PSI relative to PSII. Obviously such manipulation of the radiation environment of the leaf or plant is not practical in a remote monitoring system.

None the less an estimate of the percentage oxidation of P-700 would be a very useful measure as this is highly linearly correlated with the quantum efficiency of electron transport in vivo (Weis et al., 1987; Harbinson and Hedley, in prep.). Though it is not possible to measure the steady state pool of $P\text{-}700^+$ directly it should be possible to measure the half time for $P\text{-}700^+$ reduction and this value is fundamental in determining the degree of oxidation and hence the quantum efficiency of PSI. The oxidation of P-700 and the loss of PSI quantum efficiency is controlled by two factors. The first is the supply of excitation energy to PSI and the associated P-700 reaction centres - this generates $P\text{-}700^+$ from P-700 at a rate which is determined by irradiance and the absorbance of PSI: neither of these is under any short term 'control' (with exceptions, for example, pulvinic action may control irradiance). The second factor is the reduction of $P\text{-}700^+$ by electrons from PQH_2: this factor is determined by physological processes, principally the intrathylakoid pH (Tikhonov et al., 1984). It is the balance between the rate of creation of $P\text{-}700^+$ and rate of reduction that determines the level of oxidation of the P-700 pool. So under conditions of constant irradiance the $P\text{-}700^+$ pool is determined by the $t_{1/2}$ for $P\text{-}700^+$ reduction (Fig. 2). By implication $t_{1/2}$ is inversely correlated with the quantum efficiency of electron transport and with the potential for CO_2 fixation. The half time for $P\text{-}700^+$ reduction is a key physiologically controlled determinant of photosynthetic quantum efficiency.

Measurement of $P\text{-}700^+$ reduction remotely

Unlike the percentage oxidation of P-700 it should be relatively easy to monitor the $t_{1/2}$ for $P\text{-}700^+$ reduction remotely. The $t_{1/2}$ for $P\text{-}700^+$ reduction is a function of the intrathylakoid pH (Tikhonov et al., 1984) and the concentration of PQH_2 (Stiehl and Witt, 1969) in the thylakoid membrane. A short, single turnover (with respect to PSII) flash will have little

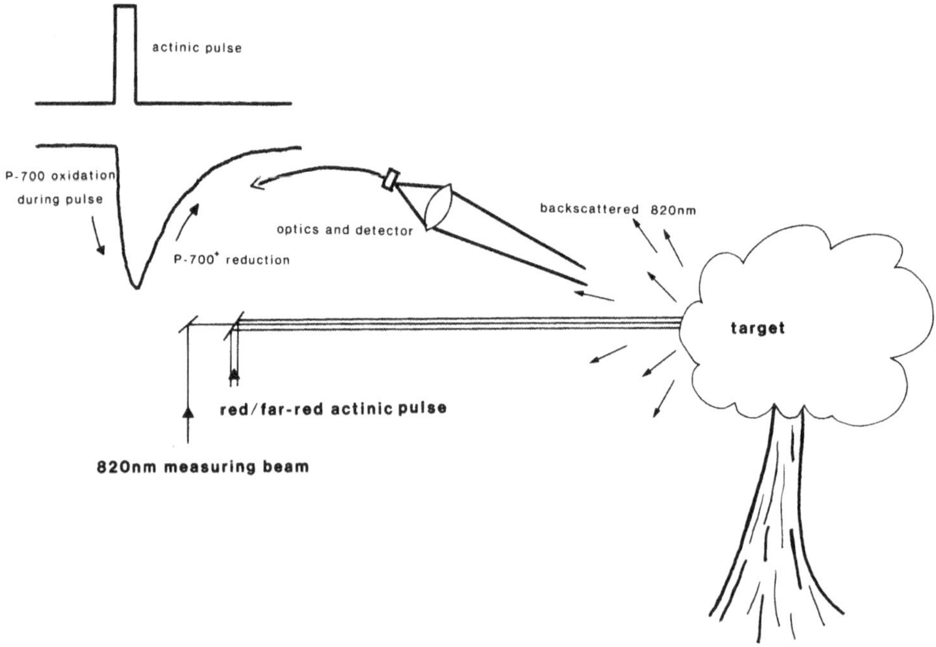

FIGURE 3. A Schematic diagram showing how the $t_{1/2}$ for $P\text{-}700^+$ reduction could be measured remotely.

effect on either of these because of the size of the PQH_2 pool and buffering capacity of the intrathylakoid space. The reduction of a transient increase in the steady-state $P\text{-}700^+$ pool produced by a short laser pulse directed at a leaf or canopy (Fig. 3) will have the same $t_{1/2}$ as the dark reduction of the $P\text{-}700^+$ pool, even though the laser flash is superimposed upon a background irradiance. If reduction of the actinic effects of the laser pulse were to be crucial, then a far-red laser (~700nm) could be used: this would principally excite PSI and generate an excess of P-700$^+$ with less disturbance to the ΔpH and redox state of the PQH_2 pool.

Monitoring of the absorbance changes associated with the generation and reduction of the excess population of P-700$^+$ could be effected using a co-axial 820nm measuring beam. This could be easily produced from any of a large number of solid-state communications lasers emitting in this spectral region. These devices by their nature are easy to modulate over a very wide bandwidth so modulation of the measuring beam with the associated ability to eliminate cross talk between the measuring beam signal and signals derived from other sources would be easy.

The range over which the system could be used would be a function of the measuring beam intensity, the f. number and focal length of the detector optics, and the noise equivalent power of the detector. Obviously there is a considerable spectrum of design options available that would allow the technique to be used at ranges of only a few centimetres upwards. The problem of leaf movement during the measurement window might restrict the application of the technique under certain conditions though the extent to which this will be a problem remains to be determined.

Application of the technique

Following the laser pulse the transient excess of P-700$^+$ produced will dark reduce until the steady-state P-700$^+$ population is restored. Our experience with decay kinetics from individual leaves suggests that this decay will be monotonic, following first order kinetics (Harbinson and Hedley, submitted). This in itself suggests that the $t_{1/2}$ for P-700$^+$ reduction is independent of irradiance under steady-state conditions and this perhaps surprising observation has been confirmed explicitly. If the leaf data can be extrapolated to the canopy then just as in individual leaves variations in the irradiance within the volume of tissue interacting with the measuring beam will not influence the $t_{1/2}$ for P-700$^+$ reduction. However whatever the character of the decay kinetics recorded the essential point is that the slower the kinetics the less the quantum efficiency of PSI, and also by implication of PSII. In our experience 'stress' such as low temperatures, photoinhibition, herbicides and low CO_2 partial pressure all act to increase the $t_{1/2}$ for P-700$^+$ reduction and therefore increase the pool of P-700$^+$ at any given irradiance.

The simple relationship that percentage oxidation of P-700 has to the quantum efficiency of PSI and to electron transport make it a potentially very useful probe of thylakoid processes in situ. We feel that the use of measurements based on kinetics rather than steady-state levels will allow the technique to be applied under conditions where the accurate control of the radiation environment required for steady-state measurements is impractical. The application of such a technique should allow non-invasive, remote probing of the performance of electron transport between PQH_2 and P-700 in vivo with all the implications this has for photosynthetic potential and quantum efficiency, and this technique could be just as valuable in controlled laboratory measurements as in the field.

REFERENCES

Bendall DA, 1982. Photosynthetic cytochromes of oxygenic organisms. Biochimica et Biophysica Acta 683: 119-151.
Borg DC, Fayer J, Felton RG and Dolphin D, 1970. The pi-cation radical of chlorophyll a. Proceedings of the National Academy of Sciences 67: 813-830.

Bradbury M and Baker NR, 1981. Analysis of the slow phases of the *in vivo* chlorophyll fluorescence induction curve. Changes in the redox state of photosystem II electron acceptors and fluorescence emission from photosystems I and II. Biochimica et Biophysica Acta 635: 542-551.
Haehnel W, 1976. The reduction kinetics of chlorophyll a_1 as an indicator for proton uptake between the light reactions in chloroplasts. Biochimica et Biophysica Acta 440: 506-521.
Haehnel W, 1984. Photosynthetic electron transport in higher plants. Annual Review of Plant Physiology 35: 659-693.
Harbinson J and Hedley CL, 1988. The dark reduction of photochemically oxidized P-700 *in vivo*. Submitted to Plant Cell and Environment.
Harbinson J and Woodward FI, 1987. The use of light-induced absorbance changes at 820nm to monitor the oxidation state of P-700 in leaves. Planta, Cell and Environment 10: 131-140.
Katoh S, Shiratori I and Takamiya A, 1962. Purification and some properties of spinach plastocyanin. The Journal of Biochemistry 51: 32-40.
Ke B, 1972. The rise time of photoreduction, difference spectrum and oxidation-reduction potential of P430. Archives of Biochemistry and Biophysics 82: 82, 183-186.
Klimov VV, Klevanik AV, Shuvalov VA and Krasnovsky AA, 1977. Reduction of phaeophytin in the primary light reduction of photosystem II. FEBS Letters 82: 183-186.
Krause GH and Weis E, 1984. Chlorophyll fluorescence as a tool in plant physiology. II. Interpretation of fluorescence signals. Photosynthesis Research 5: 139-157.
Kyle DJ, Baker NR and Arntzen CJ, 1983. Spectral characterization of Photosystem I fluorescence at room temperature using thylakoid protein phosphorylation. Photobiochemistry and Photobiophysics 5: 79-85.
Melis A, 1982. Kinetic analysis of P-700 photoconversion: Effect of secondary electron donation and plastocyanin inhibition. Archives of Biochemistry and Biophysics 217: 536-545.
Nuijs AM, Shuvalov VA, van Gorkom HJ, Plijter JJ and Duysens LNM, 1986. Picosecond absorbance - difference spectroscopy on the primary reactions and the antenna-excited states in Photosystem I particles. Biochimica et Biophysica Acta 850: 310-318.
Rühle W and Wild A, 1979a. Measurements of cytochrome f and P-700 in intact leaves of *Sinapis alba* grown under high light and low light conditions. Planta 146: 377-385.
Rühle W and Wild A, 1979b. The intensification of absorbance changes in leaves by light-dispersion. Planta 146: 551-557.
Stiehl HH and Witt, HT, 1969. Quantitative treatment of the function of plastoquinone in photosynthesis. Zeitschrift fur Naturforschung 24b: 1588-1598.
Tikhonov AW, Khomutov GB and Ruuge EK, 1984. Electron transport control in chloroplasts. Effects of magnesium ions on the electron flow between two photosystems. Photobiochemistry and Photobiophysics 8: 261-269.
Velthuys BR, 1988. Spectroscopic characterization of the acceptor state Q_A and the donor state S2 of photosystem II of spinach in the blue, red and near-infra-red. Biochimica et Biophysica Acta 933: 249-257.
Weis E, Ball JT and Berry J, 1987. Photosynthetic control of electron transport in leaves of *Phaseolus vulgaris*: evidence for regulation of photosystem 2 by the proton gradient. *In*: Progress in Photosynthesis Research Vol. II p553-556 ed. Biggins, J. Martinus Nijhoff, Dordrecht, The Netherlands.

A SIMPLE MODEL FOR LEAF OPTICAL PROPERTIES IN VISIBLE AND NEAR-INFRARED : APPLICATION TO THE ANALYSIS OF SPECTRAL SHIFTS DETERMINISM

F. BARET (1), B. ANDRIEU (2), G. GUYOT (1)

(1) INRA, Bioclimatologie, BP 91, 84140 MONTFAVET (France)
(2) INRA, Bioclimatologie, 78850 THIVERVAL-GRIGNON (France)

ABSTRACT

A simple model of leaf optical properties spectra is derived from experimental measurements of leaf characteristics. It allows to simulate reflectance and transmittance spectra of the leaf depending on it's chlorophyll concentration and leaf mesophyll structure.
This model is used to generate relationships between chlorophyll concentration, leaf mesophyll structure and spectral shifts in the red edge. Results shows that an increase in chlorophyll concentration induces spectral shifts towards longer wavebands. But the effect of leaf mesophyll structure is significant, particularly for high chlorophyll concentration.

KEY WORDS : Remote Sensing, Reflectance, Transmittance, Leaf, Chlorophyll, Model, Spectral-shift, High spectral resolution

INTRODUCTION

One of the principal goals in remote sensing of agricultural crops is to use the spectral information to estimate biological parameters that are related to the canopy characteristics and that can be used as input into growth models for yield prediction. Reflectance measurements used for these estimations have mainly been performed with broad spectral bands of 50 nm to 200 nm corresponding to actual satellites (NOAA, SPOT, LANDSAT MSS, TM...). Recently, investigation using high spectral resolution measurements were developed. These studies are focussed on the analysis of the "red-edge". This domain corresponds to the transition between visible region, which is mainly affected by pigment absorption properties, and near infrared region which is affected by leaf mesophyll structure.

The wavelength position of the inflexion point of the spectrum between 670 and 760 nm has been studied by HORLER et al. (1980, 1983) and determines the so-called "red-shift" or "blue-shift" depending on the direction of the shift. These studies were performed at leaf level and were followed by some experimental observations above canopies (FERNS, 1984 ; VANDERBILT and GRANT, 1984 ; SCHUTT et al., 1984 ; BARET et al., 1987). But all of these experimental results give only a partial understanding on the determinism of spectral shifts because of the complexity of the radiometric response of crops. Canopy reflectance depends numerous factors which are often strongly correlated in the particular conditions of one experiment. The use of reflectance model simulations can considerably improve the understanding of the determinism of spectral shifts. But they require a good description of the elementary mecanisms which that affect the reflectance and transmittance spectra of leaves. Models representing leaf optical properties were developed since the seventies. The KM and "plate models" theories derived by Allen and Richardson (1968) and Allen et al. (1969) gave a good description of optical properties of compact leaves. Ray tracing techniques were also developed (Allen et al., 1973 ; Kumar and Silva, 1973) but are time consuming and awkward to use.

H. K. Lichtenthaler (ed.) Applications of Chlorophyll Fluorescence, 345–351.

The aim of this paper is to present a simple model describing leaf reflectance and transmittance spectra in visible and near infrared domains. This model will be adjusted on experimental data and will be used to generate relationships between spectral shifts and leaf characteristics such as pigment concentration or mesophyll structure.

MATERIALS AND METHODS

Measurements are performed on wheat (cv: Fidel) leaves at different stages of senescence during the grain filling period. Whole plants are collected and brought immediately to the laboratory. Leaves are picked up to measure the reflectance and the transmittance of the upper surface in the middle of the blade. We use the Varian Cary 17 D I spectrophotometer. Spectra are monitored from 410 nm to 850 nm with 10 mm steps. The spectral resolution is around 2 nm. The sampled area which varies from 3 mm^2 to 150 mm^2 is fixed by the spectroradiometer slit width size depending on the incoming spectral energy flux. We use an integrating sphere that allows to measure the directionnal–hemispherical reflectance or transmittance. Baryum sulfate is taken as reference. The accuracy of the measurements is better than 1 %.

Pigment concentration is measured on two disks of 9 mm diameter taken as close as possible to the point where we measure reflectance and transmittance. We determine chlorophyll a, chlorophyll b and total carotenoïds according to the methodology defined by Lichtenthaler (1987).
Leaf water content, leaf thickness and leaf specific area implement those former characteristics of the leaf.

RESULTS

13 leaves were used to have a sufficient range between green active leaves and fully senescent dead leaves. Table 1 resumes the differing measurements :

leaf	Chlor a $\mu g.cm^{-2}$	Chlor.b $\mu g.cm^{-2}$	Carot. $\mu g.cm^{-2}$	Thickn. *mm	Water cont. %	spec. area ** $cm^2.mg^{-2}$
1	53.7	16.3	16.1	0.205	67.6	0.163
2	49.4	16.9	14.1	0.178	67.7	0.214
3	49.2	14.7	14.6	0.216	67.1	0.177
4	17.4	6.7	5.7	0.182	61.7	0.257
5	17.2	6.7	6.4	0.279	75.4	0.273
6	15.2	5.6	5.5	0.181	66.6	0.270
7	1.8	1.4	3.7	0.232	60.6	0.211
8	1.7	0.9	4.6	0.217	64.7	0.197
9	0.6	0.2	2.5	0.206	69.9	–
10	0.0	0.0	0.0	0.229	9.8	0.288
11	0.0	0.0	0.0	0.247	6.9	0.270
12	0.0	0.0	0.0	0.219	6.9	0.303
13	0.0	0.0	0.0	0.260	5.6	0.303

Table 1 : Pigment concentration and physical characteristicsof the leaves
* The thickness is calculated as the median of 10 measurements per blad
** The specific area is the ratio between the area and the dry weight of the leaf (samples correspond to about 400 mm^2)

Leaf water content is very stable since chlorophyll exists and decreases drastically for fully senescent leaves. Leaf thickness seems to be a poor indicator of leaf structure perhaps because of the difficulty to get a representative measurement. The veins affects the measurements, even if the use of the median instead of the mean minimizes this bias. The measurement performed with the SONY electronic micrometer is also sensitive to the pressure applied to the sensor. The specific area seems to be more reliable to leaf structure characterized by the reflectance value in the near-infrared (760 nm) as shown by Fig. 1. If we assume that the cell walls have a constant weight per unit area, an increase of the specific area of the leaf would lead to an increase of the number of cell wall interfaces inside the leaf. In consequence, according to Gates et al. (1965), the increase of interfaces will produce higher reflectance and lower transmittance. We observed those results but with some dispersion for the transmittance values.

Pigment concentrations are strongly intercorrelated (r > 0.96). This means that in natural conditions as ours, it would be difficult to separate the effects of chlorophyll a, chlorophyll b and carotenoïds on leaf spectra. We can observe a large increase in reflectance or transmittance in the visible domain when pigment concentration decreases (figure 2). Unlikely, in the near infrared domain, reflectances or transmittances plateaued around 760 nm. The different observed spectra (reflectance or transmittance) evolve within a domain limited by the optical properties of dark-green leaves (lower limit) and the optical properties of fully senescent leaves (upper limit) in which chlorophyll a and b and carotenoïds are replaced by brown pigments. Brown pigments produce the progressively decreasing absorption starting in the blue and disappearing in near infrared. These observations are the basis of the following simple model for leaf optical properties.

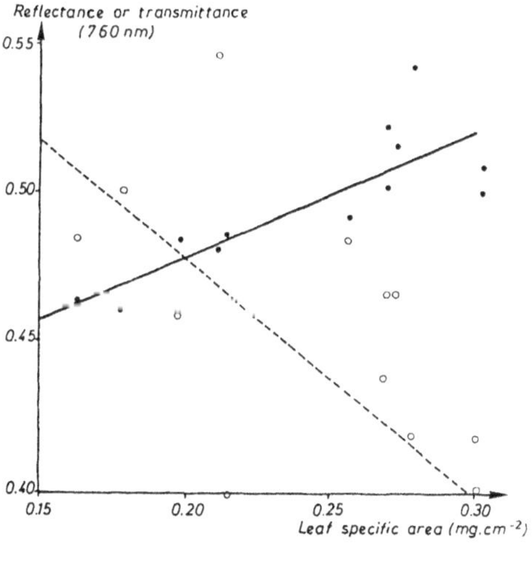

Figure 1 Relationships between leaf specific area and near infrared (760 nm) reflectance (points, solid line) or transmittance (open circle, dashed line).

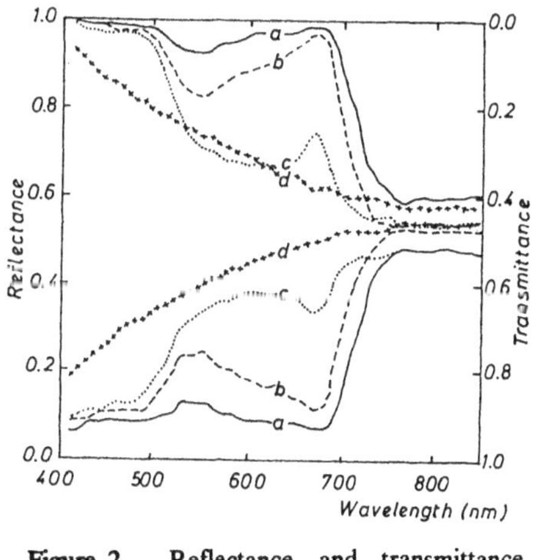

Figure 2 Reflectance and transmittance spectra of typical wheat leaves at different stages.
a, b, c, d spectra correspond respectively to leaves 2, 5, 8 and 10 of table 1.

LEAF REFLECTANCE AND TRANSMITTANCE MODELS

Because radiation extinction through an absorbing medium increases exponentially with absorbant concentration, the following simple equations (1) and (2) are proposed to describe the variations of leaf reflectance (transmittance) with pigment concentration from 410 to 850 nm :

$$r_\lambda = r_{\infty\lambda} + (r_{o\lambda} - r_{\infty\lambda})\, e^{-kr_\lambda \cdot C} \quad (1)$$

$$t_\lambda = t_{\infty\lambda} + (t_{o\lambda} - t_{\infty\lambda})\, e^{-kt_\lambda \cdot C} \quad (2)$$

r_λ (t_λ) is the reflectance (transmittance) of the leaf.
$r_{\infty\lambda}$ ($t_{\infty\lambda}$) is the infinite reflectance (transmittance) when pigment concentration reaches infinity. This parameter has been set to the mean value corresponding to leaves (1, 2 and 3).
$r_{o\lambda}$ ($t_{o\lambda}$) is the reflectance (transmittance) of dead leaves.
kr_λ (kt_λ) is the extinction coefficient ($cm^2.\mu g^{-1}$).
C is the pigment concentration ($\mu g.cm^{-2}$).
The subcript λ denotes that the parameter corresponds to the particular wavelength λ .

Because of the strong correlation between pigment concentrations, we have defined C as the concentration of chlorophylls a and b which have the most important absorption properties especially in the red domain. In the same way and because the apparition of brown pigments must be strongly correlated with the disparition of chlorophyll during natural senescence, we don't take this variable explicity into account but indirectly through chlorophyll a and b concentrations.
The value of $r_{o\lambda}$ ($t_{o\lambda}$) mainly depends on leaf structure and on brown pigment concentration. We express $r_{o\lambda}$ ($t_{o\lambda}$) by the following expression :

$$r_{o\lambda} = \frac{r_{760}}{r_{760*}} \cdot r_\lambda^* \quad (3)$$

$$t_{o\lambda} = \frac{t_{760}}{t_{760*}} \cdot t_\lambda^* \quad (4)$$

r_{760} (t_{760}) is the reflectance (transmittance) in the near infrared (760 nm) which is only affected by leaf structure.
r_{760*} (t_{760*}) is the mean reflectance (transmittance) value for dead leaves in the near infrared (leaves 10, 11, 12, 13 from table 1).
r_λ^* (t_λ^*) is the mean reflectance (transmittance) value for dead leaves (10, 11, 12, 13). Figure 3 shows the spectra r_λ^* (t_λ^*) and $r_{\infty\lambda}$ ($t_{\infty\lambda}$) which correspond to the limits into which the reflectance or transmittance will vary.

This simple model needs 2 input variables (r_{760} (t_{760}), C), 3 constants (r_{760*} (t_{760*}), r_λ^* (t_λ^*), $r_{\infty\lambda}$ ($t_{\infty\lambda}$)) and 1 adjusted parameter (kr_λ , (kt_λ)). The parameters kr_λ and kt_λ are adjusted using non linear fitting (Bouvier et al., 1985) of equations (1) and (2) on experimental data from table 1. The fitted extinction coefficients for reflectance and transmittance exhibits a very similar trend (figure 4). We can observe the well known absorption peaks for carotenoïds and chlorophylls (440 nm, 480 nm, 670 nm).

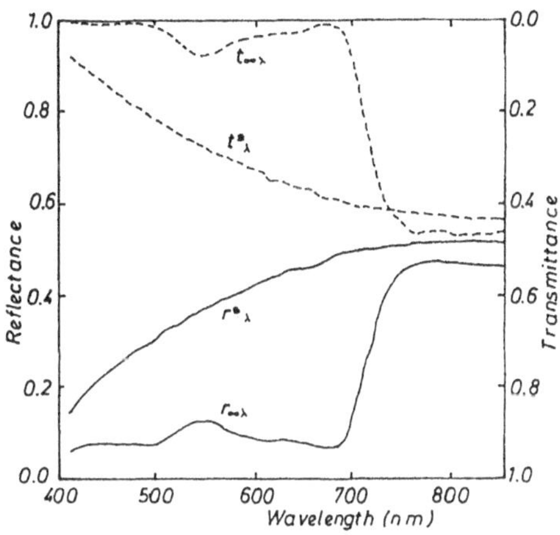

Figure 3 Mean spectra of green leaves (leaves 1, 2, 3 :$r_{\infty\lambda}$, $t_{\infty\lambda}$ and dead leaves (10, 11, 12, 13 : r_λ^*, t_λ^*).

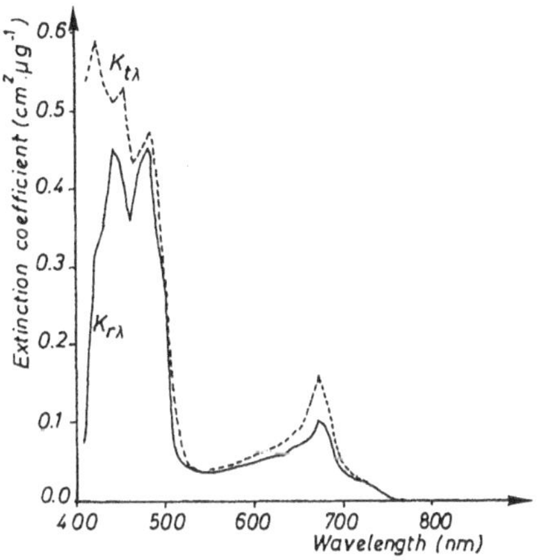

Figure 4 Extinction coefficients spectra for reflectance (solid line) and transmittance (dashed line)

Figure 5 shows that models (1) and (2) are able to give a good description of wheat leaf optical properties during natural senescence. The limits of this semi-empirical approach are due to the three simplifications introduced :

(i) The effects of each pigments (chlorophyll a, chlorophyll b, carotenoïds, brown pigments) are not explicitely introduced. This means that the model can only be applied when the correlations observed between the different pigment concentrations are strong and are the same as in experiment.

(ii) Equations (3) and (4) are empirically derived. More attention might be taken to intruduce the influence of mesophyll structure and brown pigment concentration on the upper limit.

(iii) The specular reflection is not explicitely introduced. For the reflectance, it must be considered as an additive term which is practically independant of the wavelength in the visible and near-infrared (Vanderbilt and Grant, 1983). In our case, this effect was minimized by nadir viewing measurements.

From these results, we can use the proposed models to simulate spectra of leaves with continuous and independant variations in chlorophyll concentration and leaf structure (reflectance or transmittance in the near infrared). From simulated spectra, we can evaluate the spectral shifts at leaf level and discuss about their signification.

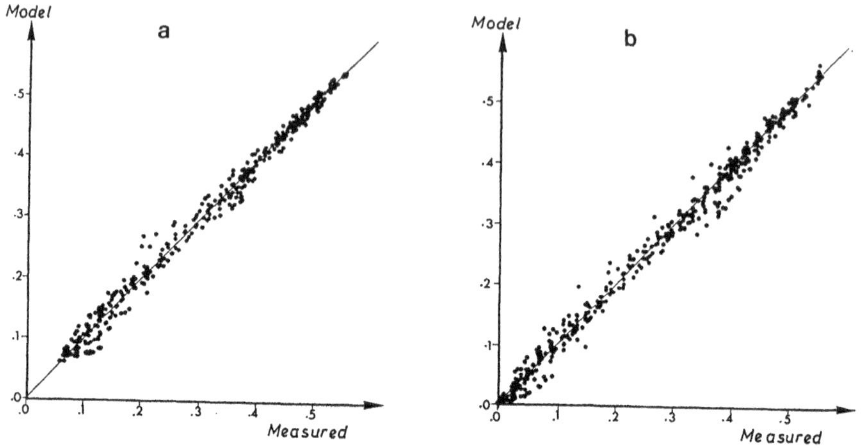

Figure 5 Comparison between reflectance (a) and transmittance (b) data simulated (equations 1 and 2) and measured (512 data)

SIGNIFICANCE OF SPECTRAL SHIFTS IN THE RED EDGE

Because of the similarly between reflectance and transmittance models, we will restrict the discussion only on leaf reflectances. Spectral shifts are classicaly determined by the position of the inflexion point in the red edge (see HORLER et al., 1983). But the accuracy will be very poor when spectral resolution decreases (or when the wavelength interval between two consecutive measurements is too large) and when the red edge has some linear feature (HORLER et al., 1983). For these reasons, we prefer to express the spectral a function of the wavelength (λ_i) corresponding to the average reflectance between the red (670 nm) and the near infrared (760 nm). This technique has been already used to characterize spectral shifts at canopy level (BARET et al., 1987). Model simulations (1) are performed with pigment concentration varying from 2 to 70 $\mu g.cm^{-2}$ and leaf reflectance at 760 nm ranging from 0.40 to 0.55.

Fig. 6 shows that chlorophyll concentration has a strong effect on spectral shift. An increase in chlorophyll content produces a shift towards longer wavelengths. An increase in near infrared reflectance, due to a variation in leaf mesophyll structure, will shift λ_i towards shorter wavelengths. This effect will be maximum for high pigment concentrations. λ_i vary between 693 nm, for low chlorophyll concentration, and 715 nm for high chloropyll concentration and high near- infrared reflectance. These results are in very good agreement with experimental results of Horler et al. (1983).

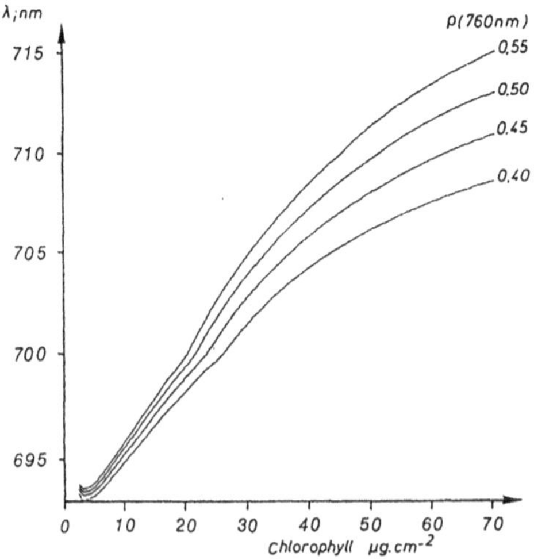

Figure 6 Effect of chlorophyll concentration and leaf reflectance in near infrared on spectral shifts in the red edge

CONCLUSION

The semi-empirical model representing the optical properties of leaves, adjusted on our experimental data, seems to be reliable in these conditions. But its application domain is limited because the particular effects of the different pigments are not specifically considered. The model will only be valid if the relationships between the different pigment concentrations are the same as in our experiment. This model must be improved in order to introduce the effects of the different pigments. If we focus our analysis on the red edge we practically have just to consider the chlorophyll a and b concentrations. As these concentrations are functionnaly well correlated (Clayton, 1980 in Horler et al., 1983) we can use the proposed models (equations 1 and 2) to describe with enough confidence the spectral feature of the red edge. Model simulations show that, at the leaf level, the observed spectral shifts are mainly due to variations in chlorophyll concentration and, in a lesser way, in leaf mesophyll structure. However, the chlorophyll fluorescence is not introduced in this model. It seems that this factor can also affect the shift of the red edge. But the fluorescence spectra which are presented in the literature, are generally expressed in relative units. For improving the model, it is necessary to have fluorescence spectra expressed in radiative energy.

REFERENCES

Allen WA, Richardson AJ, 1968. Interaction of light with a plant canopy. J. Optic. Soc. Amer. 58:1023-1028

Allen WA, Gausman HW, Richardson AJ, Thomas JR, 1969. Interaction of isotropic light with compact plant leaf. J. Optic. Soc. Amer. 59(10)

Allen WA, 1973. Willstäter-Stoll theory of leaf evaluated by ray tracing. Applied optics 12(10):2448-2453

Baret F, Champion I, Guyot G, Podaire A, 1987. Monitoring wheat canopies with high spectral resolution radiometer. Remote Sens. Environ. 22:367-378

Bouvier A, Gelis F, Huet S, Messean A,Neveu P, 1985. CS-NL. INRA biometrie, 78350 Jouy en Josas, France

Gates DM, 1965. Spectral properties of plants. Applied Optics 4(1):11-20

Grant L, 1987. Diffuse and specular characteristics of leaf reflectance. Remote Sens. Environ. 22:309-322

Horler DNH, Barber J, Barringer AR, 1980. Effect of heavy metals on the absorbance and reflectance spectra of plants. Int. J. Remote Sens. 1(2):121-136

Horler DNH, Dockray M, Barber J, 1983. The red edge of plant leaf reflectance. Int. J. Remote Sens. 4(2):273-288

Kumar R, Silva L, 1973. Reflectance model of a plant leaf. LARS information note n°022473

Lichtenthaler HK, 1987. Chlorophyll and carotenoïd: Pigments of photosynthetic membranes. in "methods in enzymology" vol 48:350-382

Schutt JB, Rowland RR, Heartly WH, 1984. A laboratory investigation of a physical mecanism for the extended infrared absorption ("red-shift") in wheat. Int. J. Remote Sens. 5(1):95-102

Vanderbilt VC, Grant L, 1983. Light polarization measurements: A method to determine the specular and diffuse light scattering properties of both leaves and plant canopy. 1st Int. Coll. Signatures Spectrales D'objets en Teledetection, Avignon 12-16 sept. 1983:55-66

Vanderbilt VC, Grant L, 1985. Geometric basis for the far red shift in crop canopies. in Int. Geosci. and Remote Sens. Symp. (IGARSS'85), Amherst, M.A., Oct. 7-9 1985:874-879

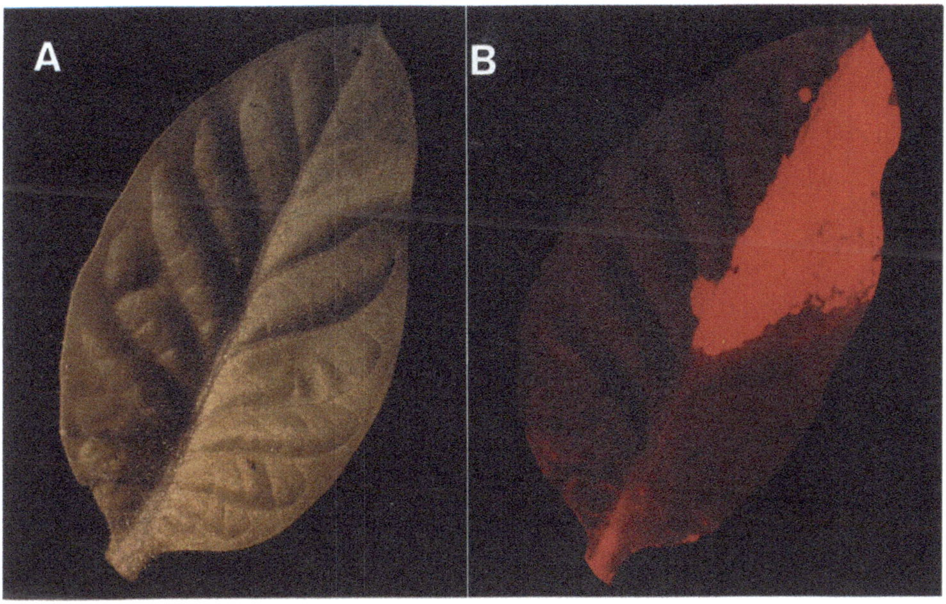

Colour plate No 1. Tobacco leaf (Nicotiana tabacum, aurea variety Su/su) without and with blue-light induced chlorophyll fluorescence. A. photographed in white light. B. taken in the low steady state of fluorescence,5 min after onset of illumination. In the bright red fluorescing upper part of the leaf photosynthesis was blocked by the herbicide diuron (DCMU). The blue excitation light was eliminated by a red cult-off filter.

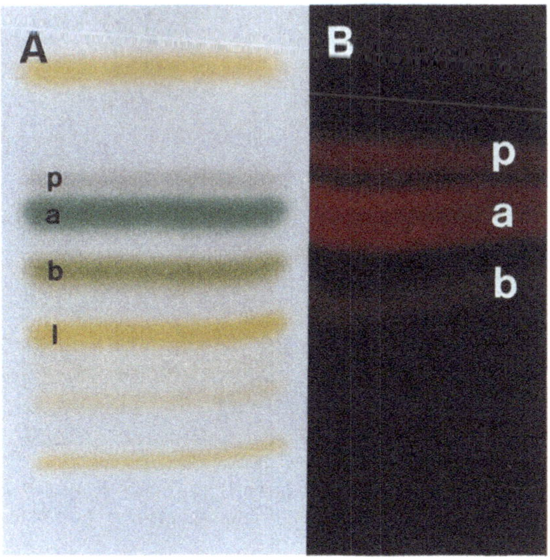

Colour plate No 2. A. Separation of photosynthetic pigments by thin layer chromatography (solvent: light petrol with isopropanol, 9 % and dioxan, 25 %). B. Blue-light induced fluorescence of pheophytin a (p), chlorophyll a (a) and chlorophyll b (b). The yellow carotenoid lutein (l) does not show red fluorescence. Photo B is enlarged.

Colour plate No 3. Section of the classified image (using maximum likelyhood) of the Mauzenberg test area which contained trees of damage class 3/4 (red) and those of damage class 2 (green). The light blue areas are clearings where dead or highly damaged trees have been felled (after Schmuck et al., 1987). The classification is based on airborne reflectance measurements.

Colour plate No 4. Damaged spruce tree (Picea abies) in the Egge Mountains (Northern Germany ca. 350 m above sea level) with yellowish discolouration of the older needle years.

Colour plate No 5. Discoloured branches and leaves of stressed forest trees (forest decline, Waldsterben) with signs of chlorophyll breakdown. A. Green and photochemically bleached needles of spruce (Picea abies), B. branch of the fir (Abies alba) with very small yellowish-green younger needle years (whereas the several year old needles are dark-green). C. branches of the hornbeam (Carpinus betulus), green control (left) and with increasing water and air-pollution stress (middle and right).

Colour plate No 6. Large-scale die-off of spruce forests (*Picea abies*, Waldsterben, forest die-back) in the Iser Mountains in Czechoslowakia (ca. 600-700 m altitude).

Colour plate No 7. View of ca. 50 year old spruces (Picea abies) of the northern Black Forest. A. location Althof (450 m above sea level, damage class 0/1), B. location Mauzenberg (700 m above sea level, mostly damage class 3/4). The photos were taken in June 1988.

Colour plate No 8. Highly damaged fir trees (Abies alba) in the northern Black Forest with signs of "storch nests" at the top of the tree.

Colour plate No 9. Appearance of branches and needles of spruces from A. the Althof site (damage class 0/1) and B. the Mauzenberg site (damage class 3/4). The highly damaged spruces (see also colour plate No 7) contain only 2 to 3 needle years, much fewer needles per needle year and a lower chlorophyll content (photos were taken in June 1988).

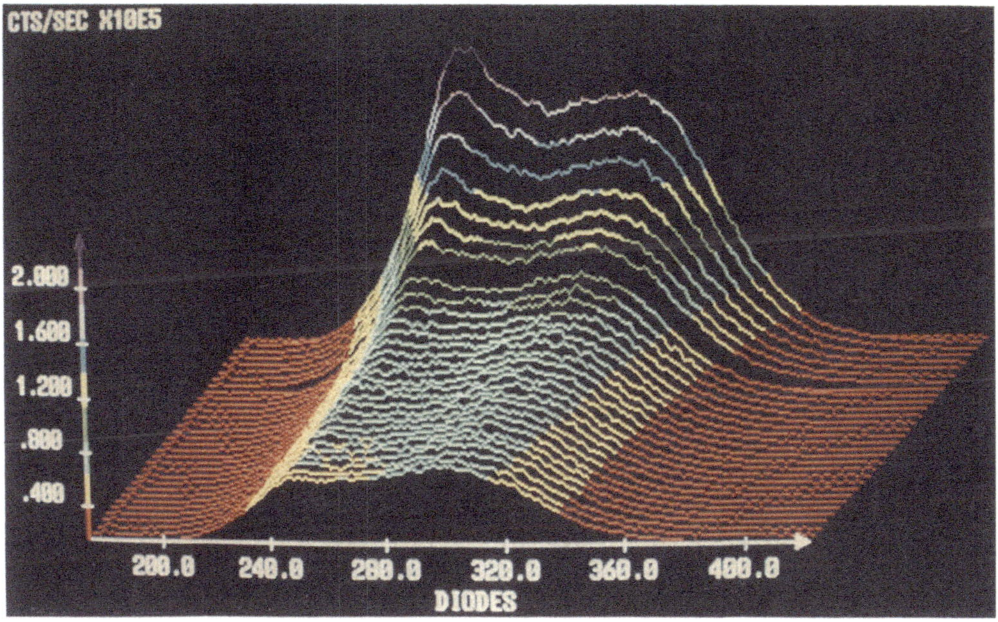

Colour plate No 10. Fluorescence emission spectra of a Phaseolus leaf taken every 200 ms during the slow decline of the induction kinetic (first spectrum at the end of the z-axis). For further details see Fig. 1 on page 80.

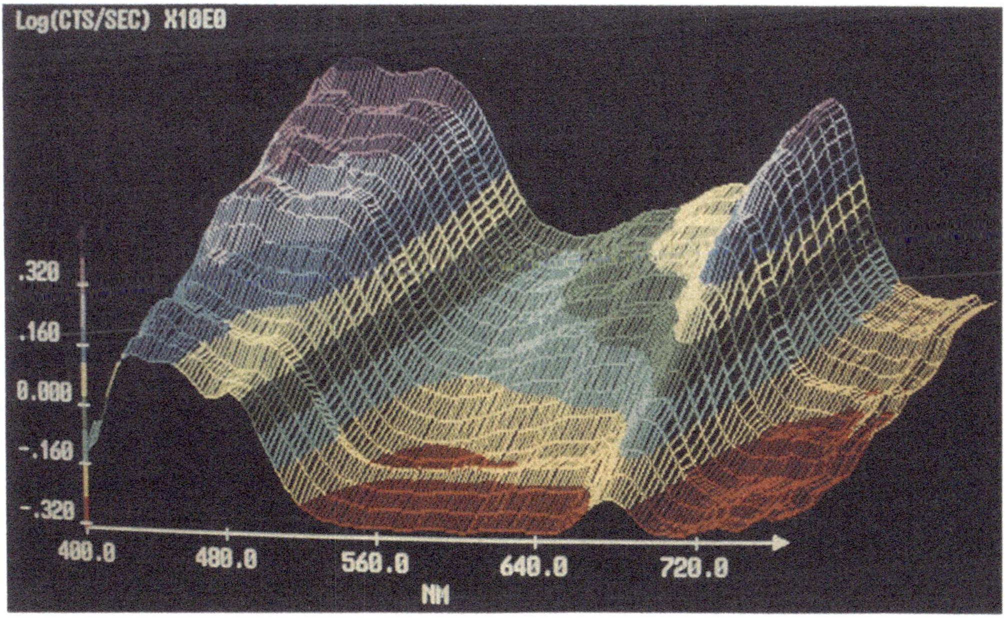

Colour plate No 11. Absorption spectra of an etiolated Phaseolus leaf taken with the VIRAF-spectrometer every 55 min during 17 h of greening in the light. For furhter details see Fig. 5 on page 330.

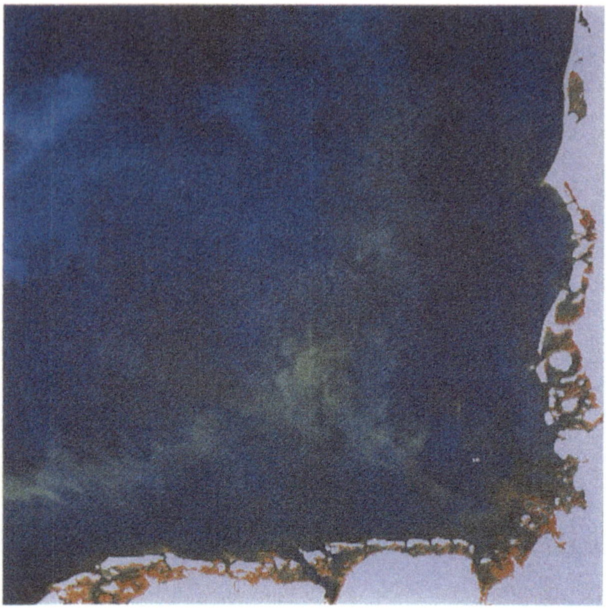

Colour plate 12: Image of the water leaving radiance of the German Bight, North Sea, as seen by the Coastal Zone Colour Scanner (CZCS) after atmospheric correction; the colour of the water is determined by suspendend matter, phytoplankton chlorophyll and Gelbstoff (s. Doerffer: Remote Sensing..)

Colour plate 13: Horizontal distribution of chlorophyll in the Fladen Ground, North Sea, as derived from sunlight induced chlorophyll fluorescemce, which was measured with a radiance spectrometer from an altitude of 600 m (s. Doerffer: Remote Sensing.., Fig. 3)

AUTHOR INDEX

SUBJECT INDEX